Security and Trends in Wireless Identification and Sensing Platform Tags:

Advancements in RFID

Pedro Peris Lopez
Carlos III University of Madrid, Spain

Julio C. Hernandez–Castro
Portsmouth University, UK

Tieyan Li
Institute for Infocomm Research, Singapore

Information Science
REFERENCE

Managing Director: Lindsay Johnston
Senior Editorial Director: Heather A. Probst
Book Production Manager: Sean Woznicki
Development Manager: Joel Gamon
Development Editor: Hannah Abelbeck
Assistant Acquisitions Editor: Kayla Wolfe
Typesetter: Lisandro Gonzalez
Cover Design: Nick Newcomer

Published in the United States of America by
        Information Science Reference (an imprint of IGI Global)
        701 E. Chocolate Avenue
        Hershey PA 17033
        Tel: 717-533-8845
        Fax: 717-533-8661
        E-mail: cust@igi-global.com
        Web site: http://www.igi-global.com

Library of Congress Cataloging-in-Publication Data

Security and trends in wireless identification and sensing platform tags: advancements in RFID / Pedro Peris Lopez, Julio C. Hernandez-Castro, and Tieyan Li, Editors.
      p. cm.
   Includes bibliographical references and index.
   Summary: "This book highlights new research regarding wireless identification and sensing platform (WISP) tags, security, and applications, serving as a reference on WISP technology and presenting recent advances in this field"--Provided by publisher.
   ISBN 978-1-4666-1990-6 (hardcover) -- ISBN 978-1-4666-1991-3 (ebook) -- ISBN 978-1-4666-1992-0 (print & perpetual access) 1. Wireless sensor networks. 2. Radio frequency identification systems. 3. Wireless sensor networks--Security measures. 4. Radio frequency identification systems--Security measures. I. Peris Lopez, Pedro, 1978- II. Hernandez-Castro, Julio C., 1972- III. Li, Tieyan.
   TK6570.I34S43 2013
   004.6--dc23
                              2012012922

British Cataloguing in Publication Data
A Cataloguing in Publication record for this book is available from the British Library.

All work contributed to this book is new, previously-unpublished material. The views expressed in this book are those of the authors, but not necessarily of the publisher.

Table of Contents

Section 1
Fundamentals

Chapter 1

Francesca Lonetti, Istituto di Scienza e Tecnologie dell'Informazione "A. Faedo",
Consiglio Nazionale delle Ricerche, Italy
Francesca Martelli, Istituto di Informatica e Telematica,
Consiglio Nazionale delle Ricerche, Italy

Chapter 2

Pablo Picazo-Sanchez, University School of Computer Sciences of Madrid (UPM), Spain
Lara Ortiz-Martin, University School of Computer Sciences of Madrid (UPM), Spain
Pedro Peris-Lopez, Carlos III University of Madrid (UC3M), Spain
Julio C. Hernandez-Castro, Portsmouth University, UK

Section 2
Security

Chapter 3

Di Ma, University of Michigan-Dearborn, USA
Nitesh Saxena, University of Alabama, Birmingham, USA

Chapter 4

Mike Burmester, Florida State University, USA
Jorge Munilla, University of Málaga, Spain

Section 3
Applications

Detailed Table of Contents

Section 1
Fundamentals

Chapter 1
Francesca Lonetti, Istituto di Scienza e Tecnologie dell'Informazione "A. Faedo",
Consiglio Nazionale delle Ricerche, Italy
Francesca Martelli, Istituto di Informatica e Telematica, Consiglio Nazionale delle Ricerche, Italy

Fast and reliable identification of multiple objects that are present at the same time is very important in many applications. A very promising technology for this purpose is Radio Frequency Identification (RFID), which is fast pervading many application fields, like public transportation and ticketing, access control, production control, animal identification, and localization of objects and people. The problem approached in this chapter is the tag identification in RFID systems. This problem occurs when several tags try to answer at the same time to a reader query. If more than one tag answers, their messages will collide on the RF communication channel, and the reader cannot identify these tags. There are two families of protocols for approaching the tag collision problem: a family of probabilistic protocols, and a family of deterministic ones. In this chapter, the authors give an overview of the most important approaches and trends for tag identification in RFID systems and provide the results of a deep comparison of the presented tag identification protocols in terms of complexity and performance.

Chapter 2
Pablo Picazo-Sanchez, University School of Computer Sciences of Madrid (UPM), Spain
Lara Ortiz-Martin, University School of Computer Sciences of Madrid (UPM), Spain
Pedro Peris-Lopez, Carlos III University of Madrid (UC3M), Spain
Julio C. Hernandez-Castro, Portsmouth University, UK

Radio Frequency Identification (RFID) is a common technology for identifying objects, animals, or people. The main form of barcode-type RFID device is known as an Electronic Product Code (EPC) and the most popular standard for passive RFID tags is Class-1 Generation-2. In this technology, the information transmitted between devices is through the air, therefore adversaries can eavesdrop these messages passed on the insecure radio channel and finally, the security of the system can be compromised. In this chapter, the authors analyze the security of EPC Class-1 Generation-2 standard, showing its security weaknesses and presenting some possible countermeasures.

Section 2
Security

Chapter 3

Recent technological advancements enrich many RFID tags with sensing capabilities. This new generation of RFID devices – supporting sensing, computation, and RFID communication - can facilitate numerous promising applications for ubiquitous sensing and computation. They also suggest new ways of providing security and privacy for RFID systems by utilizing the unique characteristics of sensor data and sensing technologies. In this chapter, the authors highlight these new possibilities and advocate the use of sensing-enabled RFID tags in security-critical applications. The purpose of this chapter is to bring awareness of these opportunities to both the research and industrial community, and to incite interests in sensing-centric security and privacy research and development for the future generation of RFID systems.

Chapter 4

Radio Frequency Identification (RFID) is a challenging wireless technology with a great potential for supporting supply and inventory management. In this chapter the authors consider a particular application in which a group of tagged items are scanned to generate a record of simultaneous presence called a grouping-proof. Grouping-proofs can be used, for instance, to guarantee that drugs are shipped (or dispensed) accompanied by their corresponding information leaflets, to couple the user's electronic passport with his/her bags, to recognize the presence of groups of individuals and/or equipment and more generally to support the security of supply and inventory systems. Although it is straightforward to design solutions when the verifier is online since it is sufficient for individual tags to authenticate themselves to the verifier, interesting security engineering challenges arise when the trusted server (or verifier) is not online during the scan activity. So, the field of grouping-proofs is very active, and many works have been published so far. This chapter details the setting for RFID grouping-proofs and discuss the threat model for such applications. The authors analyze some of the grouping-proofs proposed in the literature describing their advantages and disadvantages. Then, general guidelines for designing secure grouping-proofs are proposed. Finally, some examples of grouping-proofs that are provably secure in a strong security framework are presented.

Chapter 5

Michael Hutter, Institute for Applied Information Processing and Communications (IAIK),
Graz University of Technology, Austria
Erich Wenger, Institute for Applied Information Processing and Communications (IAIK),
Graz University of Technology, Austria
Markus Pelnar, Institute for Applied Information Processing and Communications (IAIK),
Graz University of Technology, Austria
Christian Pendl, Institute for Applied Information Processing and Communications (IAIK),
Graz University of Technology, Austria

In this chapter, the authors explore the feasibility of Elliptic Curve Cryptography (ECC) on Wireless Identification and Sensing Platforms (WISPs). ECC is a public-key based cryptographic primitive that has been widely adopted in embedded systems and Wireless Sensor Networks (WSNs). In order to demonstrate the practicability of ECC on such platforms, the authors make use of the passively powered WISP4.1DL UHF tag from Intel Research Seattle. They implemented ECC over 192-bit prime fields and over 191-bit binary extension fields and performed a Montgomery ladder scalar multiplication on WISPs with and without a dedicated hardware multiplier. The investigations show that when running at a frequency of 6.7 MHz, WISP tags that do not support a hardware multiplier need 8.3 seconds and only 1.6 seconds when a hardware multiplier is supported. The binary-field implementation needs about 2 seconds without support of a hardware multiplier. For the WISP, ECC over prime fields provides best performance when a hardware multiplier is available; binary-field based implementations are recommended otherwise. The use of ECC on WISPs allows the realization of different public-key based protocols in order to provide various cryptographic services such as confidentiality, data integrity, non-repudiation, and authentication.

Chapter 6

J. H. Kong, The University of Nottingham Malaysia, Malaysia
L.-M. Ang, Edith Cowan University, Australia
K. P. Seng, The Sunway University, Malaysia

This chapter presents a low complexity processor design for efficient and compact hardware implementation for WISP system security using the involution cipher Anubis algorithm. WISP has scarce resources in terms of hardware and memory, and it is reported that it has 32K of program and 8K of data storage, thus providing sufficient memory for design implementation. The chapter describes Minimal Instruction Set Computer (MISC) processor designs with a flexible architecture and simple hardware components for WISPs. The MISC is able to make use of a small area of the FPGA and provides security programs and features for WISPs. In this chapter, an example application, which is Anubis involution cipher algorithm, is used and proposed to be implemented onto MISC. The proposed MISC hardware architecture for Anubis can be designed and verified using the Handel-C hardware description language and implemented on a Xilinx Spartan-3 FPGA.

Section 3
Applications

Chapter 7

Enamul Hoque, University of Virginia, USA
Robert F. Dickerson, University of Virginia, USA
John A. Stankovic, University of Virginia, USA

This chapter presents a sleep monitoring system based on WISP tags. The authors show that their system accurately infers fine-grained body positions from accelerometer data collected from the WISP tags attached to the sides of a bed. Movements, duration, and bed entrances and exits are also detected by the system. The chapter presents the results of an empirical study from 10 subjects on three different mattresses in controlled experiments to show the accuracy of the inference algorithms. The authors also evaluate the accuracy of the movement detection and body position inference for six nights on one subject, and compare these results with two baseline systems. Preliminary data investigating the correlation between sleep stages from the Zeo and movement is also presented.

Chapter 8

David Parry, Auckland University of Technology, New Zealand
Anne Philpott, Auckland University of Technology, New Zealand
Alan Montefiore, Auckland University of Technology, New Zealand

A systematic literature review of published sources that discuss radio frequency identification technology, ubiquitous health care, and dosage measurement was performed. The results were then critiqued. Methods of storing data and using Radio Frequency Identification (RFID) were studied. These results were used as an aid for developing a prototype system for monitoring medication dosages in a home health care environment. The combination of an RFID technology – the Intel Wireless Sensor Platform (WISPs) and the construction of a specific pill dispensing container in this prototype demonstrated that it is possible to use RFID technology to effectively and ubiquitously monitor and track drug taking compliance. With further refinements on the dispensing unit and optimization of the software this product could be manufactured and released to home care patients to help increase compliance and reduce health related issues. This could form the heart of a modular telecare data collection system. RFID-based devices that can store data in standardized formats may allow incremental development of home telecare systems in an economical fashion.

Chapter 9

Ana V. Alejos, University of Vigo, Spain
Iñigo Cuiñas, University of Vigo, Spain
José Antonio Gay Fernández, University of Vigo, Spain
Manuel García Sánchez, University of Vigo, Spain

Traceability and embedded sensing are analyzed in this chapter by three main approaches: firstly, a Wireless Sensor Network; secondly, a Sensor Area Network; and, finally, a Wireless Identification and Sensing Platform. This chapter presents an introduction to the "RFID F2F" action, and its application to the wine sector briefly describing a wine pilot developed in Spain. The traceability system resulting of the WSN and RFID integration is sketched and concisely described. The current deployment of this pilot is commented. In a second block, this chapter introduces the accomplishment of an RFID tracking through a mesh of individual active radiofrequency (RF) barriers composed by active emitter and receiver nodes/tags that cover only small individual areas. The result is a Sensor Area Network (SAN). Finally, the authors of this chapter discuss the Wireless Identification and Sensing Platform technology. WISP chips have the capabilities of RFID tags – compliant with EPC Class-1 Generation-2 standard – but they also support embedded sensing and computing. WISP technology is shown as the next step forward in the design of pervasive devices. The chapter discusses the main features of the emerging computational RFID technologies.

In this chapter the authors present their information model implemented for one pilot developed in the "RFID from Farm to Fork" (F2F) project which looks for the extension of RFID technologies throughout the complete food chain. They describe the privacy assessment proposed by the European Union that allows the evaluation of the privacy and security impact for a RFID application under study. The main privacy risks have been identified and described by the related EU Directives concerning RFID technology. The authors describe the questionnaire elaborated by the EU to assess the privacy robustness level of a RFID application, and they showcase a real wine pilot deployed in Spain. In this chapter, the authors also examine the privacy risks in the middleware communication with both RFID reader and back-end system. The EPCIS has been the Open Source middleware solution adopted in the F2F project. For the F2F pilot deployed in the wine sector, the authors describe the privacy impact assessment questionnaire designed for this case. Finally, they discuss the threads on the RFID tags, the advantages provided by the WISP technology in this regard and its repercussion on the risk questionnaire.

Preface

We are in the era of ubiquitous computing in which the use and development of Radio Frequency Identification (RFID) is becoming more widespread. RFID systems have three main components: readers, tags, and database. An RFID tag is composed of a small microchip, limited logical functionality, and an antenna. Most common tags are passive and harvest energy from a nearby RFID reader. This energy is used both to energize the chip and send the answer back to the reader request. The tag provides a unique identifier (or an anonymized version of that), which allows the unequivocal identification of the tag holder (i.e. person, animal, or items).

The information provided by the environment may be useful to enrich the identification process and even increase the security level of the system. WISP (Wireless Identification and Sensing Platform) devices go towards this direction and Intel Seattle promotes and leads this avant-garde and challenging project.

WISP chips have the capabilities of passive RFID tags. These tags operate in the Ultra High Frequency Band (UHF) and conform to EPC Class-1 Generation-1 and Generation-2 standards. Apart from this, WISPs support sensing capabilities, being able to measure light intensity, temperature, acceleration, strain, liquid level, and so on. Moreover, these devices support on-board a 16-bit general-purpose microcontroller that provides much higher computing capabilities in comparison to those supported by simple passive RFID tags.

The explosion of using WIPS devices is conditional on the development of secure and privacy-friendly solutions such as occurred in RFID technology. Nevertheless, there are no books dealing with this matter in the literature and this book covers this gap offering the most recent advances of this research area. The book consists of 10 chapters divided into three sections. Section 1 presents WISP fundamentals, EPC Gen-1 and Gen-2 standards, and security and privacy issues linked to this technology. Section 2 includes a comprehensive collection of security proposals covering protocols, primitives, and secure-sensing solutions. Section 3 provides practical and real examples in which the automatic identification (RFID) is combined with the sensors capabilities.

Pedro Peris Lopez
Carlos III University of Madrid, Spain

Julio C. Hernandez-Castro
Portsmouth University, UK

Tieyan Li
Institute for Infocomm Research, Singapore

Acknowledgment

This book was a long project that would not have been possible without the great efforts and time invested by all the contributors. Moreover, the reviewers provided valuable comments that greatly help to improve the quality of this book. Special thanks go to Hannah Abelbeck, IGI Global, for the patient professionalism and continuous support during all the process.

Pedro Peris Lopez
Carlos III University of Madrid, Spain

Julio C. Hernandez-Castro
Portsmouth University, UK

Tieyan Li
Institute for Infocomm Research, Singapore

Section 1
Fundamentals

Chapter 1
Tag Identification Protocols in RFID Systems

Francesca Lonetti
*Istituto di Scienza e Tecnologie dell'Informazione "A. Faedo",
Consiglio Nazionale delle Ricerche, Italy*

Francesca Martelli
Istituto di Informatica e Telematica, Consiglio Nazionale delle Ricerche, Italy

ABSTRACT

Fast and reliable identification of multiple objects that are present at the same time is very important in many applications. A very promising technology for this purpose is Radio Frequency Identification (RFID), which is fast pervading many application fields, like public transportation and ticketing, access control, production control, animal identification, and localization of objects and people. The problem approached in this chapter is the tag identification in RFID systems. This problem occurs when several tags try to answer at the same time to a reader query. If more than one tag answers, their messages will collide on the RF communication channel, and the reader cannot identify these tags. There are two families of protocols for approaching the tag collision problem: a family of probabilistic protocols, and a family of deterministic ones. In this chapter, the authors give an overview of the most important approaches and trends for tag identification in RFID systems and provide the results of a deep comparison of the presented tag identification protocols in terms of complexity and performance.

INTRODUCTION

An RFID system consists of radio frequency (RF) tags attached to objects that need to be identified and one or more networked electromagnetic readers. The great appeal of RFID technology is

that it allows in-formation to be stored and read without requiring either contact or a line of sight between the tag and the reader. For this contactless feature, RFID technology is an attractive alternative to bar code in the distribution industry and supply chain, since it can hold more data. Two kinds of tags are possible in RFID systems: active or passive. Active tags have storage capa-

DOI: 10.4018/978-1-4666-1990-6.ch001

bilities and are provided with power sources for computation and transmission. The complexity and cost of mounting a power source onto a tag, the size/weight of a battery, and the difficulty of recharging it, limit the applicability of active tags: these are not practical for use with disposable consumer products, for instance. Passive tags instead rely only on RF energy induced by the electromagnetic waves emitted by the reader. In a typical communication sequence, the reader emits a continuous radio frequency wave. When a tag enters in the RF field of the reader, it receives energy from the field, for modulating the signal according to its stored data. Continuous advancements in protocols and circuit design, make the reliability and the read range of passive tags RFID systems continuously improving. Besides, their cost continues to decrease, thus leading to an increase of their applications.

In RFID systems, a reader recognizes objects through wireless communications with tags, and it must be able to identify all tags as quickly as possible. However, signals of readers or tags may collide, since a shared wireless channel is used. Collisions are divided into reader collisions and tag collisions. The reader collisions problem *(Engels et al. 2002), (Waldrop et al. 2003)* occurs when two or more readers communicate on the same frequency at the same time (reader-to-reader frequency interference), or when neighboring readers attempt to query the same tag simultaneously (multiple reader-to-tag interference). Frequency interference is avoided by having readers operating on different frequency bands. Multiple reader-to-tag interference can be avoided only by having neighboring readers operating at different times or frequencies. A simple and distributed reader anti-collision protocol is presented in *(Waldrop et al. 2003)*. Reader collisions can be easily solved, since RFID readers, that are entities with abundant memory and computation power, can detect collisions.

The problem approached in this chapter is the tag collision one, that occurs when several tags try to answer at the same time to a reader query. The reader queries the tags for their ID by broadcasting a request message. Upon receiving such message, all tags send an answer back to the reader. If only one tag answers, the reader identifies the tag. If more than one tag answer, their messages will collide on the RF communication channel, and the reader cannot identify these tags. Given the low functional power and energy constraints in each tag, it is unreasonable to assume that tags can communicate with each other directly, and that they can notice their neighboring tags or detect collisions. At the beginning the reader does not know anything about the tags. Each tag $i \in 1, ..., n$ has a unique ID string in $0, 1^k$, where k is the length of the ID string. Tags anti-collision protocols are required for identification in RFID systems, they specify the algorithms for the reader and the tags so that the reader can collect all the tag IDs. There are two families of protocols for approaching the tag collision problem *(Klair et al. 2010)*: a family of probabilistic protocols, and a family of deterministic ones. The first ones are Aloha-based protocols *(Abramson 1970)*; *(Bonuccelli et al. 2006)*; *(Cha & Kim 2006)*; *(EPC standard 2005)*; *(Lee et al. 2005)*; *(Peng et al., 2007)*; *(Schoute 1983)*; *(Vogt 2002)*; *(Wieselthier et al. 1989)*, the last ones are tree-based protocols *(Chiang et al. 2006)*; *(Choi et al. 2005)*; *(Law et al. 2000)*; *(Myung & Lee 2006)*; *(Myung et al. 2006; 2007)*; *(Zhang & Vojcic 2005)*. In addition, there are also hybrid approaches, where randomization is applied in tree schemes *(Hush & Wood 1998)*; *(Micic et al. 2005)*; *(Ryu et al., 2007)*.

A comparative evaluation of probabilistic and deterministic tag anti-collision protocols is provided in *(Choi et al.2007)*. As specified in *(Choi et al.2007)*, probabilistic tag anti-collision schemes are based on ALOHA. ALOHA is one of the basic medium access control mechanisms. In ALOHA, each tag generates a random number and waits for its transmission time according to the number chosen. If the data transmitted by a tag is not interfered by other data, the reader can identify the tag. A tag continues to do the same work after its transmission; generating a new random number and transmitting its own data after waiting for random amount of time. If during the interval two or more tags transmit, a collision occurs. In order to solve partial collision problems, transmission time is divided into discrete time intervals called slots in the slotted ALOHA. According to the number of tag transmissions in a slot, slots can be divided into three types as follow: readable slot in which exactly one tag transmits its data and the reader recognizes a tag successfully; collided slot in which more than two tags transmit their data, a tag collision occurs and the reader cannot recognize any tag; idle slot in which no tag transmits its data. More details about Aloha based protocols are in the following.

As specified in *(Choi et al.2007)*, in deterministic tag anti-collision protocols, a tag determines the point of transmission upon receiving a message from a reader. A reader divides tags into two groups. The process of dividing tags is continued until a group contains only one tag. The fact that the number of tags in that group is one means that a reader can identify the tag successfully. The dividing process of a group is continued until a reader identities all the tags.

For instance, in the query tree (QT) protocol that is a deterministic one, the reader transmits a query to tags. The query contains the prefix of the tag identification (ID) number. Every tag in the range of the reader compares the query received by the reader with its ID number and transmits its ID number to the reader in case the result of comparing is true. This scheme uses the prefix of tag ID number, to divide tags into two groups. Tags in one group transmit their ID number to the reader. Tags in the other group wait for next query of the reader. The content of the query is the identifier of each group. The reader repeats to divide tags into two groups until the number of tags in a group is one. When the number of tags in a group is one, the reader successes to identify one tag. More details about query tree protocols are in the following.

The work in *(Turner, 2003)* tries to answer to the question: Deterministic or Probabilistic, which is best? According to *(Turner, 2003)*, in a probabilistic protocol the operation and nature of the protocol does not absolutely guarantee that every tag in the reader range will be identified but the aim of a good probabilistic protocol is to maximize the probability of every tag being read. Probabilistic protocols operate by distributing tags in time slots to ensure that each tag gets an opportunity to send its identity to the reader. The more uniformly the tags are distributed, the better is the chance that an individual tag message will be heard by the reader.

A deterministic protocol will always be able to read every tag in the reader range, provided that each of the tags has a unique identity number and that all tags remain in the reader's field for the duration of the identification process. In deterministic protocols, the reader prompts any tag matching the response selection criteria. Such systems do not always require the tags to send any data to the reader as the reader can derive the tag ID by successively building up the tag ID one (or more) bits at a time.

In this chapter, we will provide an overview of the most important approaches and trends for tag identification in RFID systems. We first address the Aloha-based protocols and the most used tags estimation functions. Then, we survey the tree based protocols focusing on the query tree ones and we briefly deal with some hybrid approaches. Finally, we will propose a comparison of some of the presented protocols in terms of their complexity and performance.

WISP TECHNOLOGY

WISP (Wireless Identification and Sensing Platform) is a wireless, battery-free platform for sensing and computation that is powered and read by a standard compliant Ultra-High Frequency (UHF) RFID reader.

The WISP device includes a general-purpose fully programmable ultra-low-power 16-bit flash microcontroller and implements the bi-directional communication primitives required by the Electronic Product Code (EPC) RFID standard. To the reader, the WISP appears to be an ordinary RFID tag *(Smith et al. 2006)(WISP Wiki)*. Moreover, WISPs tags also support sensing and computing and have been used to sense quantities such as light, temperature, acceleration, strain, liquid level.

Even without their sensing capabilities, WISPs can also be used as open and programmable RFID tags: they have been used to investigate embedded security, in *(Philipose et al. 2005)* the authors show an implementation of the RC5 encryption algorithm on WISPs.

WISPs are emerging for their attractive features dealing with ubiquitous sensing. Alternative approaches to ubiquitous sensing include wired sensors that are well-suited to create purpose-built instrumented environments supporting long-term observation. This approach has the advantage that there is no battery lifetime or battery size constraint but the drawback is that it needs for wires. A second approach, favored by the Wireless Sensor Networks community, is to use battery powered devices that communicate by ordinary radio communication, often in a peer-to-peer way. One disadvantage of this approach is the size and lifetime constraints imposed by batteries. Using WISPs tags it is possible to deliberately transmit power from a large source device to the sensor platforms, which then harvest this "planted" power.

There are some applications using WISP platforms as the first accelerometer to be powered and read wirelessly in the UHF band *(Yeager et al. 2008)*. However, the WISP attractive features for ubiquitous sensing, have not yet been explored very thoroughly. WISPs present some advantages with respect to conventional RFID tags. Specifically, conventional RFID tags are fixed function devices that typically use a minimal, nonprogrammable state machine to report a hard-coded ID when energized by a reader. The Electronic Product Code (EPC) standard operates in the Ultra-High Frequency (UHF) band (915MHz in the U.S.), which has substantially improved the range and field-of-view for RFID reading over previous generations of RFID technology. The "EPC Class 1 Generation 1" specification ("Gen1") was the first UHF RFID standard to be widely deployed. The standard's broad adoption enabled a new generation of applications and interoperable products. This standard has been supplanted by a second generation of the specification, "EPC Class 1 Generation 2".

RFIDs have been used for sensing in several contexts. Conventional short range HF RFID tags with a worn RFID reader have been used for activity monitoring in eldercare scenarios. Integrating RFID tags with secondary sensors has been proposed or implemented in various contexts, and a small number of commercially available RFID sensors exist. In almost all cases, these devices are fixed-function, and simply report a unique ID and sensor data. Most of the commercially available products are "active tags," meaning that the sensor platform is battery-powered; they use the RFID channel for communication but not for getting power *(Yeager et al. 2008)*. Also, existing RFID-sensor devices are generally not programmable platforms supporting arbitrary computation. WISPs represent an alternative design philosophy that focuses on the programmability of a full microcontroller as part of a sensor-enhanced, passive RFID device.

One commercially available fully programmable microcontroller with an RFID interface is described in *(WISP Product)*. However, this device can only transmit one bit of sensor data per read event, and operates at 125kHz, which limits its range to inches.

In *(Philipose et al. 2005)*, the authors present the first WISP that implements the Gen 1 protocol on its own, without using a commercial RFID tag. This was the first WISP to use a single antenna for power harvesting, reader-to-WISP data downlink, and WISP-to-reader data uplink. Like its predecessors, this WISP also encoded the sensor data in the EPC ID, but was able to control all bits of the ID.

ALOHA-BASED PROTOCOLS

Probabilistic protocols are mainly based on Aloha protocol *(Abramson 1970)* and its variants *(Schoute 1983)*. These protocols do not eliminate collisions, but they reduce the probability of their occurrences, since tags try to answer at randomly generated distinct times.

Pure Aloha

In pure Aloha *(Abramson 1970)* protocol, each tag transmits its ID randomly and waits for the reader answer: if it receives an ACK, then it has been recognized, otherwise a collision occurred, it receives a NACK from the reader and then it waits a random backoff interval before trying to transmit its ID again.

Figure 1 shows three tags in a reader's interrogation zone.

Systems based on Pure Aloha suffer from the collision problem that reduces the throughput to

18%. To improve the performance of Pure Aloha protocols some optimizations *(Bolic et al. 2010)* presented below have been proposed trying to reduce collisions and improve the reader's read rate.

Pure Aloha with Muting

In this variant, the reader sends a mute command after identifying a tag. After receiving the mute command the identified tag stops responding to future requests. In this way, after every successful identification, the number of tags in the reader's interrogation zone reduces. Figure 2 shows the behavior of Pure Aloha with muting. Initially, there is a collision between tag 1 and tag 2 transmissions. These tags then retransmit after a random delay. When tag 3, for instance, is identified, the reader sends a mute command to the tag.

Pure Aloha with Slow Down

When a tag is identified it is instructed to increase its back off time using a slow down command. In Figure 3 we present the tags behavior if slow down is used or not. At the top of the figure, if slow down is not used, we see that tag 3's reply interferes with the transmission of tag 1 and tag 2. If slow down is used (bottom of the figure),

Figure 1. Reader and tags interactions (Bolic et al. 2010)

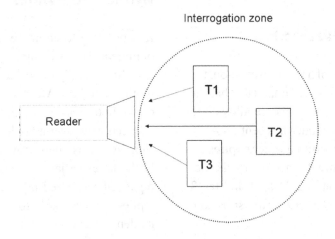

Figure 2. Pure Aloha with muting (Bolic et al. 2010).

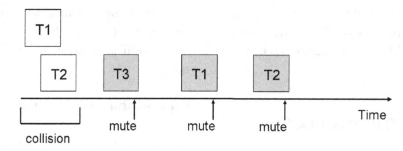

Figure 3. Pure Aloha with slow down (Bolic et al. 2010).

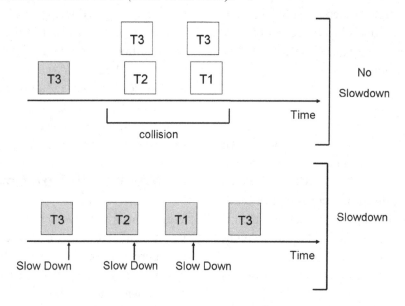

the reader instructs tag 3 to back off for a longer period of time. Consequently, the reader has a higher probability of identifying tag 1 and tag 2 successfully.

Pure Aloha with Fast Mode

After detecting the start of a tag transmission, the reader transmits a "silence" command that has the effect of muting other tags. Tags are allowed to transmit again after the reader has sent an ACK command or until their waiting timer expires.

As we show in Figure 4, once the reader detects a transmission from tag 2, tag 1 and tag 3 are silenced until the tag finishes its transmission.

Otherwise, these tags would have collided with tag 2's reply.

Hybrids Pure Aloha

It is possible to combine the above Pure Aloha optimizations to create other Pure Aloha variants. Specifically, Pure Aloha with fast mode and muting, and Pure Aloha with fast mode and slow down. In the former variant, tags are temporarily silenced whenever a tag has started its transmission. A tag is then muted after identification. In the latter variant, instead of muting tags, the replay of identified tags is slowed so that their replies are less likely to collide with those from unidentified tags.

Figure 4. Pure Aloha with fast mode (Bolic et al. 2010).

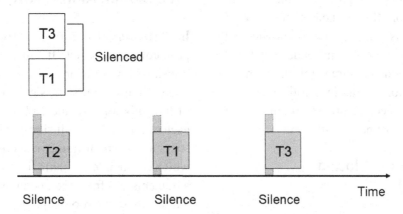

Slotted Aloha

The main problem that limits the reading rate of Pure Aloha systems is partial collisions.

In Slotted Aloha, the time is assumed to be slotted and all tags have a local clock for synchronization. A time slot is a time interval in which tags transmit their ID, and collisions can occur only at slot boundary, i.e. there are not partial collisions. Instead of replying on a continuous timeline, tags are required to respond at pre-defined slots. When a collision occurs, the involved tag waits for a random number of slots before retransmitting. The maximum throughput of Slotted Aloha protocols is around 36%. This performance gain is due to the fact that collisions only occur at the start of each slot as opposed to any time in Pure Aloha.

As in Pure Aloha, there are many optimizations *(Bolic et al. 2010):*

- **Slotted Aloha with Muting or Slow Down**. This variant has the same operating principle as Pure Aloha with muting or slow down, but it operates in a slotted manner.
- **Slotted Aloha with Early End**. A reader closes a slot if it does not detect any transmission at the beginning of a slot. Two commands are used: start-of-frame (SOF) and end-of-frame (EOF). The former is

used to start a reading cycle, and the latter is used by the reader to close an idle slot early. As a result, tags can transmit sooner, leading to a higher reading rate. Moreover, a reader is able to conserve energy as it can start receiving the next reply sooner.

- **Slotted Aloha with Early End and Muting**. After identifying a tag, the reader sends a mute command to the tag, removing the tag from contending in subsequent slots. In addition, idles slots are closed early using EOF command.
- **Slotted Aloha with Slow Down and Early End**. This variant combines slow down with the early end feature. In other words, as well as closing idle slots, identified tags are instructed to slow their replies.

Basic Framed Slotted Aloha

To further improve the Slotted Aloha performance, a variant has been introduced *(Schoute 1983),* that we call Basic Framed Slotted Aloha (BFSA): slots are arranged in frames, and a tag can select only one slot in a frame for its transmission. The frame length is set, known by all tags, and remains unchanged for the entire protocol execution. Each tag randomly selects a slot in a frame for its transmission: if there is no collision, the reader sends an acknowledgement to the tag, which becomes silent until the end of the protocol; otherwise, in case of collision or idle slot (i.e. slot not selected by any tag) no ack is sent, so all non-acknowledged tags know that the protocol is not ended yet. Read cycles are repeated for unrecognized (namely, whose transmission results in a collision) tags, until all tags have been identified. So, in the last read cycle there must be no collision.

Dynamic Framed Slotted Aloha (DFSA)

The Dynamic Framed Slotted Aloha (DFSA) *(Vogt 2002)* protocol changes the frame size dynamically. As BFSA, DFSA operates in multiple rounds, with the difference that at the end of each round the reader estimates the number of participating tags, according to the Chebyshev's estimation function (see below) for deciding a proper size of the next transmission frame.

The constraint of this protocol is that the frame size cannot be increased indefinitely as the number of tags increases, but it has an upper bound. This implies a very high number of collisions when the number of tags exceeds the maximum admitted frame size. An Advanced Framed Slotted Aloha (AFSA) protocol, analyzed in *(Lee et al. 2005)*, overcomes such a problem by dividing the unread tags into a number of groups, and interrogating each group separately. The performance of DFSA protocol is improved by the Variant Enhanced Dynamic Framed Slotted Aloha (EDFSA) protocol *(Peng et al. 2007)*, where a dynamic approach for group dividing is adopted.

The frame size affects the performance of Aloha based algorithms. A small frame size results in many collisions, and so increases the required total number of slots when the number of tags is high. In contrast, a large frame size may result in more idle time slots, when the number of tags is small. Different methods estimating the number of unread tags, so allowing the reader to choose an optimal frame size for the next read cycle, are presented in *(Cha & Kim 2005; 2006)*.

Tree Slotted Aloha (TSA)

In *(Bonuccelli et al. 2006)*, the Tree Slotted Aloha protocol is proposed. It aims at reducing tag transmission collisions by querying only those tags colliding in the same slot of a previous frame of transmissions. At the end of each frame, for each slot in which a collision occurred, the reader starts a new small frame, reserved to those tags which collided in the same time slot. In this way, a transmission frame can be viewed as a node in a tree, where the root is the initial frame; leaves represent frames where no collision occurred. The behaviour of TSA protocol is illustrated in Figure 5. To establish the size of "child" frames, TSA uses Chebyshev's estimation function in a way similar to *(Vogt 2002)*. More precisely, consider a level l in the tree, let n_i be the estimated number of transmitting tags in frame i. Then, if $c_1{}^i$ is the number of identified tags and $c_k{}^i$ is the number of collision slots in frame i, the expected number of transmitting tags in each of the $c_k{}^i$ child frames at level $l + 1$ is given by

$$\frac{n^i - c_1^i}{c_k^i}$$

Dynamic Tree Slotted Aloha (DTSA)

When the number of tags to be identified is very high (of the order of thousands) and the initial frame size is set too small with respect to the actual number of tags, the estimation function used by TSA may not define proper frame sizes. This is because the n which minimizes the Chebyshev's inequality is searched in the range $[c_1 + 2c_k, 2(c_1 + 2c_k)]$, and when $c_1 = 0$ the upper limit $4c_k$ may be insufficient to capture the real number of transmitting tags. This problem has been approached in *(Maselli et al. 2008)*. The improvement to TSA is in considering not only the outcomes of frames at a given level, but exploiting also the knowledge

Figure 5. Example of TSA protocol execution

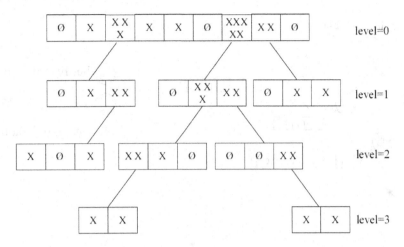

acquired during all previously completed frames. Based on the assumption that the slot allocation of each tag is independent from those ones of other tags, then the number X of tags falling in a given slot r is given by the binomial distribution

$$P\{X = r\} = \binom{n}{r}\left(\frac{1}{N}\right)^r\left(1 - \frac{1}{N}\right)^{n-r}$$

then the expected number of tags in a slot is given by

$$E[X] = \frac{n}{N}$$

At the end of level l, the size of each child frame in level $l + 1$ is computed as in TSA; when the first child frame has been completed, the knowledge on the number of tags transmitting in such a frame can be used to refine the size of the remaining frames of that level. When also the second frame ended, the reader could further refine the estimation of the number of tags going to transmit in subsequent frames, and so on. More precisely, the size of a frame i at level l is estimated as

$$S_i = \frac{1}{i-1}\sum_{j=1}^{i-1} t_j$$

where t_j is the number of tags that participated in the frame j at level l; in other words, the size of frame S_i is estimated as the mean value of tags that participated to sibling frames previous to i. As TSA proceeds in a depth-first order, at the same level, a frame is executed after all collisions in previous sibling frames have been resolved, so it is possible to recursively apply the estimation method on deeper levels of the tree.

Binary Splitting and Tree Slotted Aloha (BSTSA)

In *(La Porta et al. 2011)*, a protocol which combines the strengths of TSA with those of binary splitting protocol (BS) is proposed. BS is used to accurately estimate the number of tags, then TSA is used to identify them. In particular (see Figure 6), in the first phase, the protocol splits tags into groups by applying binary splitting, until a single-tag group is obtained, i.e. the leftmost leaf of tree is reached. When BS ends, the protocol starts identifying the right siblings on the tree, beginning from the right sibling r_i of the leftmost leaf l_i. Since in BS tags generate

Figure 6. Binary splitting and tree slotted Aloha protocol

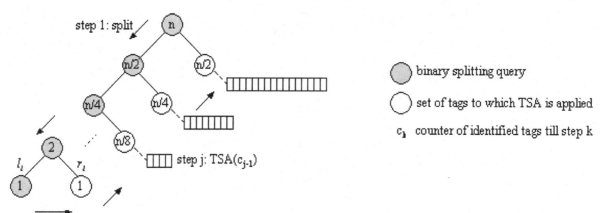

a random binary number for splitting into two groups, at each tree level, each right-side group has approximately the same size of the left-side group which approximately contains half of the tags of the parent node (say $l_i - 1$). This knowledge is used to properly size the TSA frame to identify the tags in group r_i. Once such identification is ended, the sum of the number of tags identified in the sibling child nodes r_i and l_i can be used to initialize the size of the TSA frame to identify tags in the right sibling of the parent node $r_i - 1$. The time efficiency (which is different from system efficiency, see below) of this protocol is about 75% in case of RFID networks of small size (up to 500 tags), reaching 80% for larger networks (from 500 to 5000 tags), with a gain over TSA and DTSA of about 15% *(La Porta et al. 2011)*.

Estimation Functions

Almost all protocols presented in this section use, or could use (for improving performance) a tags estimation function. We present here some of them. These functions aim at computing the real number of tags involved in the identification process, based on the outcomes of the complete frames. In particular, at each reading cycle, we obtain a triple $<c_0, c_1, c_k>$ quantifying the empty slots, slots in which exactly one tag transmitted its ID, and slots with collisions, respectively.

The simplest estimation function is given by

$$c_1 + 2c_k \tag{1}$$

since, for each collision, at least two tags have been involved *(Vogt 2002)*. This can be considered as a lower bound on the tag number.

Another estimation function is based on Chebyshev's inequality which asserts that the outcome of a random experiment involving a random variable X, is most likely somewhere near the expected value of X. Using this property it is possible to compute the distance between the effective result $< c0, c1, ck >$ and the expected result $< a0, a1, ak >$ of a reading cycle. By minimizing such a distance, defined in the following equation, it is possible to estimate the number n of tags transmitting in such a cycle.

$$\in \left(N, c_0, c_1, c_{k0}\right) = \min_n \left\| \begin{pmatrix} a_0^{N,n} \\ a_1^{N,n} \\ a_{\geq 2}^{N,n} \end{pmatrix} - \begin{pmatrix} c_0 \\ c_1 \\ c_k \end{pmatrix} \right\| \tag{2}$$

The triple $<a_0, a_1, a_k>$ entries indicate the expected number of empty slots, slots filled with one tag, and slots with collision, respectively. N

and n denote the frame size in the reading cycle and the number of tags, respectively. When the reader uses a frame size equal to N, and the number of responding tags is n, the expected value of the number of slots with r responding tags is given by

$$a_r^{N,n} = N \times \binom{n}{r}\left(\frac{1}{N}\right)^r \left(1 - \frac{1}{N}\right)^{n-r} \qquad (3)$$

Then it is needed to establish a range in which the value in equation (2) is searched. In *(Bonuccelli et al. 2006)*, n is searched in the range *[c_1 + 2c_k, ..., 2(c_1 + 2c_k)]*, i.e. between the lower bound of equation (1) and as upper bound 2(c_1 + 2c_k), since by simulation they saw that with too higher values no further accuracy is achieved in the estimation.

In *(Kodialam & Nandagopal 2006)* two estimators are developed: a zero estimator (ZE) and a collision estimator (CE), obtained by setting $r = 0$ and $r = 1$ in equation (3), respectively:

$$ZE = e^{-\left(\frac{n_0}{N}\right)} = \frac{c_0}{N}$$

$$CE = 1 - \left(1 + \frac{n_k}{N}\right)e^{-\left(\frac{n_k}{N}\right)} = \frac{c_k}{N}$$

where c_0 and c_k are used to determine n_0 and n_k. If $n_0 < n_k$ then the estimation is equal to n_0, otherwise it is n_k.

In *(EPC standard 2005)*, tag estimation is performed in the following way. All frames have size 2^Q, and after each collision slot Q is increased by a constant c, $0.1 \le c \le 0.5$. After an idle slot, Q is decremented by c, while successful slots do not change Q. Since a non integer constant is added to Q, by considering the integer part only, after a certain number of collision (idle) slots, Q will become $Q + 1$ ($Q-1$).

TREE-BASED PROTOCOLS

Aloha-based protocols can not perfectly prevent collisions. In addition, they have the serious problem that a tag may not be identified for an unlimited time (the so called *tag starvation problem*). Tree-based tag anti-collision protocols have a longer identification delay than Slotted Aloha ones, but they are able to avoid the tag starvation.

Tree-based protocols can be classified into the following categories: Binary search, Query Tree and Tree Splitting.

Binary Search Protocols

In binary search protocol *(Capetanakis 1979) (Center 2003)*, the reader performs identification by recursively splitting the set of answering tags. Each tag has a counter initially set to zero. Only tags with counter set to zero answer the reader's queries. After each tag transmission, the reader notifies the outcome of the query: collision, identification, or no-answer.

When tag collision occurs, each tag with counter set to zero, adds a random binary number to its counter. The other tags increase by one their counters. In such a way, the set of answering tags is randomly split into two subsets. After a no-collision transmission, all tags decrease their counters by one.

An improvement of the binary search protocol has been proposed in *(Zhang & Vojcic 2005)*.

In this paper, binary search protocol has been modified in order to take advantage of signal subtraction when queries are issued on a set of tags, and later on one subset of such tags. More specifically, in that protocol named BSIC, when a query leads to a collision, the received signal is stored in the reader. Then, the reader identifies the most significant bit with collision, say the *j-th*. This means that the first *j-1* bits of all answering tags are identical. Then, it issues another query for those tags with ID not larger than a threshold with the first *j-1* bits equal to those received, the *j-th* equal to 0, and all the other bits equal to 1. The new received signal is then subtracted from the previous one, for getting the signal that all the tags with ID larger than the threshold would have generated, without asking them to transmit. This saves half of the queries the usual binary search protocol performs. In *(Zhang & Vojcic 2005)*, it is formally proved that the protocol performance, in terms of number of reader or tags transmission, is always twice the number of tags to be identified. Considering the above description, tags are required to transmit only the bits from *j+1* to the less significant, instead of all the ID: the most significant bits are indeed already known.

Query Tree Protocols

The idea of splitting tags having a tree structure has been taken up into the Query Tree (QT) protocols *(Law et al. 2000)* consisting of round of queries and responses.

In each round, the reader asks the tags to answer if their ID matches a given prefix. If more than one tag answer, the reader knows that there are at least two tags having that prefix. The reader then appends symbol 0 or 1 to the prefix and continues to query for longer prefixes.

Table 1. Communication between the reader and the tags with the QT protocol

Step	Query	Response
1	0	collision
2	1	collision
3	00	collision
4	01	no response
5	10	101
6	11	110
7	000	000
8	001	001

Figure 7. The query tree for the example of Table 1

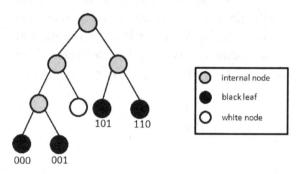

When a prefix matches a tag uniquely, the tag can be identified. Once a tag is identified, the reader starts a new round of queries with another prefix. As an example, in Table 1 we show the communication between the reader and the tags for identifying the tags with the following ID: 000, 001, 101, 110. The query tree for the example of Table 1 is presented in Figure 7. Since tags do not need additional memory except the ID, query tree protocols have the advantage to be *memoryless*, and for this they require low functional and less expensive tags. Moreover, the only computation required for each tag is to match its ID against the binary string in the query. However, since they use prefixes, their performance is sensitive to the distribution of tag IDs which a reader has to identify as we will show in the following of this chapter.

In *(Law et al. 2000)*, many improvements of QT protocol have been presented, for reducing its running time. Among them, the most important is that one that we call Query Tree Improved (QTI) *(Law et al. 2000)*, that avoids the queries that certainly will produce collisions.

Suppose that a query of prefix *"q"* results in a collision and the query of prefix *"q0"*, results in an empty slot, then the reader skips the query prefix *"q1"* and performs directly the queries *"q10"* and *"q11"*. Another improvement proposed in *(Law et al. 2000)*, is the so called "aggressive advancement" (QTAA, for short), in which every internal node of the query tree has four sons: after the query of prefix *"q"* resulting in a collision, the reader does the following queries *"q00"*, *"q01"*, *"q10"*, *"q11"*. A performance evaluation of these protocols is presented below.

In *(Chiang et al. 2006)*, a Prefix-Randomized Query Tree protocol is proposed, where tags randomly choose the prefixes rather than using their ID-based ones. The identification time of this protocol is improved with respect to QT protocol because it is no longer affected by the length and distribution of tag IDs as in QT protocol.

Knowledged Query Tree Protocol

Often, the reader of an RFID system can have an approximate knowledge of the distribution of ID of tags to be identified. In case of Class-1 Generation-2 UHF RFIDs standard *(EPC standard 2005)*, the memory containing the tag identifier is divided into four fields: one for the tag class, one for the designer identity, one for the model number, and one for the tag serial number. So, in this case, tags are logically divided in groups, that are given by the first three fields (product class, designer and model) stored in the memory. So, the IDs belonging to different models, or having different designers, are within specific, non-consecutive ranges. This naturally leads to consider IDs distributions not as a large, single, uniform one, but rather as several blocks of isolated smaller ones. In *(Bonuccelli et al. 2007a)*, a new query tree based protocol named Knowledged Query Tree (KQT) is proposed. It reduces the number of transmission collisions by taking advantage of the knowledge the reader can have about the number of tags to be identified, and the distribution of their IDs.

Specifically, the authors of *(Bonuccelli et al. 2007a)* assume that:

- Each tag $t \in \{0, ..., N-1\}$ has a unique ID string $t_{id} \in \{0, 1\}^k$, where k is the length of ID strings and N_{exp} is an estimation of the number of tags, $g(\cdot)$ is the discrete probability function of tags IDs distribution as-

sumed by the reader and $g(t_{id})$ is the probability assumed by the reader that tag t_{id} is in its reading range;

- The IDs are estimated not smaller than L and not larger than R, where L and R are the extreme points of the interval of IDs in the reader range, with $L, R \in [0..2k-1]$, $L < R$ and the IDs range $[L...R]$ is partitioned into N_{exp} nearly equiprobable intervals;

- The intervals $I = \{ I_1, I_2, ...,I_{Nexp} \}$ are denoted as $I_i = [SP_i, EPi]$, and SP_i and EP_i are such that

$$L = SP_1 \leq EP_1 < SP_2 \leq ... < SP_{Nexp} \leq EP_{Nexp} = R.$$

Notice that $SP_i + 1$ can be larger than $EP_i + 1$. In order to make such intervals equiprobable, it must be:

$$\sum_{j=SP_i}^{EP_i} g(t_j) \geq \frac{1}{N_{exp}}$$

where EP_i is the smallest value for which the above inequality holds, for each EP_i, $1 \leq i \leq N_{exp}^{1}$. In addition, it must be

$$\sum_{j \notin I} g(j) = 0$$

that is the probability of having tags out of the range $[L..R]$ is zero.

A small example explaining the above notation is depicted in Figure 8, where $k = 4$, and the assumed probability function $g(\cdot)$ is the uniform distribution. Black nodes represent tags actually present in the reader range and, together with grey nodes, represent the range of possible tags. White nodes represent tags having probability zero of being in the reader range.

In the KQT protocol, the reader works in cycles. At the beginning of the identification process, the reader uses the estimation (N_{exp}) of the number of tags to be identified, and the probability distribu-

Figure 8. Example of KQT IDs distribution

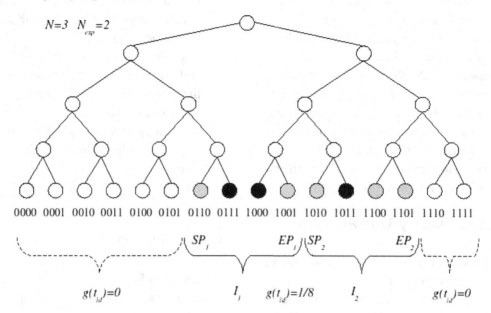

tion function $g(\cdot)$, for dividing the range of possible IDs in N_{exp} equiprobable intervals.

Then, the reader starts the identification process N_{exp} cycles long, one for each interval I_i. At cycle i the reader broadcasts SP_i, EP_i and a command *activation*. All tags having their own ID between SP_i and EP_i become active and answer with their complete ID; otherwise they remain silent. If only one tag answers, the reader identifies it, and starts the cycle $i + 1$. Instead, if a collision occurred, the reader computes

$$h = \min_{k} \left\{ k \in N \, such \, that \, 2k \geq (EP_i - SP_i + 1) \right\}$$

and broadcasts a *subtract* command followed by SP_i and h. Active tags compute the last h bits of the difference between their ID and SP_i, i. e. tag t computes

$$x_i^h = (t_{id} - SP_i)$$

and extracts the last h bits. At this point, the reader starts a procedure which is identical to the classical QT protocol, by querying prefixes at most h bits long. When all tags with IDs between SP_i and

EP_i have been identified, the reader starts the next cycle. An example of application of this procedure is shown in Figure 9, where the mini-tree is that one generated for solving collision for interval I_1 of Figure 8. If the reader assumptions on $g(\cdot)$ and on N_{exp} are exact, then there will be one tag to be identified in each of the above intervals.

Figure 9. Example of KQT reader procedure working

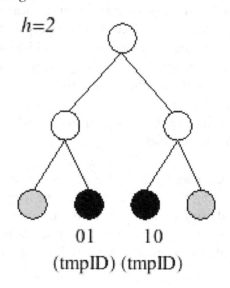

The key idea of the KQT protocol is avoiding repeated collisions when tag IDs are concentrated in particular intervals of the IDs universe. In such a case, classical QT protocols have to visit the whole universe tree, from the root down to the leaves. Instead, KQT manages collisions in a more efficient way, since it limits the number of responding tags. In addition, KQT does not degrade its performance when the number of bits assigned to each ID grows, since by dividing them in intervals, only the last h bits are considered, independently of the total length of IDs. On the other hand, with respect to classical QT protocols, in KQT tags hardware is a little more complicated, since they have to compute comparison and subtraction, but this is acceptable since the gain in the identification process is high enough, as we will show in the following.

Signal Strength Detection

A new efficient approach for tag identification in RFID system has been presented in *(Bonuccelli et al. 2009)*. This work makes the following assumptions about signals transmitted by multiple entities:

- The receiver performs the sum of the signals, with no loss of information;
- The signal corresponding to a bit equal to 0 is the complement of that corresponding to a bit equal to 1 (this is what happens, for instance, when a Manchester bit coding is used). Such signals are called "-1" and "+1", respectively. For instance, the simultaneous transmission of four tags with IDs 011010, 010101, 110001, 011111 will be received by the reader as "-2 4 0 0 0 2". From the collided tags answers, a reader can detect the exact number of transmitting tags, provided that all such tags have transmitted the same bit (say "1") in the same position (say, the second one). If the first tag has "0" in second position, instead

of a "1", the signals received by the reader would be "-2 2 0 0 0 2". This sort of "capture effect" prevents from exactly knowing the number of answering tags;
- With a high probability, when there are multiple answering tags, the received signal is not only formed by "+1" and "-1". This happens only when the number of "0" bits and that of "1" bits differ by exactly one. In this case, we have a "false positive", namely we identify one tag that is not present in our set, whose ID is given by the sum of the IDs of the colliding tags. Notice that tag IDs are 48, 96, ..., bits long, and so the probability of this "false positive" is extremely low.

The protocol, called P, works as follows. Initially, the reader asks all tags to transmit their own complete ID. From the answer, which is the sum of all IDs, the reader starts to divide the set of tags in a recursive way until all tags are identified. The partitioning is done in the following manner: the reader chooses the first bit position, say p_1, for which the answer presents the maximum absolute value, say m_1. Two cases can occur:

- $m1 = 1$: In this case, a tag ID is recognized;
- $m_1 > 1$: In this case, the reader searches the first position, say p_2, where a value, say m_2, equal

to the minimum absolute value is present in the previous answer. Then, it issues a new query by asking all tags having "1" in position p_2, with the same sign of m_2. Then, the reader repeats the above procedure with the outcome of the last query.

When a tag ID is recognized, the reader obtains another tag ID or the sum of more than one tag ID, by subtracting the recognized ID from the previous answer, so without issuing any query. More specifically, when the reader issues a query to currently active tags, some of them answer to the query, while the sum of IDs of the others is obtained by difference. By recursively applying the previous procedure, the reader always splits the responding tags into two groups: one having "+1" in a given position, the other having "-1" in the same position.

Since any two tag IDs differ in at least one bit position, the protocol converges to a complete identification of all tags.

In Figure 10, we show the execution of P protocol on a set of seven tags. When the reader issues a query, it transmits a string of bits representing the positions of all m_2 issued so far and the corresponding values; this string is initially empty (empty string is coded with Ø).

The queries issued by the reader are represented with bold arrows between father nodes and left son nodes. Arrows between father nodes and right son nodes, instead, represent IDs' sums obtained by difference between the previous and the actual answers. The execution of P protocol is equivalent to a depth first search (DFS) on the tree, where the leaves represent the IDs of the tags to be identified.

Tree Splitting Protocols

Tree Splitting protocols operate by splitting responding tags into multiple subsets using a random number generator.

Myung et al.(*Myung & Lee 2006); (Myung et al. 2006; 2007)* proposed two adaptive tag anti-collision protocols: the Adaptive Query Splitting protocol, which is an improvement of the Query Tree protocol, and the Adaptive Binary Splitting protocol, which is based on the Binary Tree protocol. For reducing collisions, the proposed protocols use information obtained from the last process of tag identification, assuming that in most object tracking and monitoring applications the set of RFID tags encountered in successive reading cycles from a reader, does not change substantially, and information from a reading cycle can be used for the next one.

In particular, the Adaptive Binary Tree Splitting protocol *(Myung et al. 2006)*, achieves a more efficient identification by reducing not only collisions but also unnecessary idle slots. In this protocol, tags that can be in the transmit or wait state, have two counters named Progressed Slot

Figure 10. Example of P protocol execution

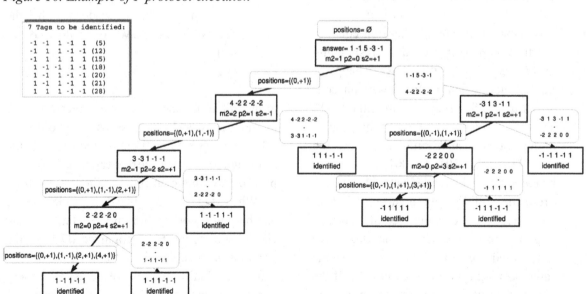

Counter (PSC) and Allocated Slot Counter (ASC) respectively. The former is incremented by one whenever the reader successfully identifies a tag, the latter specifies a tag's transmitting timeslot. A tag can transmit only when its ASC and PSC are equal. The reader informs tags about the read result of the last timeslot. If there was a collision, tags in the transmit state or collided tags select a random binary number and add it to their current ASC. For no response or idle slots, tags in the wait state decrement their ASC by one. Finally, if there was only a single response, tags in the wait state increment their PSC by one. Once all tags are identified, the reader ends the reading process using a terminating slot counter (TSC). The value of TSC is updated after each timeslot as follows: if there was a collision, the reader increments TSC by one, for an idle slot, the reader reduces TSC by one, and for a slot with a single response, TSC is left unchanged. If PCS becomes greater than TSC, the reader terminates the reading process. After all tags are identified, the reader and tags preserve their TSC and ASC value. If a new tag is added to the tag set, it is allowed to choose an ASC value ranging from zero to TSC. If two tags have the same ASC value then they are split into two subsets by generating a unique random binary number. Moreover, if a tag departs from the reader's interrogation zone, tags in the wait state decrement their ASC and TSC by one to eliminate idle slots.

HYBRID PROTOCOLS

(Alotaibi et al. 2010) proposed a Query Tree Aloha protocol, which is a combination of the QT technique and the FSA protocol. QT is used to group the responding tags according to their ID prefix, while FSA is adopted for allowing the tags matching the prefix to respond in randomly chosen time slots. Another hybrid approach is that presented in (*Al-Medhwahi et al. 2010*): in the proposed protocol, each round of identification is divided into two stages: in the first one, FSA is used by setting a frame size equal to the number of tags identified in the previous identification round; at the end of the frame, unidentified tags enter in the second stage in which identification is completed by applying binary search.

(Micic et al. 2005) also proposed an hybrid protocol: it is based on an n-ary partition of competing tags, together with the application of a modified binary protocol, where n is the number of tags to be identified, which is assumed known. More precisely, at each step, each tag transmits with probability $1/n$. If collision occurs, the modified binary protocol is applied to colliding tags, until all of them have been identified. Then, n is updated and a new cycle is started. System efficiency of this protocol is about 46%. However, in *(Micic et al. 2005),* an evaluation of the protocol behaviour when the value n used in the protocol is over/under-estimated is not given.

EPCGLOBAL

The EPCglobal Class-1 Generation-2 UHF RFID standard (*EPC standard 2008*) was developed by EPCglobal Inc. in collaboration with the Auto- ID. The goal of this project is to establish an RFID platform based on the needs of industrial supply chain management and enterprise resource planning.

The standard EPCglobal procedure for identification is a Framed Slotted Aloha based protocol. In this protocol, the reader first communicates to tags a number Q, then each tag generates a random integer number, say x, in the range $[0..2^Q-1]$ and it will transmit its identity in slot x. In the following, we give a detailed description of the inventory operation as defined in the standard EPCglobal.

In the standard, three different processes are defined:

- The SELECT operation, in which the reader establishes a subset of tags for communication;
- The INVENTORY operation, which is the identification process;
- The ACCESS operation, in which the reader can read and/or write information from/to the tag memory.

In this chapter, we are interested in the inventory process. During this operation, a tag can be in one of the following states:

- Ready: the tag is ready for participating at the inventory process; it is the initial state;
- Arbitrate: in this state, the tag is involved in the inventory process, but it does not have to transmit since its slot counter did not reach the zero value; every time a tag in this state receives a QueryRep command from the reader, it decreases its slot counter; when the counter reaches zero, the tag goes in the Replay state;
- Replay: when a tag enters this state, it has to transmit a new RN16; after that, if it receives a valid ACK from the reader, it goes in the Acknowledged state; otherwise, if it receives a non valid ACK, or a valid ACK but with a RN16 different from that it sent before, it returns to the Arbitrate state;
- Acknowledged: a tag reaches this state when the reader has identified it.

The reader (called Interrogator in the standard) can use the following commands during the inventory process:

- Query: used by the reader to initiate the inventory round; it includes the following fields:
  - Command-code which identifies the command;
  - DR to set the tag-to-reader link frequency;
  - M (cycles per symbol) to set the tag-to-reader data rate and modulation format;
  - TRext which is a flag indicating if the tag-to-reader preamble is prepended with a pilot tone;
  - Sel is a 2-bit field which is used to select the tags that must partecipate to the inventory process (all or those ones having/not having the SL flag set);
  - Session: identifies the session id for the inventory round;
  - Target selects whether Tags whose inventoried flag is A or B participate in the inventory round. Once a tag is recognized, it may change its inventoried flag from A to B (or vice versa);
  - Q sets the number of slots in the round; it is a integer number in the range (0,15);
  - Besides, a CRC-5 is added to protect the Query command and it is calculated over the first command-code bit to the last Q bit. A tag receiving a Query with an erroneous CRC-5 it shall ignore the command.
- QueryAdjust: used to modify (increase or decrease) the Q parameter; it includes the following fields:
  - Session: it has the same meaning of the corresponding parameter in the Query command; a tag receiving a QueryAdjust whose session number that does not match that one received in the Query, should ignore the command;
  - UpDn specifies whether and how tags must adjust the Q parameter; it can be:
    - 110: Increment Q (i.e. $Q = Q + 1$);
    - 000: No change;
    - 011: Decrement Q (i.e. $Q = Q - 1$).

A tag, whose Q value is 15, if it receives a QueryAdjust with UpDn = 110 it shall change Up- Dn to 000 before executing the command; likewise, a Tag, whose Q value is 0, if it receives a QueryAdjust with UpDn = 011 it shall change UpDn to 000 before executing the command. The suggested (but not mandatory) algorithm for choosing the appropriate Q value is shown in Figure 11. In the figure, Q_{fp} represents the floating-point representation of Q, and C is a floating point number, whose typical values are $0.1 < C < 0.5$: the reader uses small values of C when Q is large, and larger values of C when Q is small. Based on the number of received tag responses, the reader decides if to increase, decrease or maintain unchanged the Q value.

- QueryRep: used for inducing the tags to decrement their slot counters; after that, if slot = 0, they backscatter an RN16 to the reader. It includes only the Session field seen before. A tag receiving a QueryRep whose session field differs from that one received in the previous Query command shall ignore it. In practice, this command is used to notify the beginning of a new slot

for transmission. Upon the reception of this command, tags decrement their slot counters and, if equal to zero, they will respond with a newly generated RN16, otherwise they remain silent.

- ACK: used by the reader to acknowledge a single Tag; after the 2-bit command-code, it includes the RN16 previously backscattered by a tag. A tag receiving this command with the RN16 it has sent before, will replay with its entire ID and move from the replay state to the acknowledged state. A tag receiving an ACK with an incorrect RN16 shall return to arbitrate state without responding;

- NAK: used by the reader at any instant of the inventory round; a tag receiving a NAK should return to the arbitrate state, without any reply.

The inventory process starts with a Query command, containing the Q parameter. Upon the reception of this command, each tag generates a random value belonging to the interval $[0..2^Q-1]$ and set the counter to that value. Each received QueryRep command induces the tags to decrement

Figure 11. Reader algorithm for adjusting the Q parameter

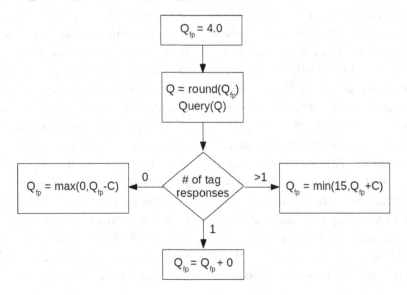

their slot counters. When the counter reaches 0, the tag will move in the Replay state and it will answer to a QueryRep command by sending the randomly generated RN16.

If only one tag answers by sending its RN16 (i.e. a 16-bit random number), the reader replies with an ACK message containing the same RN16. If the involved tag receives a wrong (or different) RN16, it returns in the Arbitrate state, otherwise it moves in the Acknowledged state and it will remain silent for the rest of the inventory process.

If more than one tag answer simultaneously, it means that two or more tags have generated the same slot counter: in such a case, they do not receive any acknowledgement from the reader, will return in the arbitrate state and will decrease their counter, namely, they set the counter to 77FF. This is done to avoid subsequent collisions among them. It can happen that, even in case of simultaneous transmissions, the reader succeeds in recognizing an RN16 at waveform level: in this case, the reader sends the Ack message with the recognized RN16; the tags that do not match the received RN16 move in the arbitrate state, while the matching tag replies to the reader with its entire ID, namely the PC code, the EPC code and a CRC-16.

The behaviour of the protocol is sketched in Figure 12. In the example, there are three tags to be identified. Initially, all tags are in the Ready state, and when the reader broadcasts the Query command, they extract the Q value and put it in the random generator to obtain a value for setting their slot counter. Suppose that tag B and tag C generate a value equal to zero, while tag A generate a value equal to 2. So, tags B and C send their generated RN16 and move in the Replay state, while tag A move in the Arbitrate state. Then the reader detects that there are multiple tags responding and starts a new slot of transmission by broadcasting a QueryRep command: at this point, tags B and C, since they did not receive the Ack command, set their slot counter to the maximum value (namely 77FF) and move in the

Arbitrate state; tag A instead, only decrements its slot counter. Given that now no tag has the slot counter equal to zero, the reader does not receive any transmission, so it broadcasts again a QueryRep command. At this point, the slot counter of tag A reaches zero, thus it transmits its RN16, and the reader sends back the Ack command with the received RN16. Tag A checks the received command and since it contains the RN16 previously transmitted, backscatters its entire ID and moves in the Acknowledged state. The reader now broadcasts a QueryAdjust command with UpDn field equal to "down" which induces the tags to decrement the Q value and generate a new slot counter. Tag A, which is in the Acknowledged state, will ignore this command, while tags B and C produce 0 and 1 values, respectively. At this point tag B is identified and successively tag C is identified.

PERFORMANCE COMPARISON

One of the main issues in RFID networks is the fast and reliable identification of all tags in the reader range. The reader issues some queries, and tags properly answer. Since the transmission medium is shared, the typical problem to be faced is to avoid or limit the number of tags transmission collisions. A performance comparison of the more important tag anti-collision protocols is very useful for pointing out the best performing features of RFID tag identification protocols, and so for designing new and better protocols. Performance for tags collision problem, is often computed as the ratio between the number of tags to be identified, and the number of queries (or time slots, in case of Slotted Aloha based protocols) performed in the whole identification process. This metric is referred as system efficiency (or S) in the figures below, *(Bonuccelli et al. 2006)*; *(Lee et al. 2005)*. This metric, however, is not very accurate, since it does not take into account the reader communications (which take time).

Figure 12. Example of EPCglobal protocol execution

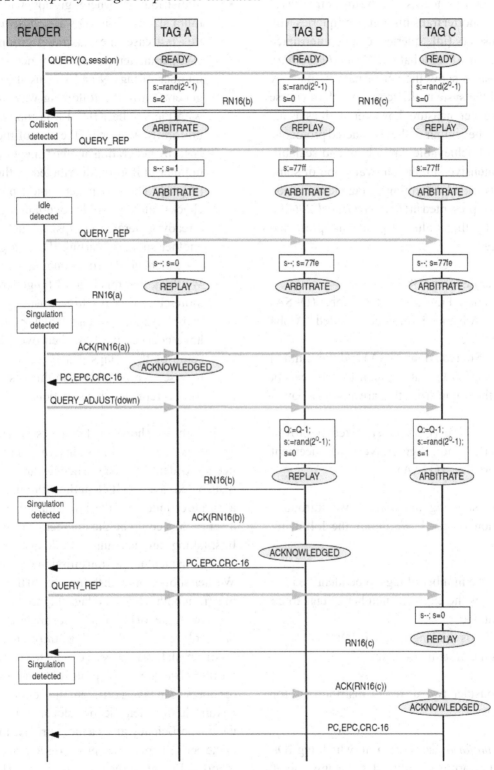

Besides, some protocols, like the query tree based ones, have shorter transmissions, and queries with no tag answers (idle queries) can be interrupted early, like in EPCGlobal (*EPC standard 2005).* We choose to show the performance values by means of the system efficiency, which of course is not an exact measure; however, it gives us the idea of the behaviour of the considered protocols, in terms of achievable throughput and scalability. An intensive simulation evaluation of some probabilistic and deterministic tag identification protocols is presented in (*Bonuccelli et al. 2007b).* Specifically the evaluated protocols, presented above, are:

- Binary Search (BS);
- Dynamic Framed Slotted Aloha (DFSA) and Advanced Framed Slotted Aloha (AFSA);
- Tree Slotted Aloha (TSA) and the Limited TSA (LTSA) that is equal to TSA except for the frame sizes, that are always a power of 2;
- Query Tree (QT), Query Tree Improved (QTI), and Aggressive Advancement Query Tree (QTAA).

Before showing the results, we introduce the notation we used. We define the following notations:

- N is the number of tags to be identified;
- N_{EXP0} is the expected number of tags to be identified;
- $BITS_{ID}$ is the length of tags ID;
- E is the system efficiency.

We consider two different distributions for the tag IDs:

- the *uniform* distribution, in which tag IDs are uniformly distributed in the universe of IDs, namely in the range $[0..2^{BITSID}-1]$

- a distribution where groups with maximum size g of tag ID's are *consecutive*: in this last case, the maximum group size g is a parameter which is a function of the number of tags N and it is used in the ID generation in the following way: together with the smallest ID of a group, which is uniformly generated, the size of the group is set by generating another integer random number uniformly distributed in the range $[0..g-1]$. The group generation procedure checks also possible overlappings with already generated groups: in such a case, another smallest starting ID for a group is generated, until no overlapping is achieved. We think that this kind of ID generation is more close to real settings: usually, a company buys a set of tags which more likely have consecutive IDs, then over the time some of these tags can be removed or a new set can be added, so this distribution tries to replicate this situation.

Figure 13 shows that the best performing protocols are QTI and TSA, in terms of total time needed to identify all tags, when IDs are uniformly distributed, and the ID length is 48 bits. A more accurate estimation of the time required to identify a set of tags could be the number of transmitted bits, taking into account that idle queries take a time equivalent to the transmission of some bits. We then show plots with the number of transmitted bits from both reader and tags. Figures 14 and 15 show the values of R_{bits} and T_{bits} for all the evaluated protocols respectively. In (*Bagnato et al. 2009*), a set of simulations has been conducted to evaluate the time efficiency of the protocols: since each slot (or query) can take a different time depending on it is an idle slot, identification slot or collision slot, the time efficiency gives a more precise measure of the performance of the protocols in practice. We report in Figure 16 the obtained results (EDFSA is the protocol we named AFSA).

Figure 13. S vs N, with $N_{EXP0} = N$, $BITS_{ID} = 48$ and uniformly distributed IDs

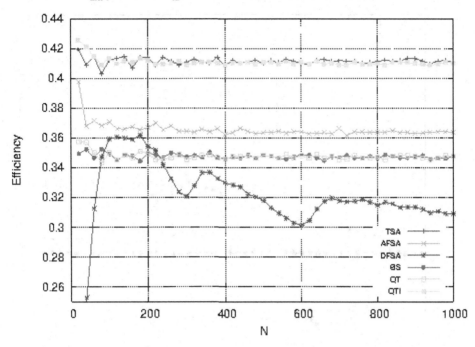

Figure 14. R_{bits} vs N with $N_{EXP0} = N$, $BITS_{ID} = 48$ and uniformly distributed IDs (Bonuccelli et al. 2007b)

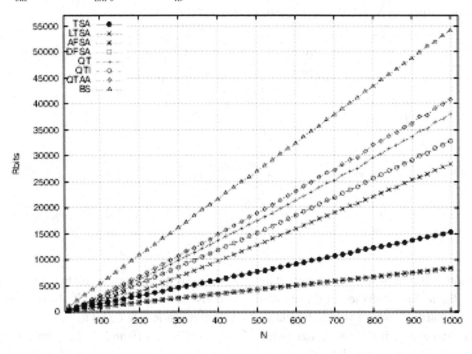

Figure 15. T_{bits} vs N, with $N_{EXP0} = N$, $BITS_{ID} = 48$ and uniformly distributed IDs (Bonuccelli et al. 2007b)

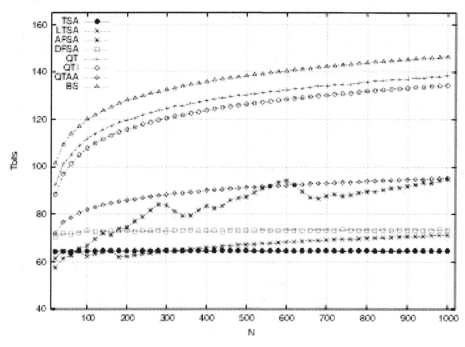

Figure 16. Time system efficiency of some analyzed protocols: in the left there are small values and in the right large values for N, respectively (Bagnato et al. 2009).

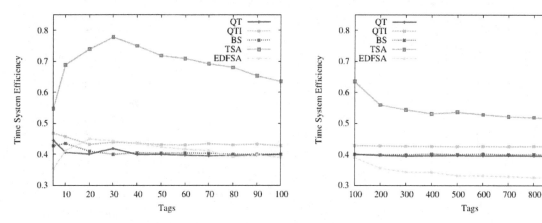

About Slotted Aloha protocols, an important parameter is the initial estimation of the number of tags to be identified, which is strictly related to the initial frame size: if the frame size is too large, there will be many idle slots, if it is too small there will be many collisions. In Figures 17 and 18, we show the behaviour of TSA and AFSA when the number of tags is underestimated and

overestimated, respectively. Here, we do not consider DFSA because it always starts with an initial frame size equal to 128.

In Figures 19 and 20 the behavior of QT, QTI and QTAA protocols are evaluated with the other distribution, namely when the tag IDs are not uniformly distributed, but there are groups of consecutive IDs. This evaluation is useful to

Figure 17. TSA and AFSA vs underestimation, with uniformly distributed IDs

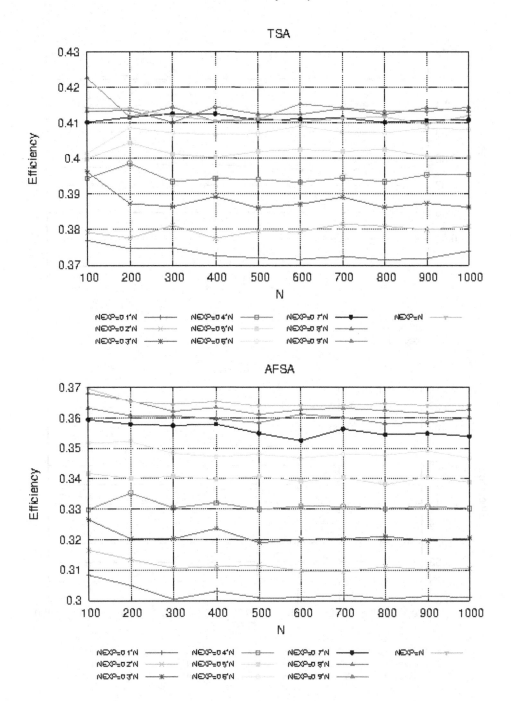

understand the behaviour of query tree protocols, since they are sensitive to the IDs distributions. From Figure 20, it can be noticed that this drawback degrades the performance when the ID length increases. Figure 21 shows the performance degradation of QTAA protocol by varying the g parameter. From this figure, we can deduce that the system efficiency increases by augmenting the

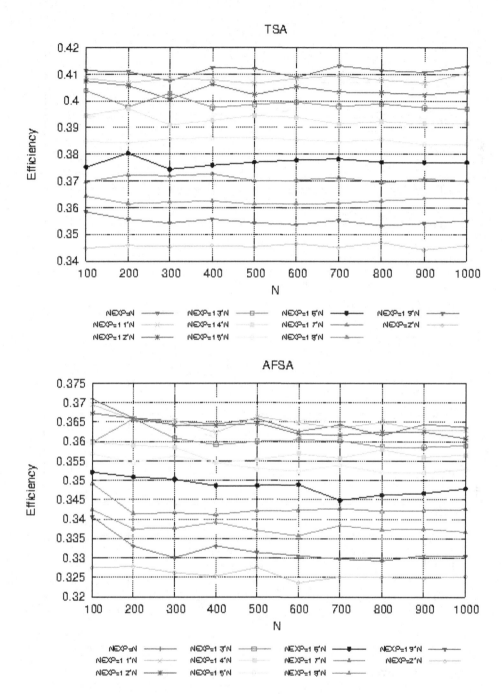

Figure 18. TSA and AFSA vs overestimation, with uniformly distributed IDs

maximum group size *g* (i.e. approaching the uniform distribution). Conversely, when there are many small groups of consecutive IDs, the performance of query tree algorithms decreases significantly. This is because the reader is forced to issue many queries, since IDs of tags belonging to the same group are different only at the last bits.

Figure 19. S vs N, with BITSID = 96 and IDs distributed in groups

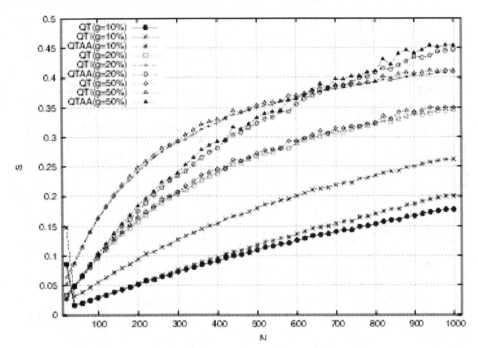

Concerning query tree protocols, the experiments performed in (*Bonuccelli et al. 2007a*) show that even a partial knowledge can highly improve the protocol performance. In particular, in Figure 22 we present the system efficiency of KQT, QTI and QTAA for different values of ID length (denoted as *k*), when the set of tags is known to be partitioned in 10 intervals (NUMint), each one of size equal to 10000 (AMP$_{int}$). For its nature, KQT performance is not influenced by the *k* values, while QT protocols degrade with increasing size of ID strings.

In Figure 23, we show the behavior of the considered protocols when the number of tags is very large.

Finally, the best performance is achieved by the protocol P presented in (*Bonuccelli et al. 2009*). Table 2 shows the average values of transmitted bits for P, BSIC, TSA, DFSA and QTI protocols, together with their confidence intervals, when the number of tags is equal to 100 and ID length is 96. Notice the very low variance of P protocol: this is because by choosing the bit position of the

minimum value (m2), the tree results to be as balanced as possible. Finally, in Figure 24, we show the average *system efficiency* of the above protocols: it is evident that the performance of P protocol is very high when the number of tags N is low, while decreases with high values of N. This happens because, with large values of N, the reader queries become longer, since more bit positions are needed (the depth of the tree increases). However, in many practical applications, the number of tags is at most few hundreds.

CONCLUSION

Tag identification is an important issue in RFID systems. In this chapter, we provided an overview of the main tag anti collision protocols focusing on their complexity and performance. Specifically, we divided them into aloha-based protocols and tree-based ones, briefly addressing also some hybrid approaches. Aloha-based protocols try to reduce collisions but they can not perfectly

Figure 20. S vs N, IDs distributed in groups with g = 10%, and different values for BID

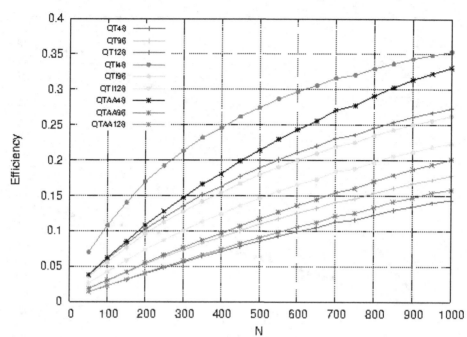

Figure 21. S vs N of QTAA, with BITSID = 96 and IDs distributed in groups

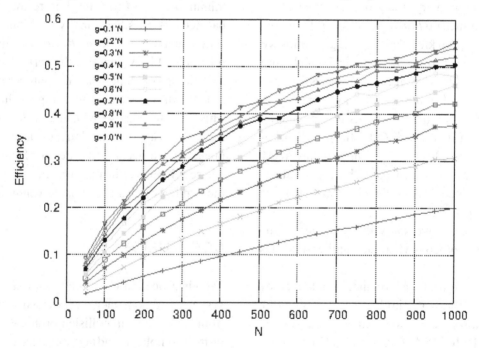

Figure 22. KQT, QTI, QTAA vs k, N = N_{exp}, NUM_{int} = 10, AMP_{int} = 10000 (Bonuccelli et al. 2007a)

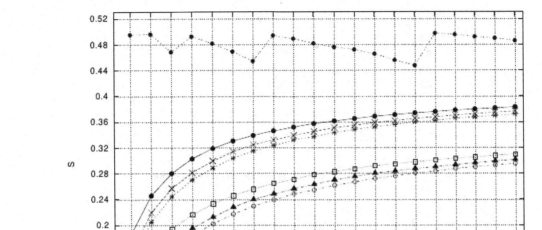

Figure 23. Efficiency of considered protocols when the number of tags N is large (BITS$_{ID}$ =96, N_{EXP}=N,g=0,1*N)

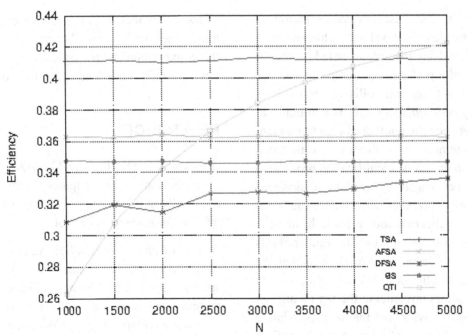

Figure 24. Efficiency of protocol P (Bonuccelli et al. 2009)

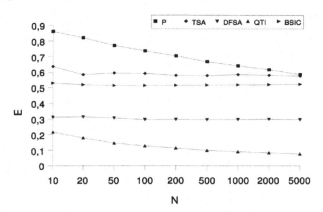

Table 2. Average values of P transmitted bits when N=100 and K=96

Protocol	95% confidence interval	min value	max value
P	14192.00$0.12	14072	14312
BSIC	18559.35$2.38	18523	18589
TSA	16370.55 $525.88	11488	24480
DFSA	32624.98$620.25	24982	39798
QTI	4562.13$361.60	69980	79966

prevent them. Tree-based protocols guarantee deterministic identification but they have a longer identification delay than aloha-based protocols. We presented a performance comparison of the more important tag anti-collision protocols in terms of *system efficiency* that is computed as the ratio between the number of tags to be identified and the number of queries (or time slots) performed in the whole identification process. In addition, we evaluated the tag anti-collision protocols according to other performance measures such as the number of reader queries, the number of bits transmitted by tags and the number of bits transmitted by the reader. We showed that the length of tag identifiers and their distribution are important factors influencing the tag identification protocol behavior, and that an approximate knowledge of the distribution of tag identifiers can highly improve the performance of query tree protocols. We also provided an overview of the WISP technology, pointing out the main advantages of this new platform with respect to the conventional RFID tags, and an exhaustive description of the EPCglobal standard.

REFERENCES

Abramson, S. (1970). The ALOHA system-Another alternative for computer communications. *Proceedings of Fall Joint Computer Conference, AFIPS Conference* 1970, Vol. 37.

Al-Medhwahi, M., Alkholidi, A., & Hamam, H. (2010). A new hybrid frame ALOHA and binary splitting algorithm for anti-collision in RFID systems. *International Conference on Software, Telecommunications and Computer Networks* (SoftCOM), (pp. 219–224).

Alotaibi, M., Bialkowski, K. S., & Postula, A. (2010). Improving the time efficiency of QTA anti-collision algorithm. *IEEE International Conference on RFID-Technology and Applications (RFID-TA)*, (pp. 205–210).

Bagnato, G., Maselli, G., Petrioli, C., & Vicari, C. (2009). *Performance analysis of anti-collision protocols for RFID systems*. IEEE 69th Vehicular Technology Conference, VTC Spring 2009, April 2009.

Bolic, M., Simplot-Ryl, D., & Stojmenovic, I. (2010). *RFID systems: Research trends and challenges*. Wiley Series in Wireless Communications and Mobile Computing, 2010.

Bonuccelli, M., Lonetti, F., & Martelli, F. (2009). Exploiting signal strength detection and collision cancellation for tag identification in RFID systems. *IEEE Symposium on Computers and Communications* (ISCC), (pp. 500–506).

Bonuccelli, M. A., Lonetti, F., & Martelli, F. (2006). Tree slotted aloha: A new protocol for tag identification in RFID networks. *Proceedings of International Symposium on World of Wireless, Mobile and Multimedia Networks, WOWMOM*, (pp. 603–608).

Bonuccelli, M. A., Lonetti, F., & Martelli, F. (2007a). Exploiting ID knowledge for tag identification in RFID networks. *Proceedings of the 4th ACM Workshop on Performance Evaluation of Wireless Ad Hoc, Sensor, and Ubiquitous Networks*, PE-WASUN, (pp. 70–77).

Bonuccelli, M. A., Lonetti, F., & Martelli, F. (2007b). Instant collision resolution for tag identification in RFID networks. *Ad Hoc Networking*, 5, 1220–1232. doi:10.1016/j.adhoc.2007.02.016

Capetanakis, J. I. (1979, September). Tree algorithms for packet broadcast channels. *IEEE Transactions on Information Theory*, IT-25, 505–515. doi:10.1109/TIT.1979.1056093

Center, M. A.-I. (2003). *Draft protocol specification for a 900 mhz class 0 radio frequency identification tag*. Retrieved from http://www.epcglobalinc.org

Cha, J., & Kim, J. (2005). Novel anti-collision algorithms for fast object identification in RFID system. *International Conference on Parallel and Distributed Systems*, Vol. 2, (pp. 63–67).

Cha, J., & Kim, J. (2006). Dynamic framed slotted ALOHA algorithms using fast tag estimation method for RFID system. *Proceedings of IEEE Consumer Communications and Networking Conference*, Vol. 2, (pp. 768–772).

Chiang, K. W., Hua, C., & Yum, T.-S. P. (2006). Prefix-randomized query-tree protocol for RFID systems. *Proceedings of IEEE International Conference on Communications*, (pp. 1653–1657).

Choi, H., Cha, J., & Kim, J. (2005). Improved bit-by-bit binary tree algorithm in ubiquitous Id System. *Advances in Multimedia Information Processing - PCM 2004, Vol. 3332 of Lecture Notes in Computer Science*, (pp. 696–703).

Choi, J., & Lee, W. (2007). Comparative evaluation of probabilistic and deterministic tag anti-collision protocols for RFID networks. *Proceeding of 2007 Conference on Emerging Direction in Embedded and Ubiquitous Computing (EUC'07), Lecture Notes in Computer Science*, (pp. 538-549).

Engels, D. W., & Sarma, S. E. (2002). The reader collision problem. In *IEEE International Conference on Systems, Man and Cybernetics*, Vol. 3, October 2002.

Hush, D. R., & Wood, C. (1998). Analysis of tree algorithms for RFID arbitration. *Proceedings of IEEE International Symposium on Information Theory*, (p. 107).

Klair, D., Chin, K.-W., & Raad, R. (2010). A survey and tutorial of RFID anti-collision protocols. *IEEE Communications Surveys Tutorials, 12*(3), 400–421. doi:10.1109/SURV.2010.031810.00037

Kodialam, M., & Nandagopal, T. (2006). Fast and reliable estimation schemes in RFID systems. *Proceedings of the 12th Annual International Conference on Mobile Computing and Networking, MobiCom,* (pp. 322–333).

La Porta, T., Maselli, G., & Petrioli, C. (2011). Anticollision protocols for single-reader RFID systems: Temporal analysis and optimization. *IEEE Transactions on Mobile Computing, 10*(2), 267–279. doi:10.1109/TMC.2010.58

Law, C., Lee, K., & Siu, K. Y. (2000). Efficient memoryless protocol for tag identification (extended abstract). *Proceedings of the 4th International Workshop on Discrete Algorithms and Methods for Mobile Computing and Communications, DIALM,* (pp. 75–84).

Lee, S., Joo, S., & Lee, C. (2005). An enhanced dynamic framed slotted aloha algorithm for RFID tag identification. *Proceedings of Mobiquitous,* (pp. 166–172).

Maselli, G., Petrioli, C., & Vicari, C. (2008). Dynamic tag estimation for optimizing tree slotted aloha in RFID networks. *Proceedings of the 11th ACM Symposium on Modeling, Analysis and Simulation of Wireless and Mobile Systems,* (pp. 315–322).

Micic, A., Nayac, A., Simplot-Ryl, D., & Stojmenovic, I. (2005). A hybrid randomized protocol for RFID tag identification. *Proceedings of 1st IEEE International Workshop on Next Generation Wireless Networks (WoNGeN).*

Myung, J., & Lee, W. (2006). Adaptive binary splitting: A RFID tag collision arbitration protocol for tag identification. *Mobile Networks and Applications, 11,* 711–722. doi:10.1007/s11036-006-7797-6

Myung, J., Lee, W., & Shih, T. (2006). An adaptive memoryless protocol for RFID tag collision arbitration. *IEEE Transactions on Multimedia, 8,* 1096–1101. doi:10.1109/TMM.2006.879817

Myung, J., Lee, W., Srivastava, J., & Shih, T. K. (2007). Tag-splitting: Adaptive collision arbitration protocols for RFID tag identification. *IEEE Transactions on Parallel and Distributed Systems, 18,* 763–775. doi:10.1109/TPDS.2007.1098

Peng, Q., Zhang, M., & Wu, W. (2007). Variant enhanced dynamic frame slotted aloha algorithm for fast object identification in RFID system. *Proceedings of IEEE International Workshop on Anti-counterfeiting, Security, Identification,* (pp. 88–91).

Philipose, M., Smith, J. R., Jiang, B., Sundara-Rajan, K., Mamishev, A., & Roy, S. (2005). Battery-free wireless identification and sensing. *IEEE Pervasive Computing / IEEE Computer Society and IEEE Communications Society, 4*(1), 37–45. doi:10.1109/MPRV.2005.7

Product, W. I. S. P. (n.d.). *MCRF200IWQ23: Passive RFID IC device.* Retrieved from http://www.datasheetcatalog.com/datasheets_pdf/M/C/R/F/MCRF202.shtml

Ryu, J., Lee, H., Seok, Y., Kwon, T., & Choi, Y. (2007). A hybrid query tree protocol for tag collision arbitration in RFID systems. *Proceedings of IEEE International Conference on Communications,* (pp. 5981–5986).

Schoute, F. C. (1983). Dynamic frame length aloha. *IEEE Transactions on Communications, 31,* 565–568. doi:10.1109/TCOM.1983.1095854

Smith, J., Sample, A., Powledge, P., Roy, S., & Mamishev, A. (2006). A wirelessly-powered platform for sensing and computation. *Proceedings of Ubicomp 2006: 8th International Conference on Ubiquitous Computing,* Orange Country, CA, USA, September 17-21 2006, (pp. 495-506).

Standard, E. P. C. (2008). *EPC radio-frequency identity protocols class-1 generation-2 UHF RFID protocol for communications at 860 MHz – 960, MHz Version 1.2.0.* Retrieved from http://www.epcglobalinc.org/standards/.

Turner, C. (2003). *Deterministic or probabilistic – Which is best? Why would you even care?* Retrieved from http://www.rfip.eu/downloads/Deterministic_or_Probabilistic.pdf

Vogt, H. (2002). Efficient object identification with passive RFID. *Proceedings of The First International Conference on Pervasive Computing,* (pp. 98–113).

Waldrop, J., Engels, D. W., & Sarma, S. E. (2003). Colorwave: A MAC for RFID reader networks. In *IEEE Wireless Communications and Networking, WCNC 2003*, Vol. 3, (pp. 1701–1704). New Orleans, Louisiana, March 2003.

Wieselthier, J. E., Ephremides, A., & Michaels, L. A. (1989). An exact analysis and performance evaluation of framed aloha with capture. *IEEE Transactions on Communications, 37*, 125–137. doi:10.1109/26.20080

WISP. (n.d.). *Wiki.* Retrieved from http://wisp.wikispaces.com/

Yeager, D. J., Sample, A. P., & Smith, J. R. (2008). WISP: A passively powered UHF RFID tag with sensing and computation . In Ahson, S., & Ilyas, M. (Eds.), *RFID handbook: Applications, technology, security, and privacy.* CRC Press.

Zhang, N., & Vojcic, B. (2005). Binary search algorithms with interference cancellation RFID systems. *Proceedings of Military Communications Conference, MILCOM,* Vol. 2, (pp. 950–955).

ENDNOTE

[1]   Note that, in order to have a total probability equal to 1, the last interval could have the sum of probabilities less than $1/N_{exp}$.

Chapter 2
Security of EPC Class-1

Pablo Picazo-Sanchez
University School of Computer Sciences of Madrid (UPM), Spain

Lara Ortiz-Martin
University School of Computer Sciences of Madrid (UPM), Spain

Pedro Peris-Lopez
Carlos III University of Madrid (UC3M), Spain

Julio C. Hernandez-Castro
Portsmouth University, UK

ABSTRACT

Radio Frequency Identification (RFID) is a common technology for identifying objects, animals, or people. The main form of barcode-type RFID device is known as an Electronic Product Code (EPC) and the most popular standard for passive RFID tags is Class-1 Generation-2. In this technology, the information transmitted between devices is through the air, therefore adversaries can eavesdrop these messages passed on the insecure radio channel and finally, the security of the system can be compromised. In this chapter, the authors analyze the security of EPC Class-1 Generation-2 standard, showing its security weaknesses and presenting some possible countermeasures.

1. INTRODUCTION

RFID is a technology that enables identification from distance (Want, 2006) and is already used for a large number of different applications, from cards accepted for building access or payments with mobile devices (Pasquet, Reynaud, & Rosenberger, 2008) to applications in sanitary environments (Benelli & Pozzebon, 2009).

There are three main components in RFID technology: tags, readers and database. Communication between a reader and a tag occurs in a packetized manner where a single packet contains a complete command from a reader and a complete response from the tag. The command and response permits half-duplex communication between a reader and a tag. The reader is connected to a database for recognizing tags' ID.

International Organization for Standardization (ISO) has several standards related to RFID technology. For instance, ISO11784 contains the

DOI: 10.4018/978-1-4666-1990-6.ch002

structure of the RFID code for animals; ISO11785 defines the air interface protocol; ISO14443 defines a set of international standards covering proximity smart cards used in payment systems; ISO18047 specifies test methods for determine the conformance of RFID tags and readers to a certain standard, and ISO18046 defines methods for testing the performance of RFID tags and readers.

The Auto-ID Center was set up in 1999 to develop the Electronic Product Code and related technologies that could be used to identify and track products through the global supply chain. In this chapter, the standard considered as the "universal" standard for Class-1 RFID tags: EPC Class-1 Generation-2, in the following EPC-C1G2, is examined (EPCGlobal, 2011).

In 2007, a classification and a description of EPC tags were made by EPCglobal (EPCglobal, 2007). This classification distinguishes between four different classes:

- **Class-1: Identity Tags:**
  Passive-backscatter tags with the following minimum features:
  ◦ An electronic product code identifier,
  ◦ A tag identifier,
  ◦ A function that renders a tag permanently non-responsive.
  ◦ Optional decommissioning or recommissioning of the tag,
  ◦ Optional password-protected access control, and
  ◦ Optional user memory.
- **Class-2: Higher-Functionality Tags**
  Passive tags with the following anticipated features above and beyond those of Class-1 tags:
  ◦ An extended tag ID,
  ◦ Extended user memory,
  ◦ Authenticated access control, and
  ◦ Additional features as will be defined in the Class-2 specification.

  These two first classes (Class-1 and Class-2) are passive. Passive tags do not have their own power source. Tags use the power received from the interrogation signal (reader) and in consequence these devices have shorter read ranges compared to active systems. In comparison to active tags, passive tags occupy smaller circuit-area and can be manufactured faster and at lower cost. Passive tags are best used on low-cost and high-quantity items, as well as on products that are very small. Because there is no battery, these tags have a long life span and function until the tag is either damaged or intentionally disabled through a kill command.

- **Class-3: Battery-Assisted passive tags (called Semi-Passive Tags in Ultra High Frequency Gen2)**
  Passive tags with the following anticipated features above and beyond those of Class-2 tags:
  ◦ A power source that may supply power to the tag and/or to its sensors, and/or
  ◦ Sensors with optional data logging.
  This second group (Class-3) are semi-passive tags. These tags use a power source (battery) to help with powering the tag when responding to the reader and to provide power to the internal memory. The use of this battery can increase the read range even though the communication technique is the same as with the passive tags (i.e., passive backscatter). However, these tags can have environmental sensors, and the data captured by theses sensors can be saved in the internal memory. Since the power source is not used for transmitting messages, it has a longer life span than a similar battery would have if it were used in an active tag. Due to size restrictions, different types of batteries can be used, such as printed batteries.
- **Class-4: Active Tags**
  Active tags with all the aforementioned features and also including the following:

- ◦ Communications via an autonomous transmitter,
- ◦ Optional user memory, and
- ◦ Optional sensors with or without data logging.

The last group (Class-4) is for active tags. Active tags typically have their own power source and can therefore achieve much longer ranges than the range provided by passive and semi-passive tags. Active tags, by definition, transmit a signal to a reader. This design also allows them to carry various sensors on board as well as support larger memory. Because of this, active tags can track any environmental changes, as well as other data, and report all this information back to the reader and the back-end system. Active tags can be used in hostile environments because they do not have to wait for the reader's request to send data. In fact, they are able to transmit by themselves.

Moreover, active tags have access to the transmitter requiring access to the power source too. Class-4 tags shall not interfere with the communications protocols used by Class-1/2/3 tags.

From the security and privacy point of view, RFID tags are one of the easiest targets for suffering several logical and physical attacks. The attackers pursue to reveal the secrets of the tags for cloning them and getting all privileges that the bearer have. The information exchanged between tags and readers should be well protected from both, confidentiality and integrity. RFID systems also need to be robust against Denial of Service (DoS), impersonation and tracking attacks. Nevertheless, nowadays it seems impossible to use classic cryptographic algorithms such as public cryptography, elliptic curves and hash functions due to the computation limits of passive tags. Current EPC tags have a couple of thousand gates, but only few hundreds can be used for security purposes.

This lack of standard cryptography in basic tags posses intriguing research challenges. In the last years, researchers have contributed with lightweight technical approaches to the problems of security and privacy linked to RFID technology.

This chapter is organized as follows. In Section 2 an introduction of EPC Class-1 is presented. Specifications of EPC Class-1 Generation-2 standard are explained in Section 3. In Section 4 the main cryptographic primitives are analysed from RFID point of view. In section 5 risk and threats are detected and classified. Lightweight cryptography is introduced and some examples of this sort of protocol are presented in Section 6. This chapter concludes with some future directions in Section 7 and some final conclusions in Section 8.

2. BACKGROUND

It is important to review EPC Class-1 Generation-1, in the following EPC-C1G1 or Gen-1, to understand why the new standard (Generation-2) was made. First generation of EPC has two specifications for tags in Ultra High Frequency Band (UHF): Class-0 and Class-1. EPCglobal defines Class-0 tags as a read-only device. Class-1 tags are defined in the EPCglobal specification for devices that are one-time programmable, but in practice commercial-available tags can be written numerous times and are regularly erased and reprogrammed in the field (M. Dobkin & Wandinger, 2005).

In Gen-1 specification, data is stored in a logical structure named Identifier Tag Memory (ITM) which is made of three well known fields: CRC, EPC and password (Figure 1).

CRC is a cyclic redundancy check where EPCs use a 16-bit CRC to detect transmission errors (i.e. more than 65,000 possible CRC values).

The Electronic Product Code (EPC) is partitioned into four different fields: version, domain manager, object class, and serial number. The header describes the structure of the remainder of the code. Domain manager is a number typically from a company or an organization. The object class is a model number. Finally, there is a serial

Figure 1. Structure of Class-1 identifier tag memory

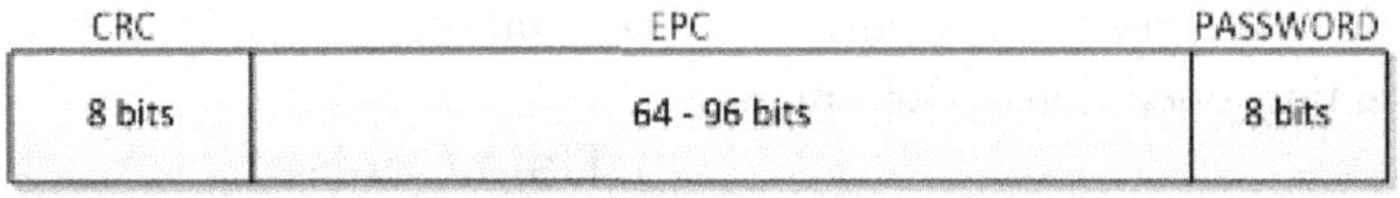

CRC	EPC	PASSWORD
8 bits	64 - 96 bits	8 bits

Table 1. Required commands

Command Name	Command Code	Tag reply
ScrollAllID	0011 0100	ScrollIDReply
ScrollID	0000 0001	ScrollIDReply
PingID	0000 1000	PingIDReply
Quiet	0000 0010	None
Talk	0001 0000	None
Kill	0000 0100	None

Table 2. Identifier programming commands

Command Name	Command Code	Tag reply
ProgramID	0011 0001	None
VerifyID	0011 1000	VerifyIDReply
LockID	0011 0001	None
EraseID	0011 0010	None

number that makes the EPC unique for a specific physical object.

Finally, the password field is an 8-bit data string used by readers to send commands and identifier-programming commands to tags. Tags must implement required commands (see Table 1) and will implement identifier-programming commands depending upon the type of ITM memory implemented on the tag (see Table 2).

Nevertheless, Gen-1 has several open issues. Here are some of them:

- A tag can be killed if the kill password is known. As seen above, password command has 8 bits and for recovering that command an attacker only has to check 256 possible combinations (brute-force at-

tack). According to the standard, this command renders the tag inactive forever.

- 16-bit CRC is the only mechanism used for the validation of the tag's ID, which means that the device can be easily validated erroneously. When the total number of the tag's readings becomes comparable in size to the total number of CRC values, it is inevitable that on occasion a noisy or spurious EPC will by chance agree with the CRC value, leading to a "phantom tag" read: an apparently valid EPC tag that does not exist physically (M. Dobkin & Wandinger, 2005).

- Class-0 and Class-1 specifications are incompatible between them and they are incompatible with ISO standards too.

- None of the specifications (Class-0 and Class-1) provide any link-level security during programming operations (Dobkin, 2007).
- Class-0 and Class-1 cannot be used globally. Class-0 tags, for instance, send out a signal at one frequency and receive a signal back at a different frequency within the UHF band. Nevertheless, this is forbidden in Europe.
- When multiple tags are present, potential collisions are solved using a bit-by-bit binary-tree search. During a tree traversal, each tag transmits the next bit of its EPC; if the reader echoes that bit, the tag continues, otherwise it transitions to a dormant state and awaits the next traversal. Therefore, the EPC number can be inferred from the reader signal that can be received at a much longer distance than the backscattered signal of the tag (Chaudhry, Thompson, & Thompson, 2005).

UHF band compared to Low Frequency (LF) or High Frequency (HF), are not as effective around metals and water. However, UHF performs well for distances greater than one meter. UHF passive tags offer a reading range of around 10 meters, and active UHF tags can be read up to 330 meters. Due to the high data rate, UHF allows for up to 1500 simultaneous tag reads per second. These devices are relatively low cost and their applications are very diverse such as baggage tracking, electronic toll collection, pallet tracking, shipping, and other supply chain applications.

2.1 Generation-1 vs Generation-2

For all of those problems (among others) described above, a new generation of RFID tags is needed. A brief summary of features of EPC-C1G1 and EPC-C1G2 is shown in Table 3. In the following, main characteristics of Gen-2 specification

are explained and compared with the previous generation.

i. Data Content

In EPC-C1G1 tags usually have 64-bits for EPC identifier and, in some cases, can have 96-bits.

In EPC-C1G2 tags require a minimum of 96-bits for the same EPC identifier. Gen-2 also allows unlimited user tag memory, but this feature is optional. The additional memory could be used to store lot codes or expiration dates, time and date stamps for transactions, input from temperature sensors, and many other additional information.

ii. Anti-Collision Protocol

In EPC-C1G1 the protocol used to avoid collisions is a bit-by-bit binary-tree search. The tag is switched to sleep mode each time is read by a reader. Multiple wakes-up and sleep cycles are necessary to ensure that all tags in a reader field are read.

In EPC-C1G2 the reader identifies an individual tag using the "Q-protocol" which is a slotted Aloha (Roberts, 1975) collision resolution. Using this procedure, tags begin broadcasting their IDs as soon as the reader field energizes them. The reader simply receives IDs, depending on chance to ensure that each tag will eventually broadcast the answer during a period when the other tags are quiet (Glover & Bhatt, 2006). The process continues until all tags in the reader field are counted.

iii. Input/Output Speed

Gen-1 provides a single generation speed; 350-450 tags/second can be read in theoretical scenario. On the other hand, Gen-2 has a maximum theoretical reading speed of around 1000 tags/second. The read speed of a tag conforming to Gen-2 is about twice than of the Gen-1 tags, with average read rates of around 450 tags per second (Peris-Lopez, Tong-Lee, & Tieyan, 2008).

Table 3. Features summary (Razaq, Luk, & Cheng, 2007)

	Field Description	Gen-1	Gen-2
General	Acceptance Level	Not a global Standard	Global standard in ISO UHF
	Arbitration	Deterministic binary tree	Probabilistic Slotted
	Anti-Collision	Binary tree algorithm with persistent state/wake states	Q-Protocol which is a variant of Slotted Aloha protocol
Communication Link	Air interface	PWM FSK and PWM PIM for Class-0 and Class-1 respectively	PIE-ASK, Miller, FM0
	Data rate	40/80 and 70/140 Kb/s for Class-0 and Class-1 respectively	40 to 640 Kb/s
	Distance	< 10m	< 10m
	Frequency range	860-930 MHz	860-960MHz
Security	Security/Password	8 and 24 bits for Class-1 and Class-0 respectively	32 bits
	Ghost reads	1.3 per 1000 for Class-0	None
	Cover-coding	No	Yes
	Data write verification	No	Yes
Enhancements	Dense reader	N/A	Yes
	Backscatter	In one format	Two formats (FM0 or Miller)
	The Q algorithm	N/A	Yes
	B symmetry	No	Yes
	Select/Persistence	No	Yes
	Write Speed (For 96 bits)	3 per second	Minimum 5 per second
	Same ID tags	Not allowed	Allowed
	Sessions	N/A	Four readers can parallel communicate with a single tag at any given time

In writing operations, EPC-C1G1 readers can write about 3 tags per second in 96 bits user's memory. Nevertheless, EPC-C1G2 can write at least, 5 tags per second in 96 bits user's memory.

iv. Tag-to-Reader Communication

Passive and semi-passive RFID tags do not use a radio transmitter; instead, they use modulation of the reflected power from the tag antenna. A principle of electromagnetic, the principle of reciprocity, states that any structure that receives a wave can also transmit a wave. The radiated wave can make its way back to the transmitting antenna, induce a voltage, and therefore, produce a signal that can be detected: a backscattered signal. Both Gen-1 and Gen-2 readers must supply power to passive tags to communicate with them and must manage spectral efficiency to avoid interferences.

EPC-C1G2 provides two new radio signaling techniques which allow to isolate tags response easily: FM0 and Miller sub-carrier.

In FM0, the tag state changes at the beginning and at the end of every symbol. In addition, a binary "0" has an additional state change in the middle of the symbol. In the left-hand side of the Figure 2, there are two different states (low and high level) for the binary "1" symbol. The right side of the figure shows the base-band signal corresponding to a series of identical binary bits

Figure 2. FM0 (EPCglobal, EPC radio-frequency protocols Class-1 generation-2 UHF RFID version 1.2.0, 2008)

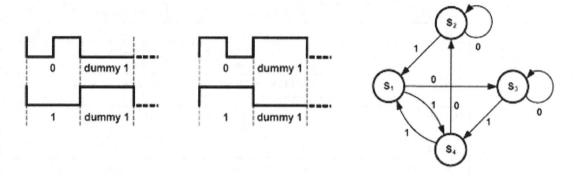

to clarify the correspondence of binary "0"s with a frequency twice as high as that of binary "1"s (Dobkin, 2007).

Miller sub-carrier (MMS) provides more flexibility for noise versus data rate tradeoffs and spectral management. In MMS, the data bits are first encoded in the opposite fashion of FM0: that is, a binary "1" is given a state transition in the middle of a symbol time and a binary "0" is not. In MMS there is no state transition at the symbol edges between consecutive 1s, or between a 1 and a 0. However, there is a state transition at a symbol edge between two consecutive 0s (Dobkin, 2007).

A example of both techniques FM0 and MMS is shown in Figure 2 and Figure 3 respectively.

v. Parallel Counting

Parallel counting means that several readers can communicate with the same tag at the same time. In EPC-C1G1 this cannot be possible because of supporting only one state. When the reader is communicating with a tag, then this tag is only linked with that reader. If another reader would want to communicate with the tag, it should have to change the link and the state and this action will loss the first reader's link.

To solve this problem, in EPC-C1G2, sessions are introduced in order to allow singulation. Sessions is a mechanism that allows tags to communicate with four readers at the same time.

Figure 3. MMS (EPCglobal, EPC radio-frequency protocols Class-1 generation-2 UHF RFID version 1.2.0, 2008)

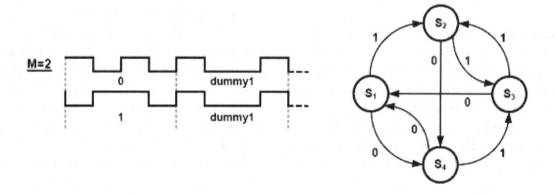

Figure 4. Gen 1 Class 0 tag's memory (Dobkin, 2007)

Page	Byte: 0	1	2	3	4	5	6	7	8	9	10	11	12	13	14
ID0	KILL PWD			NOT USED											
ID1	ID1 can't be written to								Last 12 bits of CRC + last byte seed ID1, must be programmed						
ID2	EPC												CRC		Random bits
ID3	User memory (any partition)														

vi. Tag Memory Organization

EPC-C1G1 Class 0 memory is organized into four pages, but not all of them can be written. In Figure 4 a tag's memory organization is illustrated. The ID0 page is used to record the KILL password. The ID1 page contains the random singulation code generated from a seed. The seed is the last 20 bits of the ID2 page. The last page, ID3, can be used in any fashion by the user.

EPC-C1G1 Class 1 memory is organized in 7 or 9 rows of 2 bytes each one. In Figure 5 a tag's memory organization is displayed. The CRC is stored in the first row, the EPC (most-significant-byte first) the next five or seven rows (Class 1 error check uses a sequence of 16 "0" bits after the EPC). The last row contains the kill password and the lock byte, which is set to hexadecimal "A5" to prevent further writing of the tag.

EPC-C1G2 explicitly calls for a field-writeable tag with the implication that multiple writecycles are possible. In Figure 6 a tag's memory organization is depicted. Tag's memory is made of four memory banks: bank 0 to bank 3. Bank 00 contains (at least) the 32-bit KILL and ACCESS passwords. Bank 01 contains, in addition to the tag's EPC, a 32-bit Protocol Control (PC) word, describing the length of the EPC, as well as some optional information about the tag, and the 16-bit CRC used for error checking of the EPC value. The optional Tag ID bank "10" is dedicated to manufac-turers for identifying the tags which is useful for tracking tag IC manufacturing and tracking tags inventory. Finally, optional user bank 11 is available for any application-specific data.

3. CLASS-1 GENERATION-2 SPECIFICATIONS

The EPC-C1G2 specification (EPCGlobal, 2011) defines the physical and logical requirements for RFID systems operating in the 860-960 MHz frequency range. These systems are made up of three main components, interrogators also known

Figure 5. Gen 1 Class 1 tag's memory (Dobkin, 2007)

Row	MSByte	LSByte
0	CRC	
1	EPC (MSB) →	
2	→	
3	→	
4	→	
5	→	EPC (LSB)
6	LOCK BYTE	KILL PWD

Figure 6. Gen 2 tag's memory (EPCglobal, EPC radio-frequency protocols Class-1 generation-2 UHF RFID version 1.2.0, 2008)

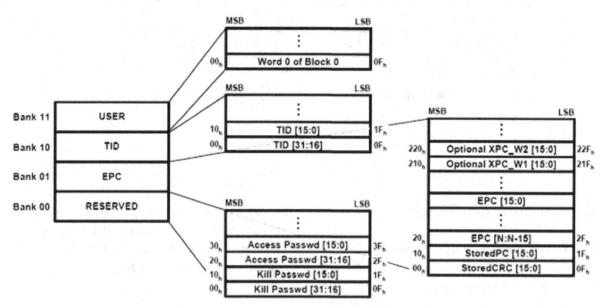

as readers, tags also known as labels and a database connected to readers.

A reader transmits information to the tags by modulating a Radio Frequency (RF) signal (860-960 MHz). As tags are passive, all of their operating energy is received from reader's RF waveform. Indeed, both information and operating energy are extracted from the signal sent by the reader.

EPC-C1G2 is a kind of systems known as Interrogator-Talk-First (ITF). In these systems, a power source initiates communication (reader) and sends queries to the labels. Tags can only answer after a message is received from the reader. The reader receives information by transmitting a continuous-wave RF signal to the tag. Communications between reader and tags are half-duplex, so the reader talks and tag listens, or vice versa but never in a simultaneous way. The tag backscatters a signal to the reader by means of the modulation of the reflection coefficient of its antenna. Backscatter radio has different requirements that conventional radio link. Among others, backscatter:

- Must supply power to passive tags.
- Must manage spectral efficiency to avoid interference.
- The need for low-complexity tags limits modulation and coding options.

3.1 Tag Identification

As shown in Figure 7, an interrogator manages tag populations using three basic operations: select, inventory and access.

a. Select. The operation of choosing a tag population for inventory and access. This operation is similar to selecting records in a database.

b. Inventory. The operation of identifying tags. Upon the exchange of several messages (inventory round), the tag sends to the reader the PC, EPC and a CRC-16 values. An inventory round operates in one and only one session at a time.

Figure 7. Managing tags population (EPCglobal, EPC radio-frequency protocols Class-1 generation-2 UHF RFID version 1.2.0, 2008)

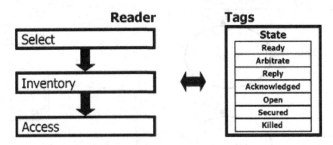

c. Access. The operation of communicating (reading from and/or writing to) with a tag, which comprises multiples commands. Tags have to be unequivocal identified before the access operation.

3.2 Tag States and Slot Counter

As defined in EPC-C1G2 standard, tags shall implement different states, as displayed in Figure 7 and Figure 8.

- Ready state. A tag enters into this state after being energized and the label remains operative (no killed). The tag shall remain this ready state until it receives an accurate query command. Tag loads a Q-bit number from its Random Number Generator (RNG), and transitions to the arbitrate state if the number is non-zero, or to the replay state if the number is zero.

- Arbitrate state. A tag in an arbitrate state shall decrement its slot counter every time it receives a query request, transitioning to the replay state and backscattering a Random Number of 16-bits (RN16) when its slot counter reaches the value 0000h.

- Replay state. A tag shall backscatter a RN16, once entering in the replay state. If the tag receives a valid acknowledge (ACK), it shall transition to acknowledge state, backscattering its PC, EPC, and

CRC-16. Otherwise, the tag remains in arbitrate state.

- Acknowledge state. A tag in an acknowledge state may transition to any state except killed state.

- Open state. After receiving a request RN command, a tag in the acknowledge state whose access password is nonzero shall transition to open state. The tag backscatters a new RN16 that both reader and tag shall use in subsequence messages. Tags in the open state can execute all access commands except lock and may transition to any state except acknowledge.

- Secured state. A tag in an acknowledge state whose access password is zero shall transition to secured state, upon receiving a request RN command. The tag backscatters a new RN16 that both reader and tag shall use in subsequent messages. A tag in the open state whose access password is nonzero shall transition to secured state, after receiving a valid access command, which includes the same handle that it previously backscattered when it transitioned from acknowledge state to the open state. Tags in secured state can execute all access command and may transition to any state except open or acknowledge.

- Killed state. Once a tag in either open state or secured state receives a kill password, it shall enter in a killed state. Kill permanently disables a tag. A tag shall notify

Figure 8. States of EPC tags (EPCGlobal, 2011)

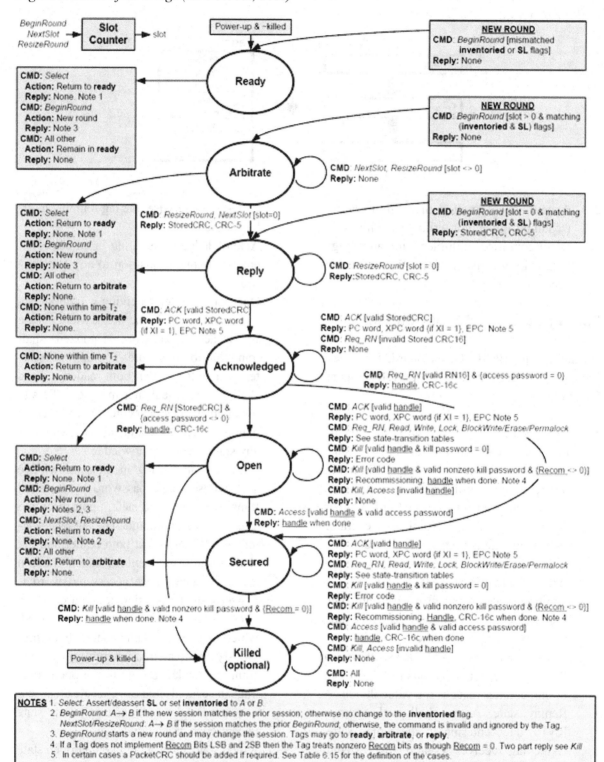

the reader that the killed operation was successfully, and shall not respond to any reader thereafter.

Tags shall implement a 15-bit slot counter. Once a query or query adjust command is received, a tag shall load into its slot counter a value between 0 and $2^{(Q-1)}$ obtained from tag's Pseudo-Random Number Generator (PRNG). Q is an integer in the range (0, 15). A query specifies Q, and a query adjust may modify Q from the prior query.

4. PRIVACY AND SECURITY

RFID systems have several advantages in comparison to other identifications systems but these systems have security problems that should be properly addressed. In EPC there are two main threats: privacy and tracking (Juels, RFID Security and Privacy: A Research Survey, 2006).

RFID tags respond to reader interrogation without alerting their owners, and these tags have an unique identifier that reveal sensitive information when are queried by readers and answer indiscriminately. Thus, when read range permits, clandestine scanning of tags is a plausible threat. This problem is been increased when this identification number is combined with any other private number of the bearer like his ID card or a constant number that identifies him to others.

Tracking or violation of location privacy is a subset of the privacy issues. This is possible because the answers provided by tags are usually predictable. In fact, most of the times, tags provide always the same identifier, which would allow a third party to easily establish an association between a given tag and her bearer. In EPC-C1G2 tags include, as seen in sections above, some static fields for the manufacturer of the object, and a product code. Therefore, a person carrying EPC tags is subject to attacks of traceability. A successful attack of an adversary might be achieved with or without recovering the secret key shared by the reader and the tag. This means that it is not enough to prove that the adversary cannot calculate the secret key in a feasible way, because she might reach her goal without it. This attack can be easily completed by an illegal reader that can silently determine what objects the bearer has: like medications bought in a pharmacy, implantable medical devices like peacemakers, passports or other ID-cards and so on.

There are two different attacks in which an adversary tries to prepare for a successful response: passive and active attacks. Passive attack is a kind of attack where the adversary eavesdrops messages from one or more runs without interfering with the communication between the parties. Active attack is a kind of attack where the adversary impersonates any of the components of the system, the reader and/or the tag, and typically replays purposefully modified messages observed in previous runs of the protocol (Vajda & Buttyán, 2003).

Although the two above-mentioned problems are the most important, EPC-C1G2 has other several vulnerabilities. In literature, authors classify these threats attending to different factors. For instance, Karygiannis et al. proposed an RFID model that divides risk into three classes (Karygicmnis, Phillips, & Tsibertzopoulos, 2006): network-based risks, business process risks and business intelligence risks. On the other hand, in (Garfinkel, Juels, & Pappu, 2005) threats are divided into those primarily affecting corporations and those primarily affecting individuals. In work (Ayoade, 2007) attacks are classified according to their purpose: data mining for profiling purposes, tracing or tracking and secret tag reading. Also in the work (Avoine & Oechslin, 2005), threats fall into two categories: information leakage and traceability. In contrast, in (Garcia-Alfaro, Barbeau, & Kranakis, 2008) threats according to a methodology proposed by the European Telecommunications Standards Institute (ETSI) are analyzed. Finally, in (Mitrokotsa, Rieback, & Tanenbaum, 2010) Mitrokotsa et al. proposed a classification in which attacks depend on the layer where each attack is taking place (physical layer, network-transport layer, application layer, strategic layer and multilayer attacks).

A. Risks and Threats

In this section, the principal threats that RFID systems have are classified according to the layer that is affected like in (Avoine & Oechslin, RFID traceability: a multilayer problem, 2005). There are five main layers: physical, network-transport, application, multilayer and strategic layer.

- Physical Layer. The physical layer in RFID communications is composed of the physical interface, the radio signals used and the RFID devices. The adversary in this layer takes advantage of the wireless nature of RFID communication, its poor physical security and its lack of resilience against physical manipulation. This layer includes attacks that permanently or temporarily disable RFID tags as well as relay attacks (Mitrokotsa, Rieback, & Tanenbaum, 2010).
  - ◦ Tag removal: RFID tags that are not embedded in items can easily be removed from an item.
  - ◦ Tag destruction: a tag may be physically destroyed (by applying pressure or tension loads, chemical exposure, abrasion caused by rough handling, etc).
  - ◦ Kill command: the KILL command is able to permanently silence an RFID tag and can be used for privacy reasons or malicious adversaries in order to sabotage RFID communications.
  - ◦ Removal/descruction reader: RFID readers can also be subject to destruction or removal that includes critical information such as cryptographic credentials.
  - ◦ Passive interference: RFID networks often operate in an unstable and noisy environments, their communication is sensitive to possible interference and collisions.
  - ◦ Acive jamming: an attacker can take advantage of the fact that an RFID tag is always listening and may cause electromagnetic jamming by continously transmiting a signal in order to prevent tags to communicate with readers.
  - ◦ Relay attack: in a relay attack, an adversary (through a device) acts as a man-in-the-middle. This device is able to intercept and modify the radio signal between the legitimate tag and reader.
- Network-Transport Layer. This layer includes all the attacks that are based on the way that RFID devices are communicated (between tags and readers).
  - ◦ Cloning Tag: an attacker can clone the tags and use these ones like if were legitimate. An example is the German passports cloning (European Digital Rights (EDRI-gram)., 2006).
  - ◦ Spoofing Tag: is a kind of cloning tag that does not physically replicate an RFID tag. The attackers employ devices that are able to emulate tags and to gain its data and privileges.
  - ◦ Impersonation: adversaries may counterfeit the identity of a legitimate reader in order to elicit sensitive information or modify data stored on RFID tags.
  - ◦ Evesdropping: attacks in which unintended recipients can intercept and read messages.
  - ◦ Counterfeiting: attacks that consist in modifying the identity of an item, generally by means of tag manipulation.
- Application Layer. This layer includes all the attacks in which the target is any information related to an RFID application.

- Unauthorized tag reading: adversaries may easily read the contents of RFID tags without leaving any trace due to not all RFID tags support protocols that authenticate reading operations.
  - Tag modification: an adversary can exploit the fact that most tags have some user writer memory and these tags are used in widespread mode for writing that tag's memory. This is dependent on the RFID standard used and the protection mechansihms employed.
- Strategic Layer. This layer includes attacks in which the target is the organization and business applications.
  - Competitive Espionage: adversaries may often have business or industrial competitors as a target.
  - Social Egineering: an adversary may even use social engineering skills to compromise an RFID system and gain unauthorized access to restricted places or information.
- Multilayer. This layer includes the attacks that belong to several layers.
  - Cover Channels: an attacker can use alternatives and non authorized communcatios channels to transmit information to the tags, due to its free memory or accessing to its data.
  - Denial of Service attacks: the normal operation of tags is interrupted by blocking its access intentionally.
  - Traffic Analyis: an eavesdropper is able to intercept messages and extract information from a communication pattern.
  - Crypto Attacks: attackers employ cryptographic attacks to break the cryptographic algorithms used and revealing or manipulating sensitive information.

- Side Channel Attakcs: it is a kind of crytographic attacks however the information that is usually exploited includes timing information, power consumption or even electromagnetic fields.
  - Replay Attacks: an adversary copies messages passed over the channel and these messages are sent later to perform impersonation attacks.

B. Countermeasures

In this subsection, we present an overview (Peris-Lopez, Hernández-Castro, Estévez-Tapiador, & Ribagorda, 2006) of the proposed solutions to overcome the security and privacy risks linked to RFID technology.

i. Kill Command

This solution was proposed by the Auto-ID Center and EPCglobal (Auto-ID Center, 2003). Each tag has a unique password, which is programmed at the time of manufacture. When tag receives a correct password, the tag will deactivate forever.

ii. Faraday Cage Approach

The privacy of objects labeled with RFID tags is protected by isolating them from any kind of electromagnetic waves using a Faraday Cage (FC), -- container made of metal mesh or foil that is impenetrable by radio signals.

iii. Active Jamming Approach

Active jamming is an alternative to the FC approach that may be done disturbing the radio channel with a device that actively broadcasts radio signals, so as to completely disrupt the communication channel, thus preventing the normal operation of RFID readers.

iv. Blocker Tag

Blocker tag scheme depends on the incorporation into tags of a modifiable bit called privacy bit. A privacy bit marks a tag as "private" and then unwanted scanning of tags mapped into the privacy zone is prevented.

v. Rewritable Memory

Each tag stores an anonymous ID and the adversary cannot know the real ID. Also, anonymous ID must be renewed frequently to solve the traceability problem.

vi. Symmetric Key Encryption

Symmetric key encryption consists in an authentication mechanism based on a simple two-way challenge-response algorithm. The cost of this approach is that tags have to support on-chip AES cipher.

vii. Public Key Encryption

There are solutions that use public-key encryption, based on the cryptographic principle of re-encryption. Nowadays, these approaches are unfeasible for low-cost tags.

viii. Schemes Based on Hash Functions

The use of schemes based on hash functions (hash lock scheme, randomized hash lock scheme and hash-chain scheme) is one of the most widely proposed solutions to solve the privacy and tracking problems that arise from RFID technology.

ix. Authentication Scheme

This solution consists of a system that authenticates readers before they can access the information in the tags in a specific system. In the mutual authentication approaches, tags are also authenticated by readers.

5. CRYPTOGRAPHIC PRIMITIVES

Standard cryptographic approaches in RFID systems require extra hardware to implement security algorithms. RFID tags are extremely resource-limited devices. For example, passive tags have no more than 2000 gates equivalent (GE) available for security (Juels & Weis, 2005). Cryptographic primitives are not available to EPC tags because these tags do not have enough GE to support standard cryptography. In the following, we introduce standard security algorithms, explain its limitations and show why some of these solutions are not workable for RFID systems.

5.1 Hash Functions

Hash functions were introduced in cryptography to provide message integrity and authentication. MD4, MD5, SHA-1, SHA-256 and MAME are nowadays some of the most widely used cryptographic hash functions. However a tiny implementation of these classic hash functions has been an impossible task. In Table 4 an analysis of these hash function can be seen and its minimum number of GE to work.

In recent years, researchers have tried to improve implementation of these primitives to reduce both power consumption and gates equivalents necessary to be supported on-chip of RFID tags. Currently, one of the smallest hash functions designed is *SPONGENT-88* with 738 GE which can do a hash operation of 88 bits and consume 990 cycles (Bogdanov, Knežević, Leander, Toz, Varici, & Verbauwhede, 2011). Between *SPONGENT-88* and the standard hash functions, there are a great variety of new proposed hash functions for RFID systems as *KECCAK*, that requires 2,520 GE (Kavun & Yalcin, 2010), *PRESENT-128*

Table 4. Hash functions comparative (Bogdanov, Leander, Paar, Poschmann, Robshaw, & Seurin, 2008)

	Hash output size	Data path size	Cycles per block	Efficiency (bps/ GE)	Logic process	GE
MD4	128	32	456	15.3	0.13µm	7350
MD5	128	32	612	10	0.13µm	8400
SHA-1	160	32	1274	4.9	0.35µm	8120
SHA-256	256	32	1128	4.2	0.35µm	10868
MAME	256	256	96	32.9	0.18µm	8100

Table 5. Ciphers comparative

	Key length	Cycles per block	Power (µW)	Logic process	GE
TEA	128	64	7.37	0.13µm	2355
AES	128	1032	4.5	0.35µm	3400
DESL	160	144	2.14	0.18µm	2309
IDEA	256	320	2.21	0.18µm	4487

(Bogdanov, Knudsen, Leander, Paar, Poschmann, & Robshaw, 2007) that requires 4,256 GE, or *QUARK* family that requires between 1,379 and 4,640 GE (Aumasson, Henzen, Meier, Naya-Plasencia, Mangard, & Standaert, 2010).

Finally, an interesting scalable universal hash function was proposed by Yüksel called WH-16 (Yüksel, 2004) in the RFID context and have a small footprint (460 gates).

5.2 Ciphers

Traditional ciphers like AES, DES, XTEA or SEA among others, excess by large the computational and storage capabilities of many constrained devices. For instance, the smallest known implementation of AES requires 3,400 GE (Feldhofer & Rechberger, A Case Against Currently Used Hash Functions in RFID Protocols, 2006) (Feldhofer, Dominikus, & Wolkerstorfer, Strong Authentication for RFID Systems Using the AES Algorithm, 2004). In Table 5 a comparative between most commonly used ciphers is presented.

However standard ciphers were not designed specifically for RFID systems (constrained device in the general context). For that reason, several lightweight ciphers have grown-up in these last years. There are two kinds of ciphers in RFID systems: block ciphers and stream ciphers. Stream ciphers encrypt individual characters of a plaintext one at a time, using an encryption transformation, which varies with time. By contrast, block ciphers tend to simultaneously encrypt groups of characters of a plaintext message using a fixed encryption transformation (Menezes, Van Oorschot, Vanstone, & Rivest, 1996).

Examples of block ciphers are: *Present* with 1,000 GE (Bogdanov, Knudsen, Leander, Paar, Poschmann, & Robshaw, 2007), *LED* with 688 GE (Guo, Peyrin, Poschmann, & Robshaw, 2011), *Katan* with 480 GE (Cannière, Dunkelman, & Knežević, 2009) or *PRINT* with 405 GE (Knudsen, Leander, Poschmann, & Robshaw, 2010).

Examples of stream ciphers are: GRAIN with 1,294 GE (Hell, Johansson, & Meier, 2007) or A2U2 with 300 GE (David, Ranasinghe, & Larsen, 2011).

5.3 Pseudo-Random Number Generators (PRNG)

In the standard of EPC, tags shall be able to generate 16-bit random or pseudo-random numbers, and shall have the ability to extract Q-bit subsets from a RN16 to preload into its slot counter. Additionally, tags should be able to temporally store at least two RN16s while powered, for example a handle and a 16-bit cover-code during password transactions. The generator (RN16) should meet the following randomness criteria:

- **Probability of a single RN16**: The probability that any RN16 drawn from the RNG has value RN16 = j for any j, shall be bounded by $0.8/2^{16}$ <P (RN16 = j) < $1.25/2^{16}$.
- **Probability of simultaneously identical sequences**: For a tag population of up to 10,000 tags, the probability that any of two or more tags simultaneously generate the same sequence of RN16s shall be less than 0.1%, regardless of when the tags are energized.
- **Probability of predicting an RN16**: An RN16 drawn from a tag's RNG 10ms after the end of T_r, shall not be predictable with a probability greater than 0.025% if the outcomes of prior draws from RNG, performed under identical conditions, are known.

There are several works in the literature that improve this pseudo-random number generator, increasing for example the number of bits that can be randomized in the random standard function. Some of these examples are *LAMED* that needs 1,566 GE (Peris-Lopez, Hernandez-Castro, Estevez-Tapiador, & Ribagorda, 2009) or *AKARI* that needs 476 GE (Martin, San Millan, Entrena, Peris-Lopez, & Hernandez-Castro, 2011). On the other hand, Che et al. proposed a new pseudorandom number generator base on a single LFSR

(Che, Deng, Tan, & Wang, 2008). This proposal was analyzed in (Melia-Segui, Garcia-Alfaro, & Herrera-Joancomarti, 2010) and the authors proposed a new scheme that is based on the use of a multiple polynomial LFSR.

6. PROTOCOLS FOR LOW-COST RFID

The support of standard and strong cryptographic primitives on low-cost tags is not a viable option nowadays by different reasons. Storage capacity (a few hundreds bits only), gates equivalents (a few thousands available for security purposes), power-consumption (tags are passively powered) are some of these reasons. Therefore, there is a room for designing lightweight protocols for low-cost RFID.

To solve this new necessity of creating these new kinds of authentication protocols that conform to low-cost RFID requirements, there are several proposed solutions: hash functions based solutions, others solutions which exclusively use non-cryptographic primitives and finally there are solutions based on the concept of human-computer authentication (Peris-Lopez P., Hernandez-Castro, Estevez-Tapiador, & Ribagorda, LMAP: A Real Lightweight Mutual Authentication Protocol for Low-cost RFID tags., 2006). In the following, these solutions are explained more thoroughly and some examples of the main representative protocols are given.

A. Hash Function Protocols

Accepting short-term limitations on low-cost tag, in (Sarma, Weis, & Engels, 2002) proposed solution that used a one-way hash function. Authors, in this design, took into account that each hash-enabled tag contains a portion of memory reserved for a "meta-ID" and operates in either an unlocked or a locked state. While unlocked, the full functionality and memory of the tag are available to anyone in the interrogation zone.

Sarma et al. proposed that the owner computes a hash value of a random key and sends it to the tag as a lock value (i.e. lock=hash(key)) in order to lock the label. Subsequently, the tag stores the lock value in the meta-ID memory location and enters into the locked state. While locked, a tag responds to all queries with the current meta-ID value and restricts all other functionalities. To unlock a tag, the owner sends the original key value to the tag. The tag then hashes this value and compares it to the lock value stored under the meta-ID. If a match is found, the tag unlocks itself. However, in this proposed solution, whenever a tag receives a query from a reader, the tag responds with its meta-ID, which is fixed. Therefore, an adversary can track the tag using the meta-ID (Choi E.Y., 2005). Many other researches provided similar authentication protocols based on hash functions inspired on Sarma et al. proposals.

D. Henrici and P. Müller introduced a simple scheme relying on one-way hash-functions that greatly enhances location privacy. The general idea is to change the ID of a tag on every read attempt in a secure manner. Authors understand secure, as a way of prevention against eavesdropping, spoofing, modification, replay, or man-in-the-middle attacks. The methods employed to accomplish the stated goals are the use of one-way hash functions, transaction identifiers to counteract reply attacks, random values generated at the back-end and different transaction numbers so that these values cannot be useful for tracking purposes. The resources demanded on tags are low because computation and data storage is mainly loaded to the back-end database (Henrici D., 2004).

In (Choi E.Y., 2005) the authors proposed an efficient authentication protocol named OHLCAP, which requires only a one-way hash function. When a reader queries to a tag, the tag and the reader authenticate each other as shown in Figure 9. Authors assert that using this scheme, leakage of information is prevented since a tag only emits its information after a successful authentication. By refreshing a message transmitted from a tag in each session, OHLCAP also provides location privacy and is secure against many attacks such as eavesdropping, traffic analysis, message interception, spoofing and replay attacks.

There are several hash-based scheme proposed such as mutual authentication scheme (Dimitriou,

Figure 9. OHLCAP protocol (Choi E.Y., 2005)

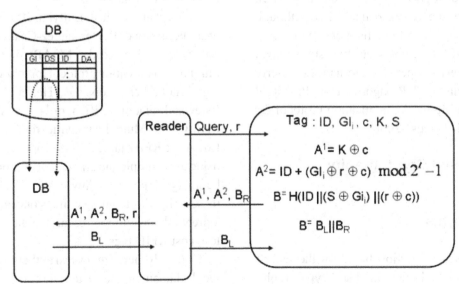

2005), which is based on the use of a secret shared, refreshed to avoid tag tracing, and a standard cryptographic hash function. Another authentication protocol, LCAP, was proposed and needs two one-way hash functions operations (Lee, Hwang, Lee, & Lim, 2005). In OSK scheme (Ohkubo, Suzuki, & Kinoshita, 2003), authors use a hash chain with two hash functions too. Moreover, J. Yang et al. in (Yang, Park, Lee, Ren, & Kim, 2005) proposed authentication protocol that combines a hash function and exclusive-or operations.

All of the above mentioned protocols apparently constitute a good and secure solution. However, in most of these proposals, no explicit algorithms are suggested and finding one is not an easy issue since standard hash functions (MD5, SHA-1, SHA-2) cannot be used (Datasheet Helion Technology, 2005). Furthermore, as well said in (Song, 2008) (Nakahara Jr, 2010) these protocols necessarily do not satisfy all the security properties claimed. For example, the OHLCAP and OSK schemes allow an adversary to track tags and the mutual authentication scheme proposed by Dimitriou is vulnerable to DoS, impersonation and tracking attacks.

However, researchers are still developing solutions based on hash function like in (Yüksel, 2004), (Song & Mitchell, 2008). OHLCAP protocol was analyzed and improved by H. Lie et.al in (Lei, Xin-me, Song-he, & Cai Zeng, 2010). Although these proposals seem to be light enough to fit in a low-cost RFID tag, the security is still an open question (Peris-Lopez P., Hernandez-Castro, Estevez-Tapiador, & Ribagorda, LMAP: A Real Lightweight Mutual Authentication Protocol for Low-cost RFID tags., 2006).

B. Minimalist Cryptography Protocols

i. Pseudonyms

Juels proposed a solution based on the use of pseudonyms (Juels, Minimalist Cryptography for Low-Cost RFID Tags, 2005). This proposal is based on using the tag's memory for storing a short list of random identifiers or pseudonyms —only authorized readers have the same random identifiers. Each time a tag is queried, it sends the next pseudonym in the list.

Pseudonym throttling is simple and practical, but has a shortcoming: the small storage capacity of RFID tags permits only a small list of pseudonyms, and hence only limited privacy protection is offered. For that reason, in the full protocol, authors allow pseudonyms in an RFID tag to be refreshed by authorized verifiers.

Although pseudonyms are commonly computed by using hash functions (and some random number), recall that the use of these primitives have not been ratified by EPC-C1G2 specification.

ii. Lightweight Protocols

Lightweight algorithms and protocols need to be designed taking into account the limitations of the tags. Motivated for this reason, several authors propose the usage of lightweight cryptography. Lightweight cryptography is a new form to encrypt private data through simple triangular operations (e.g. XOR ($\oplus$), OR ($\vee$), AND ($\wedge$) and addition mod 2^n ($+$)), and efficient non-triangular operations (mainly rotations).

Vajda et al. (Vajda & Buttyán, 2003) and Peris et al. (Peris-Lopez P., Hernandez-Castro, Estevez-Tapiador, & Ribagorda, M2AP: A Minimalist mutual authentication protocol for low-cost RFID tags, 2006; Peris-Lopez P., Hernandez-Castro, Estevez-Tapiador, & Ribagorda, LMAP: A Real Lightweight Mutual Authentication Protocol for Low-cost RFID tags, 2006) proposed protocols using these simple and efficient operations that can be surely supported on low-cost RFID. Despite of its security shortcomings, the proposal stand for a great advance in the research area of securing low-cost RFID tags.

Currently there are several authentication protocols that attempt to secure the communications

Figure 10. SASI authentication protocol (Chien, SASI: A New Ultralightweight RFID Authentication Protocol Providing Strong Authentication and Strong Integrity, 2007)

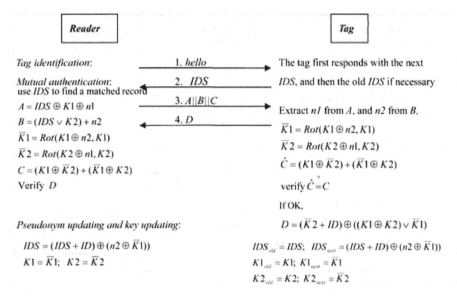

between a reader and the tags and between a reader and the associated database. In this subchapter two main protocols are presented: SASI protocol and a mutual authentication protocol conforming to EPC-C1G2 standard.

1. SASI

Chien proposed a new ultralightweight RFID authentication protocol named SASI (Chien, 2007). The protocol consist of three main stages: tag identification phase, mutual authentication phase, and pseudonym updating and key updating phase (see Figure 10).

Tag identification. Initially, the reader sends "hello" message to the tag, which first responds with its potential next IDS. If the reader could find a matched entry in the database, it starts the mutual authentication phase; otherwise, it tries again and the tag answers with its old IDS.

Mutual authentication. The reader uses IDS to find a matched record in the database. If so, it uses the matched values and two generated random integers n1 and n2 to compute the values A, B, and C. From A, B, C, the tag first extracts n1 and

n2. Then, it computes $\overline{K}1$ and $\overline{K}2$ values and verifies the value of C. If the process is successful the tag computes the response value D. Upon receiving D, the reader uses its local values to verify D and if so, the updating phase starts.

Pseudonym updating and key updating. After the reader and the tag authenticate each other, they update their local pseudonym and keys as can be seen in Figure 10.

In 2010, Avoine et al. (Avoine, Carpent, & Martin, Strong authentication and strong integrity (SASI) is not that strong, 2010) showed that eavesdropping 2^{17} sessions is enough to almost certainly disclose the full tag ID.

2. Mutual Authentication Protocol

Chien et al. proposed a mutual authentication protocol for RFID conforming to EPC-C1G2 standards in (Chien & Chen, Mutual authentication protocol for RFID conforming to EPC Class 1 Generation 2 standards, 2007). This protocol (see Figure 11) is inspired on two previous proposals (Karthikeyan & Nesterenko, 2005; Duc, Park, Lee, & Kim, 2006) and attempts to correct its security

Figure 11. Chien et al. mutual authentication protocol (Chien & Chen, Mutual authentication protocol for RFID conforming to EPC Class 1 Generation 2 standards, 2007)

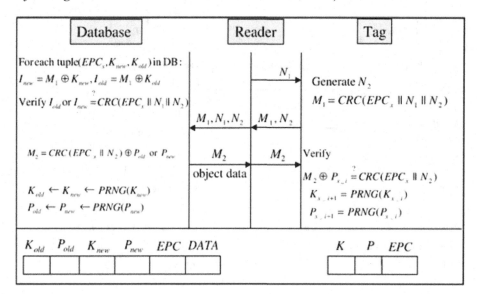

faults by the mechanisms mentioned below: 1) updating the authentication key and access key in the tag after each successful authentication to provide forward secrecy; 2) storing new and old values of the tag keys in the database to keep away from DoS attacks caused by asynchronous updates; and using random numbers to prevent from being tracked and being vulnerable againts replay attacks

In this authentication protocol, the authors use a PRNG that EPC-C1G2 tag supports on-chip 16-bit, and a 16-bit Cyclic Redundancy Code (CRC) checksum that is used to detect errors in the messages exchanged.

The scheme consists of two main phases: the authentication phase and the update phase. In the authentication phase, N_1 and N_2 randomized values are generated by the reader and the tag respectively. Tag calculates M_1 message that is used to find a matched record in the database. If both values are equal, M_1 and I_{old}/I_{new}, then the database computes M_2 message and the reader sends it to the tag. Finally, the tag verifies the correctness of M_2.

Further, in the update phase, K and P values are calculated. Database and tag update its keys as $K_{x\_i+1}=PRNG(K_{x\_i})$ and $P_{x\_i+1}=PRNG(P_{x\_i})$. The authors claimed that the updating mechanism makes the protocol resistant against replay attack. Moreover, the simultaneous maintaining of the old key and new key makes the scheme enough strong to combat DoS attack.

However, three year later, Yeh et al. demonstrated in (Yeh, Wang, Kuo, & Wang, 2010) that Chien et al.'s protocol is vulnerable to DoS and privacy attacks. In case that M_2 was intercepted, tampered or missing up to twice, the database will have no matching old authentication key and access key to complete the mutual authentication that incurs DoS attacks. Privacy attack can be done because of DATA message is transmitted in plaintext, which is very risky. Specifically the protocol only authenticates the tag and the database, and a spoofed reader can therefore access to this private information (DATA) of a tag. Moreover, each time the tag is accessed, all records kept in the database have to be computed and verified one by one to pinpoint the matching record. This process overloads the database and pulls down the overall performance of the system.

C. Human-Computer Protocols

Hoppe and Blum (Hoppe & Blum, 2001) designed a human-to-computer authentication protocol, known as HB, that took benefit of the Learning Parity with Noise (LPN) (Angluin & Laird, 1988), (Blum, Kalai, & Wasserman, 2003) problem and based on human schemes that are suitable for constrained devices like RFID tags (Kiltz, Pietrzak, Cash, Jain, & Venturi, 2011). In this sort of authentication protocols, a human wants to prove his identity to a device and the communication channel is insecure. Additionally, the human and the computer would like to reuse the secret they share in many identifications. The authors took into account that the traditional password approach was unacceptable, since a network snoop can record the password and then will be able to impersonate the user.

However HB authentication protocol, as their original authors claimed, is not secure against active attackers. For that reason, a new proposal based on HB was presented by Juels and Weis in (Juels & Weis, Authenticating Pervasive Devices With Human Protocols, 2005). That new authentication protocol, named HB+, uses the same LPN problem but in this case is applied for RFID systems that offer protection against active attacks. Yet HB+ protocol is vulnerable against active attackers as Gilbert et al. demonstrated in (Gilbert, Robshaw, & Sibert, 2005), showing a man-in-themiddle attack against all these kind of protocols.

Furthermore, the simplicity of HB+ protocol has led to the proposal of other HB-related protocols (Jr, 2010), like PUF-HB (Hammouri & Sunar, 2008), Trusted-HB+ (Bringer & Chabanne, 2008), HB++ (Bringer, Chabanne, & Dottax, HB++: a Lightweight Authentication Protocol Secure against Some Attacks, 2006), HB-MP++ (Yoon, 2009) and HB# (Ouafi, Overbeck, & Vaudenay,

2008) among others. In the following, some HB-family protocols will be explained in more detail.

i. HB Protocol

For illustrating HB protocol, imagine a bearer that holds an RFID tag and an RFID reader sharing a k-bit secret x, and the tag would like to authenticate itself to the reader. Reader selects a random challenge $a \in \{0,1\}^k$ and sends it to the tag. After that, the tag computes the binary inner-product $a \cdot x$ and sends the result back to the reader. Finally, the reader computes $a \cdot x$ value, and accepts it if the tag's parity bit is 0 (no noise).

If an attacker tries to supplant a legitimate and does not know the secret x, she will guess the correct value $a \cdot x$ half of the times. By repeating this process for n sessions, the legitimate tag can lower the probability of natively guessing the correct parity bits for all n rounds to 2^{-n}. The tag intentionally sends the wrong response with constant probability $\varsigma \in (0, \frac{1}{2})$. If the checking procedure fails at most ηn times then the tag is authenticated. Figure 12 depicts a round of the HB protocol.

HB protocol is based on two computational problems with some evidence that these computational problems are hard. These problems can be characterized as loosely based on the sparse

Figure 12. One round of the HB protocol (Juels & Weis, Authenticating Pervasive Devices with Human Protocols, 2005)

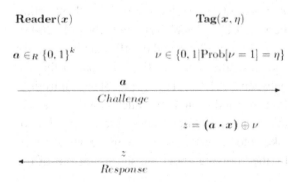

Figure 13. One round of the HB+ protocol (Juels, Minimalist Cryptography for Low-Cost RFID Tags, 2005)

Reader R		Tag T
x_T, y_T, ϵ		x_T, y_T, ϵ
$a \in \{0,1\}^k$		$b \in \{0,1\}^k$
		$\eta \in \{0,1 | \Pr(\eta = 1) = \epsilon\}$
	$\xrightarrow{a}$	$r = (a \cdot x_T) \oplus (b \cdot y_T) \oplus \eta$
	$\xleftarrow{b, r}$	
Accepts if $(a \cdot x_T) \oplus (b \cdot y_T) = r$		

subset sum problem, taken over vectors of digits, with some twists intended to allow more authentications.

ii. HB+ Protocol

HB+ protocol is a symmetric authentication protocol with a simple low-cost implementation. This protocol was proposed by Juels and Weis (Juels & Weis, Authenticating Pervasive Devices With Human Protocols, 2005) to improve the original HB protocol against active attacks. The aim of HB+ protocol is to prevent the extraction of the tag secrets by corrupt readers using adaptive (no-random) challenges.

As the authors mentioned, this scheme is no practical for humans. It requires a tag (playing the role of the human), to generate an additional k-bit random secret y for sharing with the reader (added to the k-bit random secret x value shared in HB protocol). If the tag (or human) does not generate uniformly distributed k-bit random b values, it may be possible to extract information on x or y secret values (Juels & Weis, Authenticating Pervasive Devices With Human Protocols, 2005).

In Figure 13 a round of HB+ can be seen. As shown in the protocol, the principal difference between the HB and HB+ protocol is the introduction of y and b random values in order to avoid active attacks. Unfortunately, the authors did not take into account the extra information given by the knowing about the success or fail of the

protocol and this fact is exploited during the attack proposed in (Gilbert, Robshaw, & Sibert, An Active Attack Against HB+ - A Provably Secure Lightweight Protocol., 2005).

iii. HB++ Protocol

HB++ (Bringer, Chabanne, & Dottax, HB++: a Lightweight Authentication Protocol Secure against Some Attacks, 2006) can be seen as running HB+ twice under independent secrets but with correlated challenges. Moreover, the secrets are renewed at each authentication. All the tags and readers share two functions. First, one function is introduced to link together challenges of the protocol and the other one is needed to determine secrets used for an authentication.

In Figure 14 gives a description of one round of HB++ protocol. The main differences in comparison to its antecessor are the inclusion of triangular operations and rotations, and the needed of universal hash functions (*f*). This last feature makes this protocol infeasible to adapt in EPC tags because the standard does not consider affordable the support of hash functions in the chip of the labels.

iv. HB-Family Protocols

Hammouri and Sunar proposed PUF-HB in (Hammouri & Sunar, 2008). This protocol prevents some man-in-the-middle attacks but this improvement

Figure 14. HB++ authentication protocol (Bringer, Chabanne, & Dottax, HB++: a Lightweight Authentication Protocol Secure against Some Attacks, 2006)

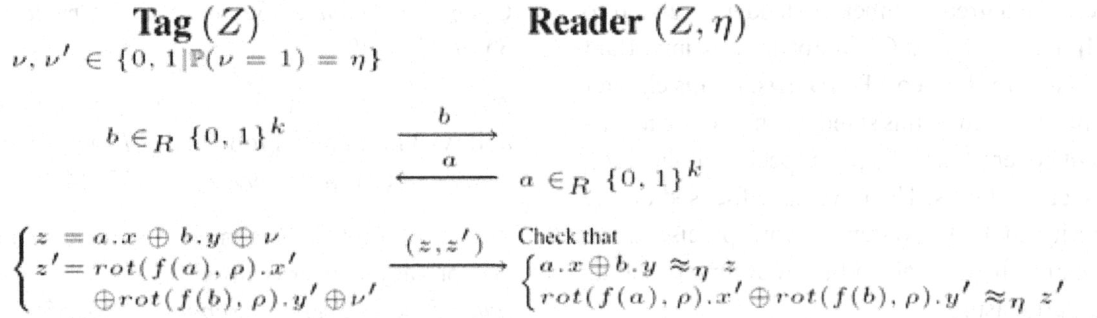

$$\textbf{Tag } (Z) \qquad\qquad\qquad \textbf{Reader } (Z, \eta)$$

$$\nu, \nu' \in \{0, 1 | \mathbb{P}(\nu = 1) = \eta\}$$

$$b \in_R \{0, 1\}^k \qquad \xrightarrow{\;\;b\;\;}$$

$$\xleftarrow{\;\;a\;\;} \qquad a \in_R \{0, 1\}^k$$

$$\begin{cases} z = a.x \oplus b.y \oplus \nu \\ z' = rot(f(a), \rho).x' \\ \quad \oplus rot(f(b), \rho).y' \oplus \nu' \end{cases} \xrightarrow{(z, z')} \begin{array}{l} \text{Check that} \\ \begin{cases} a.x \oplus b.y \approx_\eta z \\ rot(f(a), \rho).x' \oplus rot(f(b), \rho).y' \approx_\eta z' \end{cases} \end{array}$$

increases the hardware requirements for the tags (Nakahara Jr, 2010). Bringer and Chabanne in (Bringer & Chabanne, Trusted-HB: a low-cost version of HB+ secure against Man-in-The-Middle attacks, 2008) proposed Trusted-HB+, which is a protocol based on the use of hash functions. However, this protocol has several weaknesses and an adversary can break the Trusted-HB+ protocol (Nakahara Jr, 2010).

A new improvement of HB+ was proposed in (Duc & Kim, Securing HB+ against grs man-in-the-middle attack, 2007) named HB*. Other protocol called HB-MP providing more efficient performance and resistance against active attacks was proposed in (Munilla & Peinado, 2007). However both variants are insecure as shown in (Gilbert, Robshaw, & Seurin, Good variants of HB+ are hard to find, 2008).

A new protocol based on HB-MP that improves its security level against man-in-the-middle attacks was presented in (Leng, Mayes, & Markantonakis, 2008) as HB-MP+.

7. FUTURE RESEARCH DIRECTIONS

Future research directions in security and privacy of the EPC are described below. First, there is room for designing new schemes or proposing improvement protocols, which are not traceable and protect privacy of tags' holder. Furthermore,

one of the main drawbacks of many RFID applications is that tags answer indiscriminately to reader queries, compromising the privacy of tags' holder (Peris-Lopez, Orfila, Palomar, & Hernandez-Castro, 2011). In this direction, distance-bounding protocols seems an interesting solution. Several interesting proposals, in which distance bounding protocols are combined with authentication protocols, have been recently proposed. Distance bounding schemes are mainly based on measuring the delay (that enables to compute the distance) between sending out a challenge bits and receiving back the response bits. On the other hand, the use of RFID technology in Implantable Medical Devices (IMDs) starts to be a reality (Rotter, 2008). One example of this is GlucoChip (www.positiveidcorp.com) that is an implantable bio-sensing RFID microchip that measures glucose levels in the body in real time. Dr. Lee Berger has also developed a prototype of an orthopedic implant that is equipped with RFID using passive EPC Gen2 tags (Mowry, 2008). Research in security and privacy of IMDs is an important issue, in order to guarantee security and privacy protection (authentication, availability, device software and setting, device-existence privacy, device-type privacy ID privacy, measurement and log privacy, bearer privacy and data integrity) of patients that are going to use the future generations of ubiquitous IMDs (Halperin, Kohno, Heydt-Benjamin, Fu, & Maisel, 2008).

8. CONCLUSION

There are a great number of standards related to RFID technology. EPC is one of the most important standards for low-cost RFID tags. In this chapter we have reviewed this standard (including its different generations) and put a special emphasis on its security faults. Then, we have focused on the security of RFID systems. More precisely, we have presented a state of the art of threats and its countermeasures.

We hope this chapter helps to experts and non-experts to design secure RFID solutions. We are in the era in which there is a widespread use of RFID technology. In fact, RFID tags are combined with sensors (WISP devices) or even with implantable medical devices (IMDs). Nevertheless, the security of these systems cannot be discarded. That is, these new technologies provide both, many benefits and many open doors to attackers since confidential information is at stake.

REFERENCES

Angluin, D., & Laird, P. (1988). Learning from noisy examples. *Machine Learning, 2*(4), 343–370. doi:10.1007/BF00116829

Aumasson, J., Henzen, L., Meier, W., Naya-Plasencia, M., Mangard, S., & Standaert, F. (2010). Quark: A lightweight hash. *Cryptographic Hardware and Embedded Systems, CHES, 2010*, 1–15.

Auto-ID Center. (March 2003). *900 MHz class 0 radio frequency (RF) identification tag specication.*

Avoine, G., Carpent, X., & Martin, B. (2010). Strong authentication and strong integrity (SASI) is not that strong. *Proceedings of the 6th International Conference on Radio Frequency Identification: Security and Privacy Issues*, (pp. 50-64).

Avoine, G., & Oechslin, P. (2005). RFID traceability: A multilayer problem. *Lecture Notes in Computer Science, Security and Cryptology, 3570*, 125-140.

Avoine, G., & Oechslin, P. (2005). RFID traceability: A multilayer problem. In . *Proceedings of Financial Cryptography, 2005*, 125–140.

Ayoade, J. (2007). Roadmap to solving security and privacy concerns in RFID systems. *Computer Law & Security Report, 23*(6), 555–561. doi:10.1016/j.clsr.2007.09.005

Benelli, G., & Pozzebon, A. (2009). NFcare - Possible applications of NFC technology in sanitary. *Proceedings of the Second International Conference on Health* (pp. 58-65). Porto, Portugal: INSTICC Press.

Blum, A., Kalai, A., & Wasserman, H. (2003). Noise-tolerant learning, the parity problem, and the statistical query model. *Journal of the ACM, 50*(4), 506–519. doi:10.1145/792538.792543

Bogdanov, A., Knežević, M., Leander, G., Toz, D., Varici, K., & Verbauwhede, I. (2011). SPONGENT: A lightweight hash function. *In Proceedings of the 13th International Conference on Cryptographic Hardware and Embedded Systems (CHES'11)*, (pp. 312-325).

Bogdanov, A., Knudsen, A., Leander, L., Paar, G., Poschmann, C., & Robshaw, A. (2007). PRESENT: An ultra-lightweight block cipher. In *Proceedings of the 9th international workshop on Cryptographic Hardware and Embedded Systems (CHES '07)*, (pp. 450-466).

Bogdanov, A., Leander, G., Paar, C., Poschmann, A., Robshaw, M., & Seurin, Y. (2008). Hash functions and RFID tags: Mind the gap. In *Proceeding sof the 10th international workshop on Cryptographic Hardware and Embedded Systems (CHES '08)*, (pp. 283-299).

Bringer, J., & Chabanne, H. (2008). *Trusted-HB: A low-cost version of HB+ secure against man-in-the-middle attacks.* Cryptology ePrint Archive, Report 2008/042.

Bringer, J., Chabanne, H., & Dottax, E. (2006). HB++: A lightweight authentication protocol secure against some attacks. *Second International Workshop on Security, Privacy and Trust in Pervasive and Ubiquitous Computing (SecPerU'06)*, (pp. 28-33).

Cannière, C., Dunkelman, O., & Knežević, M. (2009). KATAN and KTANTAN -- A family of small and efficient hardware-oriented block ciphers. *Proceedings of the 11th International Workshop on Cryptographic Hardware and Embedded (CHES '09)*, (pp. 272-288).

Chaudhry, N., Thompson, D., & Thompson, C. (2005). *RFID technical tutorial and threat modeling.* Arkansas.

Che, W., Deng, H., Tan, W., & Wang, J. (2008). A random number generator for application in RFID tags . In Ranasinghe, D. C., & Cole, P. C. (Eds.), *Networked RFID Systems and Lightweight Cryptography, 2008 (* (pp. 279–287). doi:10.1007/978-3-540-71641-9_16

Chien, H. (2007). SASI: A new ultralightweight RFID authentication protocol providing strong authentication and strong integrity. *IEEE Transactions on Dependable and Secure Computing*, *4*(4), 337–340. doi:10.1109/TDSC.2007.70226

Chien, H., & Chen, C. (2007). Mutual authentication protocol for RFID conforming to EPC Class 1 Generation 2 standards. *Computer Standards & Interfaces*, *29*(2), 254–259. doi:10.1016/j.csi.2006.04.004

Choi, E. Y., Lee, S. M., & Lee, D. H. (2005). Efficient RFID authentication protocol for ubiquitous computing environment. In *Proceedings of SECUBIQ'05.*

Datasheet Helion Technology. (2005). *MD5, SHA-1, SHA-256 hash core for Asic.* Retrieved from http://www.heliontech.com

David, M., Ranasinghe, D., & Larsen, T. (2011). A2U2: A stream cipher for printed electronics RFID tags. *2011 IEEE International Conference on RFID (RFID)*, (pp. 176 -183).

Dimitriou, T. (2005). A lightweight RFID protocol to protect against traceability and cloning attacks. *Proceedings of SECURECOMM'05.*

Dobkin, D. (2007). *The RF in RFID: Passive UHF RFID in practice.* Newton, MA: Newnes.

Dobkin, D. M., & Wandinger, T. (2005). A radio-oriented introduction to RFID protocols, tags and applications. *High Frequency Electronics*, August, (pp. 32–46).

Duc, D., & Kim, K. (January de 2007). Securing HB+ against grs man-in-the-middle attack. *Insitute of Electronics, Information and Comunication Engineers, Syposium on Crytography and Information Security*, (pp. 23-26).

Duc, D., Park, J., Lee, H., & Kim, K. (2006). Enhancing security of EPC global Gen-2 RFID tag against traceability and cloning. In *The Proceedings of the 2006 Symposium on Cryptography and Information Security (SCIS'06)*, (pp. 17–20).

EPCglobal. (2007). *Tag class definitions.* Retrieved 2011, from http://www.gs1.org/docs/epcglobal/TagClassDefinitions_1_0-whitepaper-20071101.pdf

EPCglobal. (2008). *EPC radio-frecuency protocols class-1 Generation-2 UHF RFID Version 1.2.0.* Retrieved 2011, from http://www.gs1.org/gsmp/kc/epcglobal/uhfc1g2/uhfc1g2_1_2_0-standard-20080511.pdf

EPCGlobal. (2011). *EPC™ radio-frequency identity protocols EPC class-1 HF RFID air interface protocol for communications at 13.56 MHz Version 2.0.3.*

European Digital Rights (EDRI-gram). (2006). *Cloning an electronic passport* (pp. 4–16).

Feldhofer, M., Dominikus, S., & Wolkerstorfer, J. (2004). Strong authentication for RFID systems using the AES algorithm . In Joye, M., & Quisquater, J.-J. (Eds.), *CHES 2004* (pp. 357–370). doi:10.1007/978-3-540-28632-5_26

Feldhofer, M., & Rechberger, C. (2006). A case against currently used hash functions in RFID protocols. *First International Workshop on Information Security (IS'06)*, (pp. 372-381).

Garcia-Alfaro, J., Barbeau, M., & Kranakis, E. (2008). Security threats on EPC based RFID systems. *Fifth International Conference on Information Technology: New Generations, ITNG 2008*, (pp. 1242 -1244).

Garfinkel, S., Juels, A., & Pappu, R. (2005). *RFID privacy: An overview of problems and proposed solutions. Security Privacy* (pp. 34–43). IEEE.

Gilbert, H., Robshaw, M., & Seurin, Y. (2008). Good variants of HB+ are hard to find (G. Tsudik, Ed.) *LNCS, 5143*, (pp. 156-170).

Gilbert, H., Robshaw, M., & Seurin, Y. (2008). *HB#: Increasing the security and efficiency of HB+. En Advances in Cryptology – EUROCRYPT 2008 (Vol. 4965*, pp. 361–378). Berlin, Germany: Springer.

Gilbert, H., Robshaw, M., & Sibert, H. (2005). An active attack against HB+ - A provably secure lightweight protocol. *Electronics Letters, 41*, 1169–1170. doi:10.1049/el:20052622

Glover, B., & Bhatt, H. (2006). *RFID essentials*. O'Reilly Media, Inc.

Guo, J., Peyrin, T., Poschmann, A., & Robshaw, M. (2011). The LED block cipher. In *Proceedings of the 13th International Conference on Cryptographic Hardware and Embedded Systems (CHES'11)*, (pp. 326-341).

Halperin, D., Kohno, T., Heydt-Benjamin, T., Fu, K., & Maisel, W. (2008, January-March). Security and privacy for implantable medical devices. *IEEE Pervasive Computing / IEEE Computer Society and IEEE Communications Society, 7*(1), 30–39. doi:10.1109/MPRV.2008.16

Hammouri, G., & Sunar, B. (2008). PUF-HB: A tamper-resilient HB based authentication protocol. *Proceedings of the 6th International Conference on Applied Cryptography and Network Security*, (pp. 346-365).

Hell, M., Johansson, T., & Meier, W. (2007). Grain - A stream cipher for constrained environments. *International Journal of Wireless and Mobile Computing, 2*(1), 86–93. doi:10.1504/IJWMC.2007.013798

Henrici, D., & Muller, P. (2004). Hash-based enhancement of location privacy for radio radiofrequency identification devices using varying identifiers. *Proceedings of PERCOMW'04, Second IEEE Annual Conference on Pervasive Computing and Communications Workshops*, (pp. 149–153).

Hoppe, N. J., & Blum, M. (2001). Secure human identification protocols. *Conference on the Theory and Application of Cryptology and Information Security International* (pp. 52-66). ASIACRYPT 2001.

Juels, A. (2005). Minimalist cryptography for low-cost RFID tags. *Security in Communication Networks, LNCS, 3352*, 149–164. doi:10.1007/978-3-540-30598-9_11

Juels, A. (2006). RFID security and privacy: A research survey. *IEEE Journal on Selected Areas in Communications, 24*(2), 381–394. doi:10.1109/JSAC.2005.861395

Juels, A., & Weis, S. (2005). Authenticating pervasive devices with human protocols . In Shoup, V. (Ed.), *CRYPTO 2005, LNCS 3126* (pp. 293–198). doi:10.1007/11535218_18

Karthikeyan, S., & Nesterenko, M. (2005). RFID security without extensive cryptography. *Proceedings of the 3rd ACM Workshop on Security of Ad Hoc and Sensor Networks*, (pp. 63-67).

Karygicmnis, A., Phillips, T., & Tsibertzopoulos, A. (2006). RFID security: A taxonomy of risk. *First International Conference on Communications and Networking in China. ChinaCom '06*, (pp. 1-8).

Kavun, E., & Yalcin, T. (2010). A lightweight implementation of Keccak hash function for radio-frequency identification applications. In *Proceedings of the 6th International Conference on Radio Frequency Identification: Security and Privacy Issues (RFIDSec'10)*, (pp. 258-269).

Kiltz, E., Pietrzak, K., Cash, D., Jain, A., & Venturi, D. (2011). Efficient authentication from hard learning problems. In *Proceedings of the 30th Annual international conference on Theory and Applications of Cryptographic Techniques: Advances in Cryptology* (pp. 7-26). Tallinn, Estonia: Springer-Verlag.

Knudsen, L., Leander, G., Poschmann, A., & Robshaw, M. (2010). PRINTcipher: A block cipher for IC-printing. *Proceedings of the 12th International Conference on Cryptographic Hardware and Embedded Systems (CHES '10)*, (pp. 16-32).

Lei, H., Xin-Me, L., Song-He, Y., & Zeng-Yu, C. (2010). A one-way hash based low-cost authentication protocol with forward security in RFID system. In *Informatics in Control, Automation and Robotics (CAR), 2010 2nd International Asia Conference on* (Vol. 2, pp. 269 -272).

Leng, X., Mayes, K., & Markantonakis, K. (2008). HB-M+ protocol: An improvement on the HB-MP protocol. *IEEE International Conference on RFID*, (pp. 118-124).

Martin, H., San Millan, E., Entrena, L., Hernandez-Castro, J., & Peris-Lopez, P. (2011). AKARI-X: A pseudorandom number generator for secure lightweight systems. *2011 IEEE 17th International On-Line Testing Symposium (IOLTS)*, (pp. 228 -233).

Melia-Segui, J., Garcia-Alfaro, J., & Herrera-Joancomarti, J. (2010). Analysis and improvement of a pseudorandom number generator for EPC Gen2 tags. *Proceedings of the 14th International Conference on Financial Cryptograpy and Data Security (FC '10)*, (pp. 34-46).

Menezes, A., Van Oorschot, P., Vanstone, S., & Rivest, R. (1996). *Handbook of applied cryptography*. Boca Raton, FL: CRC Press, Inc.

Mitrokotsa, A., Rieback, M., & Tanenbaum, A. (2010). Classifying RFID attacks and defenses. *Information Systems Frontiers*, 12(5), 491–505. doi:10.1007/s10796-009-9210-z

Mowry, M. (2008). *A survey of RFID in the medical industry*. Retrieved from http://faculty.ist.psu.edu/xu/papers/jips.pdf

Munilla, J., & Peinado, A. (2007). HB-MP: A further step in the hb-family of lightweight authentication protocols. *Computer Networks*, 51, 2262–2267. doi:10.1016/j.comnet.2007.01.011

Nakahara, J. Jr. (2010). *WG2 - Lightweight cryptographic algorithms*. European Network of Excellence in Cryptology II.

Ohkubo, M., Suzuki, K., & Kinoshita, S. (2003). *Cryptographic approach to "privacy friendly" tags*. RFID Privacy Workshop.

Ouafi, K., Overbeck, R., & Vaudenay, S. (2008). On the security of HB# against a man-in-the-middle attack . In Pieprzyk, J. (Ed.), *Advances in Cryptology -- Asiacrypt 2008* (pp. 108–124). doi:10.1007/978-3-540-89255-7_8

Pasquet, M., Reynaud, J., & Rosenberger, C. (2008). Secure payment with NFC mobile phone in the SmartTouch project. *International Symposium on Collaborative Technologies and Systems, CTS 2008*, (pp. 121-126).

Peris-Lopez, P., Hernandez-Castro, J., Estevez-Tapiador, J., & Ribagorda, A. (2006). *LMAP: A real lightweight mutual authentication protocol for low-cost RFID tags*. Workshop on RFID Security (RFIDSec'06).

Peris-Lopez, P., Hernandez-Castro, J., Estevez-Tapiador, J., & Ribagorda, A. (2006). M2AP: A Minimalist mutual authentication protocol for low-cost RFID tags. *International Conference on Ubiquitous Intelligence and Computing (UIC'06)*, (pp. 912-923).

Peris-Lopez, P., Hernández-Castro, J., Estévez-Tapiador, J., & Ribagorda, A. (2006). RFID systems: A survey on security threats and proposed solutions. *Personal Wireless Communicaitons*, *4217*, 159–170. doi:10.1007/11872153_14

Peris-Lopez, P., Hernandez-Castro, J., Estevez-Tapiador, J., & Ribagorda, A. (2009). LAMED - A PRNG for EPC class-1 generation-2 RFID specification. *Computer Standards & Interfaces*, *31*(1), 88–97. doi:10.1016/j.csi.2007.11.013

Peris-Lopez, P., Orfila, A., Palomar, E., & Hernandez-Castro, J. (2011). A secure dustance-based RFID identification protocol with an off-line back-end database. *Personal and Ubiquitous Computing*, *16*(3), 1–15.

Peris-Lopez, P., Tong-Lee, L., & Tieyan, L. (2008). Providing stronger authentication at a low cost to RFID tags operating under the EPCglobal framework. *IEEE/IFIP International Conference on Embedded and Ubiquitous Computing '08* (pp. 159 -166).

Razaq, A., Luk, W., & Cheng, L. (2007). Privacy and security problems in RFID. *IEEE International Workshop on Anti-counterfeiting, Security, Identification* (pp. 402 -405).

Roberts, L. G. (1975). ALOHA packet system with and without slots and capture. *SIGCOMM Computer Communications Review*, *5*(2), 28–42. doi:10.1145/1024916.1024920

Rotter, P. (2008, April-June). A framework for assessing RFID system security and privacy risks. *IEEE Pervasive Computing / IEEE Computer Society and IEEE Communications Society*, *7*(2), 70–77. doi:10.1109/MPRV.2008.22

Sarma, S. E., Weis, S. A., & Engels, D. W. (2002). RFID systems and security and privacy implications. In *Proceedings of CHES'02, 2523*, (pp. 454–470).

Song, B. A. (2008). RFID authentication protocol for low-cost tags. In *Proceedings of the First ACM Conference on Wireless Network Security* (pp. 140-147). Alexandria, VA, USA.

Vajda, I., & Buttyán, L. (2003). Lightweight authentication protocols for low-cost RFID tags. In *Proceedings of UBICOMP'03*.

Want, R. (2006). An introduction to RFID technology. *Pervasive Computing*, *5*(1), 25–33. doi:10.1109/MPRV.2006.2

Yang, J., Park, J., Lee, H., Ren, K., & Kim, K. (2005). *Mutual authentication protocol for low-cost RFID*. Ecrypt Workshop on RFID and Lightweight Crypto.

Yeh, T., Wang, Y., Kuo, T., & Wang, S. (2010). Securing RFID systems conforming to EPC class 1 generation 2 standard. *Expert Systems with Applications*, *37*(12), 7678–7683. doi:10.1016/j.eswa.2010.04.074

Yoon, B. (2009). *HB-MP++ protocol: An ultra light-weight authentication protocol for RFID system*.

Yüksel, K. (2004). *Universal hashing for ultra-low-power cryptographic hardware applications*.

KEY TERMS AND DEFINITIONS

Active Tag: Term for a radio frequency transponder powered by a battery, which may be replaceable or sealed within the device. These kinds of RFID tags send back information to the reader using their own battery and can be read from 100 meters or more.

ALOHA: A probabilistic anti-collision algorithm that involves a node transmitting a data packet after receiving a data packet. If a collision occurs, a node becomes overfilled and transmits the packet again after a random delay. The interrogator keeps transmitting until the collision does not happen.

Authentication Protocol: It is a series of rules or steps for recognizing the identity of both reader and tag for sending private data in a secure way.

Backscatter Modulation: It is a method of communication between passive tags and readers. In this process a tag responds to a reader signal or field by modulating and reradiating the response signal at the same carrier frequency. The reflected signal is modulated to transmit data.

Cyclic Redundancy Check (CRC): It is an error detection algorithm that checks data stored on RFID tags to be sure that have not been corrupted or some of it lost.

Electronic Product Code (EPC): A serial, created by the Auto-ID Center, which complements bar codes. The EPC has digits to identify the manufacturer, product category, and individual item.

Lightweight Cryptography: It is a new form to encrypt private data through triangular operations like XOR ($\oplus$), OR ($\vee$), AND ($\wedge$), addiction mod 2^n ($+$) and non-triangular operations such as rotations. This kind of cryptography only needs a few equivalents gates for encoding data.

Passive Tag: It is an RFID device without its own power source and transmitter. Passive tag reacts to a specific inductively coupled or radiated electromagnetic field. This energy is converted by the antenna into electricity than can power up the microchip in the tag.

Reader: A device used to communicate with RFID tags. The reader has one or more antennas, which emit radio waves and receive signals back from the tag. Moreover, readers may be connected to a host computer or industrial controller with a database for managing RFID tags.

Reader-Talks-First: A means by which a passive UHF reader communicates with tags in its read field. The reader sends energy to the tags but the tags sit idle until the reader requests them to respond. The reader is able to find tags by asking all tags with a serial number (EPC).

RFID: A method of identifying unique items using radio waves. Typically, a reader communicates with a tag, which holds digital information in a microchip.

Semi-passive Tag: It is similar to active tag, except that the battery is used to run the microchip's circuitry but not to broadcast a signal to the reader. Some of semi-passive tags can be slept until a reader does not wake up it by a signal, which conserves battery life.

Singulation: A means by which an RFID reader identifies a tag with a specific serial number from a number of tags in its field. There are different methods of singulation, but the most common is the "tree walking", which involves asking all tags with a serial number that starts with either a 1 or 0 to respond. If more than one responds, the reader might ask for all tag with a serial number that starts with 01 to respond, and so on.

Slotted Antenna: An antenna that consists only of a narrow slot cut into an electrical conductor connected to the tag. Slotted antennas exhibit the same orientation sensitivity as dipoles.

Section 2
Security

Chapter 3
Towards Sensing–Enabled RFID Security and Privacy

Di Ma
University of Michigan-Dearborn, USA

Nitesh Saxena
University of Alabama, Birmingham, USA

ABSTRACT

Recent technological advancements enrich many RFID tags with sensing capabilities. This new generation of RFID devices – supporting sensing, computation, and RFID communication - can facilitate numerous promising applications for ubiquitous sensing and computation. They also suggest new ways of providing security and privacy for RFID systems by utilizing the unique characteristics of sensor data and sensing technologies. In this chapter, the authors highlight these new possibilities and advocate the use of sensing-enabled RFID tags in security-critical applications. The purpose of this chapter is to bring awareness of these opportunities to both the research and industrial community, and to incite interests in sensing-centric security and privacy research and development for the future generation of RFID systems.

INTRODUCTION

Passive RFID (Radio Frequency Identification) tags are miniaturized devices that enable automated identification in numerous applications and circumstances (e.g., access cards, contactless credit cards, e-passports and medical implants). Unfortunately, RFID tags are plagued with a wide variety of security and privacy vulnerabilities due to the weaknesses of the underlying wireless

communication radio communication. They often store sensitive information and usually respond promiscuously to any read requests. This renders the tag-specific information easily subject to eavesdropping and unauthorized reading, which further allow owner tracking, cloning or impersonation. RFID tags are also susceptible to different forms of relay attacks. They are also likely to get lost or stolen, in which case the services they provide are themselves endangered.

Providing security and privacy services for RFID tags presents a unique and formidable set

DOI: 10.4018/978-1-4666-1990-6.ch003

of challenges. This is due to the constraints of these tags in terms of computation, memory and power resources. The problem is exacerbated by the very strict and somewhat unusual requirements of RFID applications (originally geared for automation) in terms of usability. Consequently, currently deployed or proposed solutions often fail to meet these constraints and requirements. Many security vulnerabilities remain unsolved. This motivates the need for new security and privacy mechanisms suitable for different RFID applications in terms of not only *efficiency* and *security*, but also *usability*.

Recent technological advancements make the fusion of RFID and sensors viable and enable many RFID tags with low-cost sensing capabilities. Intel's Wireless Identification and Sensing Platform (WISP) is a representative sensing-enabled tag which extends RFID beyond simple identification to in-depth sensing and computation (Smith, 2006; Sample, 2007). It uses an ultra-low power 16-bit general-purpose micro-controller for sensing, computation and RFID communication. Such sensor nodes (with RFID communication interface) can be powered by harvesting Radio Frequency (RF) power from a reader. Harvested energy is stored on a capacitor that can sustain virtually unlimited charging cycles, enabling an RFID sensor to have a potentially very long lifespan and few or no requirements vis-a-vis maintenance. Also, being battery-less, these devices can have smaller form factor, which allows them to be used for sensing and computation in places where a battery-powered device cannot be placed. This new generation of RFID+sensor devices can facilitate new application domains where long life and small size (as well as, possibly, deployment in inaccessible locations) are important. They have seen use in studies on a variety of topics, from critical infrastructure health monitoring to human behavior monitoring. They also suggest new ways of providing security and privacy services by leveraging the unique characteristics of sensor data and sensing technologies for current RFID

systems: *First*, a sensor-enabled RFID tag can acquire useful contextual information about its environment (or its owner, or the tag itself) so as to make informed decisions (for better security and privacy). *Second*, sensor-based communication channels can be established for tags to interface with its reader, or with other smarter devices for enhanced and finer-grained security. *Third*, the variability and unpredictability of certain sensed data across time, space and individuals allow extraction of randomness – an essential requirement for many cryptographic functions and protocols.

In this chapter, we highlight these potential opportunities brought by the new sensing capabilities of RFIDs, and discuss in details how different sensing technologies can be leveraged to provide efficient and user-friendly security and privacy solutions. The use of sensors can be quite straightforward in some cases and pretty complicated in some other cases. Therefore further study and investigation are needed. The purpose of this chapter is to bring awareness of these opportunities to both the research and industrial community, and to incite interests in sensing-centric RFID security and privacy research and development for the future generation of RFID systems.

The organization of the chapter is as follows. We first review security and privacy threats and existing countermeasures in the current RFID systems, and identify security and privacy primitives that need further research. We then review and discuss existing and on-going research on sensing-centric security and privacy solutions for RFIDs organized into three categories: context recognition, location-limited communication, and randomness extraction. We next discuss possible attacks targeting the sensing-centric mechanisms. Future work is then discussed, and concluding remarks are outlined finally.

BACKGROUND

RFID Systems, Threats, and Countermeasures

A typical RFID system consists of three entities: tags, readers and/or back-end servers. Tags, also called transponders, are miniaturized wireless radio devices that store information about their corresponding subject. Readers, also known as interrogators, broadcasts queries to tags in their radio transmission ranges for information contained in tags and tags reply with such information. The queried information is then sent to the server (which may co-exist with the reader) for further processing and the processing result is used to perform proper actions (such as updating inventory, opening gate, charging toll or approving payment). Prominent RFID applications include supply chain management (inventory management), e-passports, credit cards, driver's licenses, vehicle systems (toll collection or automobile key), access cards (building or parking, public transport), and medical implants.

In many RFID applications, tags store sensitive and/or personally identifiable information. For example, a US e-passport stores the name, nationality, date of birth, digital photograph, and (optionally) fingerprint of its user. Similarly, a contactless credit card stores the credit card account number. When stored on a tag, such information is subject to clandestine *eavesdropping* and *unauthorized reading*. The information (might simply be plain identifier) gleaned from a RFID tag can also be used to *track* the owner of the tag, or be utilized to *clone* the tag so that an adversary can impersonate the tag's owner.

Moreover, RFID tags are susceptible to different types of *relay* attacks. One class of these attacks is referred to as *"ghost-and-leech"* (Kfir, 2005). In this attack, an adversary, called a "ghost," relays the information surreptitiously read from a legitimate RFID tag to a colluding entity known as a "leech." The leech can then relay the received information to a corresponding legitimate reader. This way a ghost and leech pair can succeed in impersonating a legitimate RFID tag without actually possessing the device. A more severe form of a relay attack, usually against payment cards, is called *"reader-and-leech"*; it involves a malicious reader using which the owner intends to make a transaction (Drimer, 2007). In this attack, the malicious reader, serving the role of a ghost and colluding with the leech, can fool the owner of the card into approving a transaction which she did not intend to make (e.g., paying for a diamond purchase made by the adversary while the owner only intending to pay for food). We note that addressing this problem requires *secure transaction verification*, i.e., validation that the tag is indeed authorizing the intended payment amount. The feasibility of executing relay attacks has been demonstrated on many RFID deployments, including the Chip-and-PIN credit card system (Drimer, 2007), RFID-assisted voting system (Oren, 2009), and keyless entry and start car key system (Francillon, 2011).

Furthermore, RFID tags may get lost or stolen. Such loss or theft leads to a complete compromise of the services provided by these tags. For example, a lost-and-found or stolen wallet containing Alice's access card would allow an adversary (e.g., a potential thief) with unauthorized entry into Alice's office building.

Numerous countermeasures addressing several of the aforementioned threats have been proposed. These, however, do not satisfy the requirements of the underlying RFID applications in terms of (one or more of) efficiency, security and usability. Existing solutions can be roughly divided into two categories -- non-cryptographic and cryptographic. Non-cryptographic approaches address the problem of unauthorized reading by means of *selective unlocking* of tags (Juels, 2003; Juels, 2005-1; Rieback, 2005). That is, tags are made to respond selectively, rather than promiscuously. However, these approaches require the presence of a special-purpose hardware and/or explicit user involvement (necessitating changes to the existing usage model). Consequently, this undermines the overall usability and acceptability of these approaches, and reduces the likelihood of their deployment.

Many cryptographic approaches, on the other hand, are beyond the reach of most RFID tags because of their complexities. For example, distance-bounding protocols aimed at thwarting relay attacks (Drimer, 20076; Francillon, 2011) are currently infeasible due to their extreme sensitivity to tag-side processing delays and need for a reader-side high precision clock. There has been a growing interest in the research community to design lightweight cryptographic mechanisms. These mechanisms can potentially be used to secure tag-reader communication, thereby preventing eavesdropping and cloning attacks. However, to be able to use these protocols (or any other cryptographic protocols for that matter), two prerequisites must first be satisfied. First, the tag and reader need to be *securely associated* or "paired" with each other, especially in scenarios where establishing pre-shared keys or a trusted infrastructure is infeasible. One such compelling scenario is granting emergency access to an implanted tag. Second, a viable means of *true random number generation (RNG)* functionality is required. Unfortunately, the current state-of-the-art fails to meet these prerequisites.

We observe that there has been, surprisingly, no prior research on protecting RFID tags in the event of tags' loss or theft. Such a protection can be achieved by means of *user(-to-tag) authentication*.

Prior Work

The discussion above motivates the need for new security and privacy mechanisms on five fundamental primitives: (1) *Selective Unlocking*, (2) *Secure Transaction Verification*, (3) *Secure Association*, (4) *Random Number Generation*, and (5) *User Authentication*. A summary of various threats these primitives protect against is provided in Table 1. In practice, several of these primitives need to be deployed together as per the requirements of a given RFID application in terms of efficiency, security and usability. We note that secure association and RNG are independent primitives, but they are to be used in conjunction with cryptographic mechanisms. Selective unlocking cannot prevent eavesdropping, tracking or cloning because the adversary can launch these attacks whenever the tag is unlocked by its owner. It can also not provide protection in the event of loss/theft because the adversary, in possession of the tag, can unlock the tag just like a legitimate owner; user authentication is the only primitive that is effective here. Similarly, secure transaction verification is the only primitive that can address reader-and-leech attacks. The traditional cryptographic protocols are ineffective against relay attacks because relay attackers can simply forward the cryptographic information back and forth between a valid tag and a valid reader. In the following paragraphs, we review the existing work related to the different primitives. We recall that there has been no prior research on user authentication.

Table 1. RFID defense primitives and threats they can protect against

	Unauthorized Reading	Eavesdropping	Tracking	Cloning	Ghost-and-Leech	Reader-and-Leech	Loss or Theft
Selective Unlocking	Yes	No	No	No	Yes	No	No
Secure Transaction Verification	Yes	No	No	No	Yes	Yes	No
Crypto + RNG + Secure Association	No	Yes	Yes	Yes	No	No	No
User Authentication	Yes	No	No	No	Yes	No	Yes

Hardware-based Selective Unlocking: Hardware-based selective unlocking schemes have been proposed previously. These include: Blocker Tag (Juels, 2003), RFID Enhancer Proxy (Juels, 2005-1), RFID Guardian (Rieback, 2005) and Vibrate-to-Unlock (Asokan, 2011). All of these approaches, however, require the users to carry an auxiliary device (a blocker tag in Juels (2003), a mobile phone in Saxena (2011), and a PDA like special-purpose RFID-enabled device in Juels (2005-1) and Rieback (2005)); such an auxiliary device may not be available at the time of accessing RFID tags, and users may not be willing to carry these devices always. A Faraday cage can also be used to prevent an RFID tag from responding promiscuously by shielding its transmission. However, a special-purpose cage (a foil envelope or a wallet) would be needed and the tag would need to be removed from the cage in order to be read. This greatly decreases the usability of such solutions as users are not willing to put up with changes to traditional usage model given that RFID devices were meant to make life easier for people. Moreover, building a true Faraday Cage that shields all communication is known to be a significant challenge. For example, a crumpled sleeve is shown to be ineffective for shielding purposes (Koscher, 2009).

Cryptographic Protocols: Cryptographic reader-to-tag and tag-to-reader authentication protocols could be used to protect against unauthorized reading and cloning, respectively. The latter can also be made privacy-preserving to prevent owner tracking (see, e.g., Dimitriou, 2005; Halevi, 2009; Molnar, 2004). Due to their complexities, many of these protocols are still unworkable even on high-end tags, and there exists another line of research on designing lightweight protocols (e.g., Bringer, 2006; Gilbert, 2008; Jules, 2005-3; Katz, 2006).

Distance Bounding Protocols: These protocols have been used to thwart relay attacks. A distance bounding protocol is a cryptographic challenge-response authentication protocol. Hence, it requires pre-exist secure association (such as shared keys) between tags and readers as other cryptographic protocols. Besides authentication, a distance bounding protocol also allows the verifier to measure an upper-bound of its distance from the prover. Using this protocol, a valid RFID reader can verify whether the valid tag is within a close proximity thereby detecting ghost-and-leech and reader-and-leech relay attacks. (We stress here that normal "non-distance-bounding" cryptographic authentication protocols have NO help in defending against relay attacks.) The upper-bound calculated by an RF distance bounding protocol, however, is very sensitive to processing delay (the time used to generate the response) at the prover side. This is because a slight delay (of the orders of a few nanoseconds) may result in a significant error in distance bounding. Because of this strict delay requirement, even XOR- or comparison-based distance bounding protocols (Brands, 1993; Hancke, 2005) are not suitable for RF distance bounding since signal conversion and modulation itself can lead to significant delays. By eliminating the necessity for signal conversion and modulation, a very recent protocol, based on signal reflection and channel selection, achieves a processing time of less than 1 *ns* at the prover side (Rasmussen, 2010). However, it requires specialized hardware at the prover side due to the need for channel selection. This renders existing protocols currently infeasible for even high-end devices, such as RFID tags.

Secure Transaction Verification: Addressing the problem of reader-and-leech relay attacks requires secure transaction verification. This can trivially be achieved by showing the transaction details (e.g., the amount of transaction) on the tag itself, assuming the tag possesses a low-cost display (Nithyanand, 2010), and having the user validate the details. This approach, however, is problematic because it requires explicit user involvement and necessitates that the tags are taken out from their containers (wallets or purses) while making the transaction. Another proposed solution for secure transaction verification is based on distance bounding protocols. However, as discussed below, these protocols are currently infeasible for many RFID tags.

Secure Association: The only prior work on RFID secure association is in the context of emergency access control to implantable medical devices (Halperin, 2008; Rasmussen, 2009). In Halperin (2008), the use of a vibrational or low-frequency audio channel is suggested to establish the shared key between an implant and a reader. Basically, the tag, equipped with a piezo, selects a random key and transmits it to the reader, equipped with a microphone, in the form of beeping. However, in a recent work (Halevi, 2010), it has been demonstrated that this approach is insecure. In particular, audio emanations associated with this channel can be eavesdropped upon using a traditional microphone (from a short distance) and a parabolic microphone (from a longer distance). The scheme of Rasmussen (2009), on the other hand, is based on distance bounding, which is currently infeasible as discussed above.

Randomness Extraction: RNG is beyond the capacity of today's average RFID tags. A recent proposal (Holcomb, 2007) involves the use of onboard RAM as the source of "true" randomness. However, our own prior work demonstrates (Saxena, 2009) the limitations of this approach. First, this method must compete with other system functionalities for use of memory. Thus, the amount of uninitialized RAM available for utilization as a randomness generator may be severely restricted. Second, RAM is subject to *data remanence*; there is a time period after losing power, during which stored data remains intact on the memory. This means that after a portion of memory has been used for entropy collection once, it will require a relatively extended period of time without power before it can be re-used. In a usable RFID based security application, which requires multiple (or long) random numbers, this may lead to unacceptably high delays. RNG based on hardware and internal noise (Tormanen, 2010)

has also been suggested. However, these methods require special and essentially single-purpose hardware, which severely limit their applicability.

SENSING-CENTRIC SECURITY AND PRIVACY: PRINCIPLES AND FEASIBILITY

Principles

To advocate the sensing-centric approach for RFID security and privacy, we first answer the question: *Why is it beneficial to use sensors for RFID security and privacy?*

The physical environment offers a rich set of attributes that are unique in space, time, and to individual objects. These attributes -- such as temperature, sound, light, acceleration or magnetic field -- reflect either the current condition of a tag's surrounding environment or the condition of the tag (or its owner) itself. They provide useful contextual information that can be leveraged to make informed decisions about the tag (or its owner). More specifically, this contextual information can be used in two ways. *First*, it can be used to add "contextual intelligence" to the tags using which they can *selectively respond* to reader interrogations, thus raising the bar even for sophisticated adversaries without affecting the RFID usage model, i.e., without imposing additional user burden. That is, rather than responding promiscuously to queries from any readers, a tag can leverage upon "context recognition" and will only communicate when it makes sense to do so. For example, an office building access card, equipped with a location sensor, can remain locked unless it is near the (fixed) entrance of the building. Similarly, a toll card, enabled with a speed sensor, can be designed to remain in a locked state except when the vehicle is travelling at a designated speed near a toll booth (such as 10~15 mph). *Second*, contextual information can be used by the server to assist in *secure transaction verification*. For example, a bank server will deny a $2000 transaction when it detects the tag (RFID credit card) is currently located in a restaurant whose normal transaction is usually less than $50.

Sensing capabilities also enable the tag to establish location-limited -- audio, visual or tactile -- communication with its reader. Being location-limited or proximity-based, these channels (unlike the primary radio channel) are (better) resilient to the man-in-the-middle attacks, and can therefore be leveraged for *secure association* between the tag and reader. Location-limited channels also allow the tag to communicate with other "smarter," personal and trusted devices (e.g., the owner's mobile phone) without the latter requiring an RF communication interface. This is attractive because computationally intensive or fine-grained sensing tasks can now be outsourced from the RFID tag to the smart device. Most beneficially, this functionality allows the auxiliary device to be used as an authentication token for the purpose of *user authentication*, providing protection in the event of tag's loss or theft.

Generating random numbers of sufficient quality for cryptographic applications is a fundamental task. While modern general purpose computers have several techniques available for the generation of high quality random numbers, this requirement is beyond the capacity of today's RFID tag. Sensing data can change in space, time, and across different objects, and many physical attributes de-correlate rapidly in space and time. This serves as a natural entropy collection process and onboard sensors can be used to collect physical parameters of different forms for use as a source of randomness.

In summary, sensor data and sensing technologies can be utilized for security and privacy purposes in three fundamentally different categories: 1. context recognition; 2. location-limited communication; 3. randomness extraction. The five primitives identified in the last section can be built on the functionalities provided by each category (as shown in Figure 1).

Feasibility

We argue the feasibility of the sensing-centric approach for RFID security and privacy in terms of technology, economy, and power budget.

Technology

Two technological advancements in industry empower and support the research on the use of sensors for RFID security and privacy. First, embedding sensors on RFID tags is a technology trend. Various types of sensors have been incor-

Figure 1. Sensing-based approach for RFID security and privacy primitives

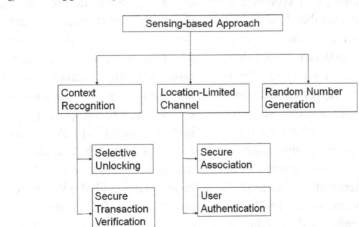

porated to many RFID tags (Holleman, 2008; Ruhanen, 2008; 100). Intel's Wireless Identification and Sensing Platform (WISP) (Smith, 2006; Sample, 2007; Sample, 2009) is a representative example of a sensor-enabled tag. It is compliant with the Electronic Product Code (EPC) protocol like an ordinary tag. What differentiates it from an ordinary tag is its inclusion of an onboard Texas Instruments MSP430F2131 micro-controller and sensors such as accelerometer, voltage sensors. Furthermore, it also has an extensible hardware architecture which allows for integration of new sensors. Second, more unconventional types of sensors have been invented recently to measure new environmental attributes and offer a richer set of information about the devices to which they are attached (Adee, 2010; Ruhanen, 2008).

Economy

Security comes at a cost. So a basic question this raises in regard to the sensing-centric approach is: "whether the cost of sensor-enabled tags is acceptable?" The cost of RFID tags is dependent on several factors such as capabilities of the tag (computation, memory), packaging of the tag (e.g., encased in plastic or embedded in a label), and the volume of tags produced. High-end RFID tags, e.g., those available on e-passports or some of the access cards that are capable of performing certain cryptographic computations like AES or RSA encryption, cost around $5, whereas low-end inventory tags that do not support any (cryptographic) computation cost only about $0.20 (Wagner, 2005). (We emphasize that our proposal generally targets high-end RFID tags that open up a wide array of applications and generally require higher level of security and privacy. Inventory tags, at least for the time being, are not within the scope of this article.) The current cost of WISP tags – equipped with a thermometer and an accelerometer – assembled from discrete components cost roughly $25 but it is expected that this number will be reduced

closer to $1 once they are mass manufactured (Buettner, 2009). This cost is certainly acceptable for high-end tags and does not affect their business model. Incorporating sensors on tags – i.e., increasing the capabilities of tags – may raise the price of tags initially. However, in the long run, following Moore's law, although advances in process technology and mass production will not reduce the cost of small chips such as RFID tags indefinitely because of the per-die cost, they should enable tags with more capabilities (such as sensing, increased computation and memory) at the same cost of today's tags (Czeskis, 2008).

Power Budget

In order to realize sensing-centric mechanisms, it is also important to understand the underlying power requirements since onboard sensors would need to draw power from the reader to work. We discuss the requirements for sensors to work under the power budget of WISPs, instead of generic RFID tags. WISP is powered by the conversion of induced RF power from the reader into DC voltage (1.8V) in wireless mode. The micro-controller (MCU) MSP430 on the WISP draws approximately 1 mA running at full speed (200A per MHz). For a sensor to be integrated with and work on the WISP platform, we have to take into account the following considerations. First, the sensor must allow for a supply voltage of 1.8V. Second, the time and current assumption necessary to make a measurement/sensing (power-on time + settling time) must be small. Third, the power required for additional circuitry essential for interfacing the sensor must be taken into consideration along with the overall power consumed. A rough approximation of the energy budget available is 1 mA for 1 ms. That is, as long as the product of the current and settling time of the sensor is less than 1mA*1ms, the sensor can be integrated with the WISP and work under the power budget. Otherwise, a storage capacitance has to be added to the WISP to support additional power consumption request. Apparently, not all types of sensor can be supported by the power budget of WISP. However, low power sensors which meet the above requirements can work within the induced power on the WISP, including those previously implemented on WISP (rectified voltage, light level, temperature, and acceleration). Several recent works have successfully integrated additional sensors on the WISP platform, including capacitive sensor (Yeager, 2008), and neural sensor (Holleman, 2008). Some other sensors that meet the above requirements include: Honeywell's HMC1053 3-D magnetometer, LOCOSYS Technology's LS20031 GPS receiver, Servoo Corporation's MS5607 pressure sensor, and ST's MP34DB01 audio sensor (microphone).

Besides choosing the appropriate sensors, it is also important to design efficient context recognition algorithms so that they can run on the WISP platform. Prior work (Czeskis, 2008; Saxena, 2010; Asokan, 2011) already demonstrate the practicality of developing different context recognition techniques, similar in requirements to the techniques proposed in this document, on the WISPs. Individual operations can be further optimized to reduce power consumption. As an alternative approach, one can also use the checkpoint strategy proposed in Salajegheh (2009) which allows a tag to perform demanding computations despite limited energy and interruptions of power that lead to complete loss of the contents of RAM. In short, the idea is for an interrupted tag to backup its RAM state just before it loses power (e.g., when the reader becomes out of range of the tag). When the reader comes in close enough proximity of the tag, the tag can retrieve its backed up state and resume the unfinished operations, without having to re-start them from the very beginning.

CONTEXT RECOGNITION

In this section, we explore context recognition – i.e., the use of contextual information retrieved by (or provided to) tag's on-board sensors – for

enhanced security and privacy of RFID tags. Specifically, we first examine how context recognition can be used for selective unlocking so as to provide improved protection against unauthorized reading and ghost-and-leech relay attacks. Secondly, we suggest ways in which context recognition can be used as a basis for secure transaction verification. This allows the tag to validate the amount of a transaction before authorizing it, thereby addressing the problem of malicious readers and reader-and-leech attacks.

The design of context recognition for RFID tags poses several challenges. First, the resource constraints of RFID tags hamper the complexity of the algorithms that can be used to judge what activity a tag is currently undergoing. Another obstacle is the lack of ways in which users can interact with their tags. RFID tags, being geared for automation, were designed to be as transparent as possible to their users, and as such lack any input or output interfaces such as buttons and displays. Moreover, many users are typically not in direct contact with their tags because they prefer to keep them inside other objects, such as wallets or purses. For example, it is a common practice to swipe one's wallet containing the tag against the reader rather than taking the tag out from the wallet and directly swiping the tag.

Selective Unlocking

There are two recently proposed selective unlocking schemes, both working with accelerometer-equipped RFID tags. We first review these two schemes and discuss their merits and demerits. We then outline some selected potential schemes based on other conventional sensor types such as magnetometer and location sensor (GPS). For each mechanism, we also suggest specific application(s) that could benefit from it. We note that, although each proposed mechanism can work in a stand-alone fashion, different mechanisms can also be employed together to provide even stronger security and better usability.

Secret Handshakes

"Secret Handshakes" is a recently proposed interesting selective unlocking method that is based on gesture recognition (Czeskis, 2008). In order to unlock an accelerometer-equipped RFID tag using Secret Handshakes, a user needs move or shake the tag (or its container) in a particular pattern. The tag will only engage in wireless communications when it internally detects these (pre-defined) gesture patterns. For example, the user might be required to move the tag parallel with the surface of the RFID reader's antenna in a circular manner.

To recognize a particular gesture pattern, or secret handshake as called in the paper, secret handshake templates are first created and stored on the WISP. Cross-correlation of the accelerometer data against a template is then computed. If the cross-correlation value exceeds some thresholds, the corresponding gesture is recognized as the movement of the tag. The accelerometer fingerprints of gesture templates used in secret handshakes should significantly differ from those which are produced during everyday activities (such as sitting, eating, walking, running, and jumping) in order to reduce false rate. A number of unlocking patterns (such as circle, triangle, 1.5-wave) were studied and shown to exhibit low error rates.

A central drawback to Secret Handshakes, however, is that a unique movement pattern is required for each tag to be unlocked. This requires subtle changes to the expected RFID usage model. While a standard, insecure RFID setup only requires users to bring their RFID tags within range of a reader, the "Secret Handshakes" approach requires that users consciously move the tag in a certain pattern. This undermines the usability of this approach. Moreover, it does not apply to applications where tags are fixed (such as a toll card usually fixed to windshield of card) or simply inaccessible (such as an implantable medical device).

Motion Detection

"Motion Detection" is another selective unlocking scheme based on accelerometer reading (Saxena, 2010). It keeps in mind the goal of not incorporating any usage model changes. In Motion Detection, a tag would respond only when it is in motion, instead of doing so promiscuously. In other words, if the device is still, it remains silent. This approach hinges on the straightforward observation that accessing a personal mobile RFID tag fundamentally involves moving it in some manner (e.g., swiping an access card in front of the reader). Although Motion Detection does not require any changes to the traditional usage model and raises the bar required for some common attacks to succeed, it is not capable of discerning whether the device is in motion due to a particular gesture or because its owner is in motion. Thus daily activities, such as walking, running or jumping, can also unlock the tag. Hence, the false unlocking rate of this approach is high, meaning there is a high chance that a tag gets unlocked when it actually should have been locked.

As we have seen above, Secret Handshakes nevertheless changes the usage model and is not suitable for applications where tags cannot be moved by hands (such as toll cards which are usually fixed on the windows or bumpers of vehicles in order to allow successful payment deduction) while Motion Detection has high false unlocking rate. This motivates the design of selective unlocking schemes based on other sensor modalities that focus not only on security and privacy, but also on usability. We suggest the following possible alternative ways to do it.

Magnetic-field triggered proximity sensing. The requirement for an RFID tag and a reader being in close physical proximity is common in most RFID applications. For example, while making a payment, a user typically needs to bring her contactless credit card (or its container) closer to the reader for transaction processing. This requirement can therefore serve as an effective means to establish a valid context for selective unlocking. In other words, the tag can get unlocked whenever it senses that it is in close proximity of a reader; otherwise it remains locked.

One possible approach to do RFID proximity sensing could be using magnetometers through magnetic field strengths. That is, on the tag side, it requires a scalar magnetometer which measures the total strength of the magnetic field the sensor/tag is subjected to. On the reader side, it requires only a physical attachment in the form of a magnet. More specifically, when the tag is brought close to the reader, as in the traditional usage model, the tag's on-board magnetometer would sense the magnetic field and the tag would get unlocked if the strength of the magnetic field is above some pre-defined threshold (i.e., matching the level of the magnetic field produced by the magnet attached to the reader).

The size and sensitivity of magnetometers can be very different. For our purpose, a small magnetometer matching the size of an RFID tag with a reasonable level of sensitivity is needed. Tiny, inexpensive atomic magnetometers about the size of a fat grain of rice have been reported. The most sensitive types of atomic magnetometers can detect fields of the order of a femtotesla (= 10^{15} Tesla) – about one-fifty-billionth the strength of Earth's magnetic field. We also note that iron and steel can cause shielding effects on magnetic fields. Other materials such as wood, Plexiglas, Styrofoam, brass, copper, aluminum, leather or paper, however, have almost no effect on shielding magnetic fields. This means that a magnetometer can work even when encased in many objects, such as wallets, purses or backpacks. This suggests that a magnetometer-equipped tag would not need to be removed from its container while accessing the tag, and therefore this magnetic-field triggered proximity sensing approach does not necessitate any changes to the existing RFID usage model.

The security of this selective unlocking scheme based on magnetic-field triggered proximity sensing relies on the inability of an adversary to induce a magnetic field of certain strength from a distance. If an adversary intends to unlock a tag, it can simply be in very close proximity of the tag, just like a legitimate reader. However, being near, increases the chances of the adversary being detected. To remain surreptitious, the adversary is therefore forced to generate a stronger magnetic field from an undetectable distance. Our preliminary investigation shows this attack does not seem feasible. According to Biot Savart's law, the magnetic field strength decreases inversely proportional to the square of the distance from the location of magnetic source (a current owing object or a permanent magnet). So, it is practically impossible to induce a strong magnetic field from a distance and therefore the magnetic-field triggered proximity sensing approach can effectively thwart adversaries who are not in close proximity of the tag.

Posture recognition based selective unlocking. In certain RFID applications, a specific posture of the tag owner may serve as a valid context. One class of such applications involves implanted medical devices (IMDs). Under legitimate IMD access, we can assume that the patient is lying down on his or her back. Thus, access to the IMD will be granted only when the patient's body is such a pre-defined unique posture. This will prevent an attacker from controlling the IMD in many common scenarios, such as while standing just behind the patient in public. Yet another class of applications that can benefit from posture based contexts involves the Passive Keyless Entry and Start (PKES) system. In such applications, a driver needs to move into the car and sit down on the driver's seat before the engine can be started automatically (while the key resides in the driver's pockets). Thus, getting into the car and sitting on the driver seat can be considered necessary posture sequences that need to be performed to unlock the car key. In turn, this will unlock the door and start the engine. Such an unlocking mechanism will prevent an adversary from launching attacks in scenarios whereby the driver is not entering the car and then sitting on the car seat. Since posture formations are human activities performed by users unconsciously, posture recognition can provide a finer-grained non-obtrusive unlocking mechanism without purposeful or conscious user involvement.

We can classify postures into two primary types: posture and posture transition. Posture means a static posture status that a user can maintain for certain duration, such as lying, sitting, standing and walking. Posture transition subsumes different human movements, such as "stand-to-sit", "sit-to-stand", "sit-to-lie", "lie-to-sit", and so on. Posture transitions capture the dynamics of human movement and usually only last for a short duration. We analyze the features of these two posture types and realize that most of the postures and some of the posture transitions can be simply detected by measuring direction changes or status changes in sagittal and transverse planes. In case of posture recognition, consider, for example, an

IMD – such as a pacemaker implanted into the patient's chest area – equipped with a 3-axes accelerometer. As the IMD is fixed to the human body, it remains static relative to the body system but has different orientations in the earth coordinate system (magnetic north and gravity) due to human body movement. Thus, we can detect such movements by simply monitoring its relative orientation change in the earth coordinate system. For example, when the patient is in the "sitting" position, the Z axis of the accelerometer points to the sky and the X-Y plane is parallel to the earth surface. When the patient lies down, the Z axis now should be parallel to the earth surface while one of the X or Y axis should point to the sky. Thus, by simply monitoring the change of directions of axes, we can tell whether a patient is lying or not. We note that mobile devices also commonly use such detection techniques based on accelerometer axis direction change to perform screen rotation functions.

In contrast, posture transition recognition is similar to gesture recognition to a certain extent. Similar to the gesture recognition schemes (like the Secret Handshakes scheme we discussed previously), in a posture transition recognition, user movement can be recorded by motion sensors such as accelerometers and the captured motion data is then compared with a reference posture template which has been recorded by performing the corresponding movement in a reference coordinate system. A match between the captured data and the reference template implies that the user has exhibited a certain posture transition defined by the reference template. However, there is one primary difference between gesture recognition and posture transition recognition, i.e., device tilt. In (hand) gesture recognition systems, users are assumed to be aware of their hand activities. So gestures are performed in a more-or-less controlled way without tilting the tag so that the effect of tilt can be greatly minimized or ignored. However, in posture transition recognition, as we do not require any explicit user involvement, the tag, placed inside a human body in the form of an IMD or into the pockets in the form of a car key, can be tilted due to the movement of human body or the device positioning itself. The reference template is usually collected in a reference coordinate system. However, once a device is tilted, movement data collected from the device is no longer in the reference coordinate system and the corresponding posture will not be detected correctly. It is therefore critical to detect the tag's orientation in order to rotate the data vector back to the reference coordinate system for correct recognition.

Current systems for full orientation estimation, such as the one in Apple iPad2, usually use a set of sensor modalities – typically including gyroscopes, accelerometers and magnetometers – to estimate device orientation. Gyroscopes are used to determine accurately angular changes while the other sensors are used to compensate the integration drift of the gyroscopes and keep this estimate drift free. However, a typical gyroscope requires about 5~10 times more power than magnetometer and accelerometer together. Moreover, its comparably larger form factor also makes gyroscope not commonly available in a tiny single package MEMS chip. Considering the resource constrained RFID platforms, it might be necessary to restrict from using gyroscopes, and instead focus on using accelerometers and/ or magnetometers for device orientation and posture estimation. As integrated accelerometers and magnetometers are commercially available in tiny packages, an RFID tag with such sensors can be flat and less obtrusive for the user, which makes them very attractive to be used in IMDs or smart car keys. There exist several attempts to use either accelerometers or magnetometers; however, it has been shown that neither of the two sensors is good enough alone to estimate full orientation. On the other hand, orientation estimation schemes that use both accelerometers and magnetometers show very promising results (Huyghe, 2009; Yun, 2008). Further study is needed to check whether these schemes (based on both accelerometer and magnetometer) are efficient enough to be applied on the RFID platform.

Location and speed based selective unlocking. In certain RFID applications, location or location-related information (such as speed) can serve as a legitimate access context. For example, an access card to an office building needs to only respond to reader queries when it is near the entrance of the building; a credit card should only work in authorized retail stores; toll cards usually only communicate with toll readers in certain fixed locations (toll booths) or when the car travels at a certain speed. Hence, location or location-specific information can serve as a good means to establish a legitimate usage context. That is, a tag is unlocked only when it is in an appropriate (pre-specified) location. It is suitable for applications where reader location is fixed and well-known in advance. If a tag is equipped with a GPS receiver, location (as well as speed) information

can be obtained through the GPS receiver. This is a pretty straightforward approach.

A pre-requisite in a location-aware selective unlocking scheme is that a tag needs to store a list of legitimate locations beforehand. Upon each interrogation from a reader, the tag obtains its current location information from its on-board GPS sensor, and compares it with the list of legitimate locations and decides whether to switch to the unlocked state or not. Due to limited on-board storage (e.g., the WISP has an 8KB of flash memory) and passive nature of tags, the list of legitimate locations must be short. Otherwise, testing whether the current location is within the legitimate list may cause unbearable delay and affect the performance of the underlying access system. Moreover, the list of legitimate locations should not change frequently because otherwise users will have to do extra work to securely update the list on their tags. Thus, selective unlocking based on pure location information is more suitable for applications where tags only need to talk with one or a few readers, such as building access cards. It may not be suitable for credit card applications as there is a long list of legitimate retailer stores, and store closing and new store opening occur on a frequent basis.

Selective unlocking based on pure location information presents similar problems for toll systems as for the credit card systems because toll cards will need to store a long list of toll booth locations. We notice that vehicles mounted with RFID toll tags are usually required to travel at a certain speed when they approach a toll booth. For example, three out of eight toll lanes on the Port Authority's New Jersey-Staten Island Outer Bridge Crossing permit 25 mph speeds for E-ZPass drivers; the Tappan Zee Bridge toll plaza and New Rochelle plaza, NY has 20mph roll-through speed; Dallas North Toll way has roll-through lanes allowing speeds up to 30 mph. Hence, "speed" can be used as a valid context to design selective unlocking mechanisms for toll cards. That is, a toll card remains in a locked state except when the vehicle is traveling at a designated speed near a toll booth (such as 25-35 mph in the Dallas North Toll Way case). GPS sensors can be used to estimate speed either directly from the instantaneous Doppler-speed or indirectly from positional data differences and the corresponding time differences.

For better protection against attacks, the speed and location can also be used together as a valid context for unlocking of toll cards. Here, the adversary will only be able to unlock the tag if both the valid location and speed criteria are satisfied.

Secure Transaction Verification

A highly difficult problem arises in situations when the reader itself, with which the tag (or the tag's user) engages in a transaction, is malicious. For example, in the context of an RFID credit card, a malicious reader can fool the user into approving for a transaction whose cost is much more than what she intended to pay. That is, the reader terminal would still display the actual (intended) amount to the user, while the tag will be sent a request for a higher amount. Perhaps more seriously, such a malicious reader can also collude with a leech and can succeed in purchasing an item much costlier than what the user intended to buy (Drimer, 2007). As discussed previously, addressing this problem requires *secure transaction verification*, i.e., validation that the tag is indeed authorizing the intended payment amount. Note that selective unlocking is ineffective for this purpose because the tag will anyway be unlocked in the presence of a valid (payment) context.

A display-equipped RFID tag can easily enable secure transaction verification (Drimer, 2007). This, however, necessitates user involvement and is prone to human errors. Distance bounding protocols have also been suggested as a countermeasure to the reader-and-leech attacks. However, these protocols are currently infeasible (as reviewed in Prior Work). In this section, we set out to explore

the possibility of using contextual information to design secure transaction verification schemes.

Contextual information can be used to assist server decision making. In an RFID system, queried information from a tag is usually sent by the reader to a back-end server for further processing and this processing result is used to make (usually security-critical) the final decision regarding the transaction (such as opening gate, charging toll, approving payment, etc.). Contextual information from a tag can help the server make more informed decisions. In the following, we suggest two possible mechanisms using context recognition as a basis for secure transaction verification to provide protection against malicious readers.

Numeric Digit Recognition. One possible approach is for the user to indicate to the tag the intended amount of transaction (instead of the tag displaying this to the user, which requires direct access to the tag). Use of touch sensors (Sample, 2009) or on-board buttons is not feasible for this purpose as they would also require direct tag access; buttons will also hamper tag's form factor. Secret Handshakes (Czeskis, 2008) could be extended though. The user could create numeric patterns depicting the amount by moving her accelerometer-enabled tag (or wallet containing the tag). For example, user can create a '5' and then two '0's up in the air to indicate a transaction worth $500. This method, however, has the same shortcomings as Secret Handshakes – it requires explicit user involvement and has usability implications. Another, potentially more user-friendly, solution is to have the user speak-out the amount of transaction (e.g., digit-by-digit), which the tag can record using an on-board microphone and decode. This method requires some form of numeric speech (digit) recognition. A simplistic, yet robust (error-tolerant) digit-based speech recognition system that can be implemented within the resource constraints of a typical RFID payment token is needed for further study.

Sensor-Centric Colocation. Location sensing could also be used to provide improved resilience, specifically, to reader-and-leech attacks. Note that under such attacks, the valid tag and the valid reader would usually not be in close proximity (e.g., the tag is at a restaurant, while the reader is at a jewelry shop (Drimer, 2007)). This is unlike normal circumstances whereby the two entities would be at the same location, physically near to each other. Thus, a difference between the locations of the tag and that of the reader would imply the presence of such attacks. Specifically, the tag (credit card) detects its current location and sends this location information encrypted with the key that it pre-shares with its issuing bank; the bank will then compare the tag's location with that of the (jewelry) merchant and reject the transaction if the two mismatch. For example, a bank server can deny a $2000 transaction when it detects that the tag (e.g., an RFID credit card) - equipped with a location sensor (GPS) - is currently in a restaurant whose normal transaction is usually less than $200 or so. We note that such a solution can be deployed, with minor changes on the side of the issuer bank, under the current payment infrastructure, where cards share individual keys with their issuer banks.

Clearly, location information can be directly obtained through the use of GPS sensors. However, one serious disadvantage with the GPS-based approach is the reliance on the GPS infrastructure. Another issue which needs to consider is the possible delay due to GPS initialization. A GPS can have either a cold start or hot start. The hot start occurs when the GPS device remembers its last calculated position and the satellites in view, the almanac (i.e., the information about all the satellites in the constellation) used, the UTC Time, and makes an attempt to lock onto the same satellites and calculate a new position based upon the previous information. This is the quickest GPS lock but it only works if the receiver is generally in the same location as it was when the GPS was last turned off. The cold start is when the GPS device dumps all the information, attempts to locate satellites and then calculates a GPS lock.

This takes the longest because there is no known or pre-existing information. For example, the GPS receiver module LS20031 from LOCOSYS Technologies can normally acquire a fix from a cold start in 35 seconds, and acquire a hot-start fix in less than 2 seconds. Hence, this approach may only apply to applications which can tolerate such delay.

This motivates the need to design a "localized" approach to location sensing, that does not require any additional infrastructure besides the RFID. One idea is to make use of (multiple) environmental sensors (such as microphone, thermometer, or magnetometer, and perhaps odor and gas sensors) as a means to derive the location-specific information. This is based on the assumption that certain ambient information, extracted by the tag and reader at around the same time (i.e., at the time of transaction), will be highly correlated if the two devices are in close physical proximity. Said differently, if a certain sensor attached to the tag and the same type of sensor attached to the reader report mismatching ambient information, this will indicate that the tag and reader are (most likely) not at the same location or close to each other.

One possibility is to investigate the use of audio sensors (microphones) for accomplishing the aforementioned approach to transaction verification: both the tag and the reader are equipped with a microphone. This choice is motivated by the intuition that the audio data captured at two different locations at a given time may be different to some extent. Whenever a transaction is initiated, a short-term data acquired by these microphones on the two devices will be transmitted to the bank server. The server will then perform correlation analysis over the two audio signals and determine the outcome of the transaction based on the degree of correlation between the signals. As an example, under a normal scenario, when both the valid tag and valid reader are at the restaurant, the data captured by their respective microphones is likely to be highly correlated, in which case the transaction will be accepted. In contrast, under an attack, when the valid tag is at the restaurant but the valid reader is at a jewelry store, the audio data is not likely to be correlated, in which case the transaction will be rejected.

LOCATION-LIMITED COMMUNICATION

Sensing capabilities can enable the tag to establish location-limited communication channels (audio, visual or tactile) with other entities. Being location-limited or proximity-based, these channels are (better) resilient to the man-in-the-middle attacks unlike the primary radio channel, and can therefore be leveraged to accomplish two fundamental security tasks. In this section, we show LLC channels can be used to accomplish two fundamental security tasks. First, LLC can be used for secure association between the tag and reader. Second, LLC allows the tag to communicate with other "smarter" devices without the latter requiring an RF communication interface. This facilitates outsourcing security tasks to the smart device and employing the smart device as an authentication token.

Secure Association

In many situations, one would need to perform an "on-the-fly" shared key agreement (or "pairing") paring.

The method from Blinking Lights (Saxena, 2006) could be applied in a relatively straightforward manner to establish pairing between an LED-equipped tag and a reader connected to a web cam. The user/administrator would first capture the tag's blinking LED using the camera; the reader would indicate the result of pairing (success or failure) and accordingly the user indicates the result on the tag, e.g. by pressing a touch sensor (Sample, 2009). Alternatively, a tag equipped with a microphone or photo-sensor may be used to transmit the authenticator in the other

direction (from reader to tag), obviating the need for pressing a button on the tag.

The above methods are workable in many pre-deployment pairing scenarios, because the tags are directly accessible (i.e., not enclosed in objects or humans) and forming LLC channels is relatively easy. However, post-deployment pairing can be complicated. A particularly challenging operation is pairing of an implanted medical tag. It is difficult to establish direct LLC channels between such a tag and the reader because the former is not physically reachable. Thus, forming visual and tactile channels is impossible, and establishing reader-to-tag audio channels is also difficult because a microphone may not be functional inside a human body. As a result, pairing of an implanted tag after post-deployment is a hard problem which requires further study.

User Authentication

Location-limited channels also allow the tag to communicate with other "smarter" personal and trusted devices (e.g., the owner's mobile phone) without the latter requiring an RF communication interface (unlike some prior solutions, such as NFC phones). This is attractive because computationally intensive or fine-grained sensing tasks can now be outsourced from the RFID tag to the smart device. Most beneficially, this functionality allows the auxiliary device to be used as an authentication token for the purpose of user authentication, providing protection in the event of tag's loss or theft. As an example, imagine Alice goes shopping carrying a contactless credit card. The card is in a default locked state and does not respond to read requests. When ready for checkout, Alice unlocks the credit card by authenticating to it. Once the transaction completes, the card again gets locked.

In this following, we first discuss Vibrate-to-Unlock, a user-to-tag authentication scheme via a vibrating mobile phone. We then suggest the possibility of a strong authentication scheme which executes a challenge-response protocol between the tag and the phone over the audio LLC.

PIN-based authentication using a vibrational LLC. "Vibrate-to-Unlock" (VtU) is a mobile phone assisted user authentication scheme which authenticates a user to multiple of her personal RFID tags through her mobile phone (Asokan, 2011). Through user authentication, a user can control when and where her RFID tags can be accessed, thus preventing attacks such as unauthorized reading, relay attacks, device lost or stolen.

VtU works by using a mobile phone as an authentication token that stores tag specific shared PINs that authenticate to multiple RFID tags. It requires a phone with the capability to vibrate and an RFID tag equipped with an accelerometer. Tags are by default in locked state and do not respond to any reader request. Authentication is achieved when a user touches her vibrating phone with an RFID tag or its container, such as a wallet. The phone's vibrations are used to encode a PIN shared with the tag. The accelerometer on the tag senses the vibrations and decodes the PIN. If the extracted PIN matches the one stored on the tag within some error tolerance, the tag authenticates the user and sends the response to the RFID reader. Otherwise, the tag maintains in locked state.

Compared with existing solutions, VtU offers several advantages. First, a double layer of protection is provided. To access a tag's service an adversary would need access to the tag as well as its user's phone. This provides improved resilience in the event of loss or theft of tags. Second, the phone acts as a "master key" that allows users to authenticate to multiple tags. Critically, unlocking one tag does not unlock other tags stored in the same container (e.g., a wallet). Third, since each tag can only be unlocked by a unique PIN stored on the phone, unauthorized reading and relay attacks are completely eliminated. In other words, false tag unlocking is not possible. Finally, VtU is automated and transparent to users and does not impose any usability constraints. In particular, users do not need to memorize any PINs. However,

it requires the carrying-on of a mobile phone by the user. As mobile phones have become an integral part of users' lives and are almost constantly available to users due to their desire to remain socially connected, it justifies the VtU approach.

Strong authentication based on the audio LLC. Audio LLC could also be used for (strong) user authentication. Here, the tag simply sends a challenge to the phone, the phone responds by encrypting the challenge using the pre-shared key, and the phone (or the user) gets authenticated only if the decryption of the response equals the challenge. In this approach, an audio codec that can be used for the purpose of challenge-response is needed.

RANDOMNESS EXTRACTION

Generating random numbers of sufficient quality for cryptographic applications is a fundamental task. While modern general purpose computers have several techniques available for the generation of high quality random numbers, this requirement is beyond the capacity of today's RFID tag. In this section, we focus on random number generation through using onboard sensors to collect physical parameters of different forms for use a source of randomness. Physical parameters collected by sensors can be human-specific or human-related. They can also just be ambient environment parameters, such as temperature, or magnetic field. They can change in space, time, and across different objects, and many physical attributes de-correlate rapidly in space and time. This serves as a natural entropy collection process and onboard sensors can be used to collect physical parameters of different forms for use as a source of randomness. In the following, we review the work of random number generation based on accelerometer. We note other sensors, such as magnetometers, thermometers, microphones, and their combinations thereof may also serve a good source of randomness.

Accelerometer-based Random Number Generation

The work of Voris (2011) investigates the use of accelerometers for *true random number generation (RNG)*. The choice of accelerometer is based on the premise that accelerometers are highly perceptive sensors, even capable of picking up minute vibrations from afar.

While the naturally occurring phenomena sensors capture are unpredictable, they necessarily contain some bias rather than being distributed uniformly. In order to establish that accelerometer reading is a good source of randomness, the min-entropy of the accelerometer output is experimentally estimated. Min-entropy is a mathematical property of a distribution. It is equal to the probability of the most likely element being drawn from a distribution X. Phrased somewhat differently, if a distribution X has a min-entropy of k, the likelihood of drawing any single element x from X does not exceed $1/2^k$ for all $x \in X$. Min-entropy is an important measurement of a distribution because it captures the amount of randomness a distribution is capable of supporting. Despite the fact that elements of X are n bits in length, due to the bias of the distribution, X may not contain enough entropy to actually support the extraction of n unbiased bits. Only k "strongly" random bits can be derived from a distribution that has a min-entropy of k regardless of the distribution's element length n. Experiments show that even when the tag is stationary, its accelerometer output can yield a sufficient min-entropy with a value of 3.4. That is, a stationary accelerometer can support the creation of 3.4 random bits per one 30-bit accelerometer sample. A high quality 128-bit random number can be generated in about 1.5 seconds when the sensor/tag is in a stationary state and much faster when it is mobile.

More interesting to know is that the entropy of a stationary accelerometer cannot be reduced in the presence of a variety of environmental changes or even under adversarial manipulations. That means accelerometers are resistant to changing environments, benign or adversary. So the best approach an attacker could take to interfere with the amount of min-entropy generated by an accelerometer would be doing nothing -- since anything else will only serve to increase the min-entropy of the readings rather than reduce it.

EVALUATION OF SENSING-CENTRIC ATTACKS

The security of sensing-enabled defense mechanisms clearly depends on the (in)capability of an adversary to either directly control the sensors or manipulate the environment in which the sensors operate. In this section, we discuss possible attacks targeting the sensing-centric mechanisms that may arise. A systematic evaluation of sensing-centric attacks is needed with respect to adversarial manipulation of sensors and sensing capabilities. First, we need to consider indirect control of sensors by means of a malicious reader, given that reader is what powers up the sensors. Additionally, we need to consider malicious manipulation of sensor's environment in order to compromise the security of the underlying mechanism. It is clear that tampering with the localized physical environment is a difficult task, for example, when compared with tampering the wireless radio environment (a property which is a foundation for our proposal). At the same time, it is still important to understand the level of security provided by sensing-centric mechanisms against localized attackers and to identify the mechanisms which remain most resistant in the face of such attackers.

Manipulation via Malicious Reader

RFID tags and associated sensors are utterly dependent on reader transmissions for energy. A malicious entity that gains control of an RFID reader could thus trivially perform a denial-of-service (DoS) attack by simply refusing to supply enough power for the sensor to operate. Rather than a DoS attacker, a cleverer opponent may attempt to manipulate onboard tag sensors by subtly adjusting reader parameters. One such attribute is the rate at which a reader issues requests to tags. If an RFID protocol requires that a tag samples its sensor data each time it wakes up, an attacker could manipulate the rate at which samples are taken by changing the frequency at which a reader issues queries. This may have undesirable consequences from a security perspective. Sensor readings taken at different periods may contain more or less entropy, for instance. The experiments with accelerometer-based random number generation (discussed previously) show that the sampling rate has a very minor impact on the unpredictability of its readings, but this effect may be more pronounced for other types of sensors. An attacker could also perhaps time queries so that sensors take environmental readings only when it is advantageous for them.

Along the same lines, an adversary could modify the signal strength of a RFID reader's transmissions in order to change the amount of power that is made available to tags. Since some tag hardware requires more power to operate than others, this could potentially alter the behavior of sensing hardware. A sensor may not operate correctly, and its output may be less accurate or more predictable when it is supplied with less power than its designers intended.

Furthermore, RFID tags are programmed to blindly obey whatever commands they receive. An opponent may therefore be capable of undermining sensor based security mechanisms by querying a tag for its sensor values. For example, consider a transponder that constructs random numbers by collecting sensor readings, saving them in memory, and hashing the stored values. An attacker who is in command of an RFID reader could issue a read instruction to learn the contents of the device's memory, including the raw sensor data. This in-

formation could possibly provide the malicious entity with an advantage in predicting the output of the random number generator. Similarly, an adversary could issue a write command to replace the acquired sensor information with values of his or her choosing.

Environmental Manipulation

Sensing-centric mechanisms that are based on sensor data extracted from the environment are subject to environmental manipulation. The question this raises is: whether it is possible for the adversary to control the environment in such a way that compromises the security of the mechanism? We discuss a few possibilities of such an adversarial control vis-a-vis some of the approaches we discussed previously.

Let us first consider the selective unlocking approach based on proximity sensing. If an adversary intends to unlock a tag, it can simply be in very close proximity of the tag, just like a legitimate reader. However, being near, increases the chances of the adversary being detected. To remain surreptitious, the adversary is therefore forced to generate a stronger magnetic field from an undetectable distance. This attack, however, is not feasible due to the physical characteristics of the magnetic field strength. Biot Savart's law can be used to predict the strength of the field as follows:

$$B = \frac{\mu_0 I}{4\pi} \int \frac{d\ell \times \hat{r}}{r^2} \qquad (1)$$

(Here, I is the current flowing through a magnetic source, vector $d\ell$ is the direction of the current, μ_0 is the magnetic constant, r is the distance between the magnetic source and the location at which the magnetic field is being calculated, and $\hat{r}$ is a unit vector in the direction of $\boldsymbol{r}$).

As per Equation (1), the magnetic field strength generally goes down drastically with distance. As one special case of this equation, when the magnetic source is an infinitely long straight wire running a current I, it suggests that it will take a wire carrying a large amount of current (500 amperes) in order to generate a magnetic field strength of just 100 *microtesla* even from a distance of 1 m. A current of 500 *amperes* will be impossible to induce even for a sophisticated attacker (as a reference, a current of about 1 ampere can cause electrocution). In another case of the Biot Savart's Law, when the magnetic source is a permanent magnet or a dipole, the magnetic field strength decreases inversely proportional to the cube of the distance. It suggests that if an attacker wants to generate a field with a strength of, for example, 100 *microtesla* at just 1 m away, it would need to generate a field with a strength of roughly 500 *Tesla* at the source (magnet) (1 *gauss* = 10^{-4} *Tesla*). Note that 500 *Tesla* is an extremely large number; given an MRI's electromagnet is only 3 *Tesla*. Also, the strongest possible lab magnet known today is only 10 *Tesla*. Based on the above analysis, we can conclude that it is not possible to induce strong magnetic field from a distance and therefore our magnetic-field triggered proximity sensing approach can thwart adversaries who are not in close proximity of the tag.

Location-based defenses rely on the GPS infrastructure and thus may be prone to GPS associated vulnerabilities such as spoofing and jamming. Successful spoofing experiments on standard receivers have been reported (Papadimitratos, 2008), indicating commercial-off-the-shelf receivers do not detect such attacks. In the context of location-aware selective unlocking, the adversary can falsely unlock the tag if it can spoof the GPS signals coming from the satellites and feed in false location information to the GPS receiver (e.g., corresponding to a toll booth location even though the car/card is at a different location). Similarly, in the context of location-aware transaction verification, the adversary can, for example, fool the valid tag into thinking that it (the tag) is at a jewelry shop even though it is in a restaurant. Of existing GPS

attack countermeasures, the one that is most suitable for the RFID setting is the scheme proposed in Papadimitratos (2008). This scheme does not require any special hardware and does not rely on any cryptography. Instead, a GPS receiver in this scheme is augmented with inertial sensors (e.g., speedometers or accelerometers). The receiver can measure the discrepancy between its own predicated value (through inertial sensors) and measurements (through received GPS signals) in order to detect spoofing and replay attacks. The scheme is applicable to any mobile RFID tag setting, such as a toll card. However, this approach detects only the inertial abnormalities but not the location abnormalities. Thus, it only applies to situations where GPS receivers are mobile.

Recently, a very interesting work on the requirements to successfully mount GPS spoofing attack has been reported (Tippenhauer, 2011). The authors show that it is easy for an attacker to spoof any number of individual receivers. However, the attacker is restricted to only a few transmission locations when spoofing a group of receivers - even when they are stationary - while preserving their constellation (or mutual distances). Moreover, conducting spoofing attack on a group even becomes impossible if the group can hide the exact positioning of at least one GPS receiver from the attacker (e.g., by keeping it mobile on a vehicle) since in such case the attacker cannot adapt to its position. This suggests a cooperative detection scheme where multiple GPS receivers can work together to detect GPS spoofing attacks by also checking their mutual distances. Although it is still hard to foresee this countermeasure can be applied in current RFID application settings, it does state that a network of GPS receivers (or GPS-enabled devices) can be setup on the field to monitor GPS signals when it is necessary and when spoofing attack is a real menace.

When using sensor-centric random number generation, an adversary's goal will be to supply environmental data that has very low (ideally, zero) entropy. For example, the output distribution of an audio sensor (microphone) is influenced by the sounds produced by the users in close proximity, or in the environment itself, and if an adversary can feed in constant audio inputs or loud noises to the microphone, the system can be enforced into a zero-entropy state. This suggests that a microphone may not be suitable to serve as a means of secure entropy collection. On the contrary, as mentioned previously, the accelerometer-based approach is quite promising in this regard. Accelerometers, unlike other sensors, are resistant to a variety of environmental variations and even to adversarial manipulation. Specifically, most benign or adversarial changes affecting accelerometer can only serve to *increase* the entropy it provides. The best approach an attacker could take to interfering with the amount of entropy would be to place the accelerometer equipped device in as stable an environment as possible.

DISCUSSION AND CONCLUSION

The current technological advancement makes the fusion of sensor and RFID viable. The enrichment of RFID with sensing capabilities opens up many promising applications for ubiquitous sensing and computing. It also suggests new ways to provide enhanced security and privacy for RFID systems relying on their new sensing capabilities. In this book chapter, we highlighted these new possibilities by reviewing and discussing existing and ongoing/potential research. We outlined that this sensing-based approach for RFID security and privacy is not only possible but also feasible both technologically and financially.

We note that this sensing-centric approach alone may not provide absolute security due to the possibility of errors associated with context recognition; however, it raises the bar even for sophisticated adversaries without affecting the RFID usage model. In addition, although the techniques discussed can work in a stand-alone fashion, they can also be used with other security mechanisms, such as cryptographic based schemes, to provide stronger cross-layer security

protection according to different security needs in various applications. Moreover, many of the ideas and techniques discussed will be applicable in the realm of other wireless (or wired) devices equipped with sensors. Because sensors serve as a bridge between the physical and the digital world, the proposed sensing-centric mechanisms will be instrumental towards providing dependability, security and privacy for complex Cyber-Physical Systems.

We stress that the overall goal of utilizing sensing technologies is to produce solutions that not only consider efficiency and security, but also usability. For each potential solution, wherever applicable, it should be evaluated by means of usability studies. Moreover, more and more unconventional types of sensors are being invented to measure new environmental attributes that offer a richer set of information about the devices they are attached to or the environment in which they operate. Thus, any sensing-centric research in this domain should be aware of these new developments related to sensors and should be able to incorporate these in order to provide stronger and better security and privacy guarantee.

REFERENCES

Asokan, N., Saxena, N., Uddin, M., & Voris, J. (2011). *Vibrate-to-unlock: Mobile phone assisted user authentication to multiple personal RFID tags.* In International Conference on Pervasive Computing and Communications, 2011.

Brands, S., & Chaum, D. (1993). *Distance-bounding protocols.* In Advances in Cryptology - EUROCRYPT, International Conference on the Theory and Applications of Cryptographic Techniques, 1993.

Bringer, J., Chabanne, H., & Dottax, E. (2006). HB++: A lightweight authentication protocol secure against some attacks . In *Security*. Privacy and Trust in Pervasive and Ubiquitous Computing. doi:10.1109/SECPERU.2006.10

Buettner, M., Prasad, R., Philipose, M., & Wetherall, D. (2009). *Recognizing daily activities with RFID-based sensors.* In International Conference on Ubiquitous Computing, 2009.

Czeskis, A., Koscher, K., Smith, J., & Kohno, T. (2008). *RFIDs and secret handshakes: Defending against Ghost-and-Leech attacks and unauthorized reads with context-aware communications.* In ACM Conference on Computer and Communications Security, 2008.

Dimitriou, T. (2005). A lightweight RFID protocol to protect against traceability and cloning attacks. In *IEEE Conference on Security and Privacy in Communication Networks, 2005.*

Drimer, S., & Murdoch, S. (2007). *Keep your enemies close: Distance bounding against smartcard relay attacks.* In 16th USENIX Security Symposium, 2007.

Francillon, A., Danev, B., & Capkun, S. (2011). *Relay attacks on passive keyless entry and start systems in modern cars.* In 18th Annual Network and Distributed System Security Symposium, 2011.

Gilbert, H., Robshaw, M., & Seurin, Y. (2008). *HB#: Increasing the security and efficiency of HB+.* In Advances in Cryptology - EUROCRYPT, International Conference on the Theory and Applications of Cryptographic Techniques, 2008.

Halevi, T., & Saxena, N. (2010). *On pairing constrained wireless devices based on secrecy of auxiliary channels: The case of acoustic eavesdropping.* In ACM Conference on Computer and Communications Security, 2010.

Halevi, T., Saxena, N., & Halevi, S. (2009). *Using HB family of protocols for privacy-preserving authentication of RFID tags in a population.* In Workshop on RFID Security, 2009.

Halperin, D., Heydt-Benjamin, T., Ransford, B., Clark, S., Defend, B., & Morgan, W. … Maisel, W. (2008). *Pacemakers and implantable cardiac defibrillators: Software radio attacks and zero-power defenses.* In IEEE Symposium on Security and Privacy, 2008.

Hancke, G., & Kuhn, M. (2005). *An RFID distance bounding protocol.* In the 1st International Conference on Security and Privacy for Emerging Areas in Communications Networks, 2005.

Holcomb, D., Burleson, W., & Fu, K. (2007). Initial SRAM state as a fingerprint and source of true random numbers for RFID tags. In *Conference on RFID Security, 2007.*

Holleman, J., Yeager, D., Prasad, R., Smith, J., & Otis, B. (2008). NeuralWISP: An energy-harvesting wireless neural interface with 1-m range. In *Biomedical Circuits and Systems Conference, 2008.* Benoit Huyghe, B., & Doutreloigne, J. (2009). *3D orientation tracking based on unscented Kalman filtering of accelerometer and magnetometer data.* In IEEE Sensors Application Symposium, 2009.

Juels, A., Rivest, R., & Szydlo, M. (2003). *The blocker tag: Selective blocking of RFID tags for consumer privacy.* In ACM Conference on Computer and Communications Security, 2003.

Juels, A., Syverson, P., & Bailey, D. (2005-1). *High-power proxies for enhancing RFID privacy and utility.* In Privacy Enhancing Technologies, 2005.

Juels, A., & Weis, S. (2005-3). Authenticating pervasive devices with human protocols. In *International Cryptology Conference, 2005.*

Katz, J., & Shin, J. (2006). Parallel and concurrent security of the HB and HB+ protocols. In *Advances in Cryptology - EUROCRYPT, International Conference on the Theory and Applications of Cryptographic Techniques, 2006.*

Kfir, Z., & Wool, A. (2005). Picking virtual pockets using relay attacks on contactless smartcard. In *Security and Privacy for Emerging Areas in Communications Networks, 2005.*

Koscher, K., Juels, A., Brajkovic, V., & Kohno, T. (2009). EPC RFID tag security weaknesses and defenses: Passport cards, enhanced drivers licenses, and beyond. In *ACM Conference on Computer and Communications Security, 2009.*

Molnar, D., & Wagner, D. (2004). *Privacy and security in library RFID: Issues, practices, and architectures.* In ACM Computer and Communications Security, 2004.

Nithyanand, R., Tsudik, G., & Uzun, E. (2010). *Readers behaving badly: Reader revocation in PKI-based RFID systems.* In European Symposium on Research in Computer Security, 2010.

Oren, Y., & Wool, A. (2009). *Relay attacks on RFID-based electronic voting systems.* Cryptology ePrint Archive, Report 2009/422. Retrieved from http://eprint.iacr.org/2009/422

Papadimitratos, P., & Jovanovic, A. (2008). *GNSS-based positioning: Attacks and countermeasures.* In IEEE Military Communications Conference, 2008.

Rasmussen, K., & Capkun, S. (2010). *Realization of RF distance bounding.* In the USENIX Security Symposium, 2010.

Ruhanen, A., et al. (2008). *Sensor-enabled RFID tag handbook.* Retrieved January 2008 from http://www.bridge-project.eu/data/File/BRIDGE_WP01_RFID_tag_handbook.pdf

Salajegheh, M., Clark, S., Ransford, B., Fu, K., & Juels, A. (2009). *CCCP: Secure remote storage for computational RFIDs.* In 18th USENIX Security Symposium, August, 2009.

Sample, A., Yeager, D., Powledge, P., & Smith, J. (2007). *Design of a passively-powered, programmable sensing platform for UHF RFID systems.* In IEEE International Conference on RFID, 2007.

Sample, A., Yeager, D., & Smith, J. (2009). *A capacitive touch interface for passive RFID tags.* In IEEE International Conference on RFID, 2009.

Saxena, N., Ekberg, J., Kostiainen, K., & Asokan, N. (2006). *Secure device pairing based on a visual channel* (short paper). In IEEE Symposium on Security and Privacy, 2006.

Saxena, N., & Voris, J. (2009). *We can remember it for you wholesale: Implications of data remanence on the use of RAM for true random number generation on RFID tags.* In Conference on RFID Security, 2009.

Saxena, N., & Voris, J. (2010). *Still and silent: Motion detection for enhanced RFID security and privacy without changing the usage model.* In Workshop on RFID Security, 2010.

Tippenhauer, N. O., Popper, C., Rasmussen, K. B., & Capkun, S. (2011). *On the requirements for successful GPS spoofing attacks.* In ACM Conference on Computer and Communication Security (CCS'11), October 2011.

Tormanen, T. (2010). *Analog IC design 2010: Lecture 9 - Noise.* Retrieved from http://framtiden.eit.lth.se/fileadmin/eit/courses/eti063/lectures2010/AnalogIC_F9.pdf

Voris, J., Saxena, N., & Halevi, T. (2011). *Accelerometers and randomness: Perfect together.* In ACM Conference on Wireless Network Security, 2011. Smith, J., Powledge, P., Roy, S., & Mamishev, A. (2006). *A wirelessly-powered platform for sensing and computation.* In 8th International Conference on Ubiquitous Computing, 2006.

Wagner, D. (2005). *Privacy in pervasive computing: What can technologists do?* Invited talk at the 1st International Conference on Security and Privacy in Communication Networks, 2005. Retrieved from http://www.cs.berkeley.edu/~daw/talks/SECCOM05.ppt

Yeager, D., Prasad, R., Wetherall, D., Powledge, P., & Smith, J. (2008). Wirelessly-charged UHF tags for sensor data collection. In *IEEE International Conference on RFID*, 2008.

Yun, X., Bachmann, E., & McGhee, R. (2008). A simplified Quaternion-based algorithm for orientation estimation from earth gravity and magnetic field measurements. *IEEE Transactions on Instrumentation and Measurement, 57*(3).

ADDITIONAL READING

Halevi, T., Saxena, N., & Halevi, S. (2010). Tree-based HB protocols for privacy-preserving authentication of RFID tags. *Journal of Computer Security (JCS) – Special Issue on RFID System Security*, 2010.

Heydt-Benjamin, T., Bailey, D., Fu, K., Juels, A., & O'Hare, T. (2007). Vulnerabilities in first-generation RFID-enabled credit cards. In *Financial Cryptography, 2007.*

Holcomb, D., Burleson, W., & Fu, K. (2009). Power-up SRAM state as an identifying fingerprint and source of true random numbers. *IEEE Transactions on Computers, 58*(9), 1198–1210. doi:10.1109/TC.2008.212

Juels, A. (2006). RFID security and privacy: A research survey. *IEEE Journal on Selected Areas in Communications, 24*(2), 381–394. doi:10.1109/JSAC.2005.861395

Juels, A., Molnar, D., & Wagner, D. (2005). *Security and privacy issues in E-passports*. In Security and Privacy for Emerging Areas in Communications Networks, 2005.

Rasmussen, K., Castelluccia, C., Heydt-Benjamin, T., & Capkun, S. (2009). *Proximity-based access control for implantable medical devices*. In ACM Conference on Computer and Communications Security, 2009.

Rieback, M., Crispo, B., & Tanenbaum, A. (2005). *RFID guardian: A battery-powered mobile device for RFID privacy management*. In Australasian Conference on Information Security and Privacy, 2005.

Chapter 4
RFID Grouping–Proofs

Mike Burmester
Florida State University, USA

Jorge Munilla
University of Málaga, Spain

ABSTRACT

Radio Frequency Identification (RFID) is a challenging wireless technology with a great potential for supporting supply and inventory management. In this chapter the authors consider a particular application in which a group of tagged items are scanned to generate a record of simultaneous presence called a grouping-proof. Grouping-proofs can be used, for instance, to guarantee that drugs are shipped (or dispensed) accompanied by their corresponding information leaflets, to couple the user's electronic passport with his/her bags, to recognize the presence of groups of individuals and/or equipment and more generally to support the security of supply and inventory systems. Although it is straightforward to design solutions when the verifier is online since it is sufficient for individual tags to authenticate themselves to the verifier, interesting security engineering challenges arise when the trusted server (or verifier) is not online during the scan activity. So, the field of grouping-proofs is very active, and many works have been published so far. This chapter details the setting for RFID grouping-proofs and discuss the threat model for such applications. The authors analyze some of the grouping-proofs proposed in the literature describing their advantages and disadvantages. Then, general guidelines for designing secure grouping-proofs are proposed. Finally, some examples of grouping-proofs that are provably secure in a strong security framework are presented.

1. INTRODUCTION

The low cost and high convenience value of RFID tags gives them the potential for massive deployment. Accordingly, they have found increased adoption in manufacturing, inventory control, healthcare domain and counterfeit prevention.

An RFID deployment involves tags, readers, and a Verifier (backend Server). Tags are wireless transponders that typically have no power of their own and respond only when they are in an electromagnetic field, while RFID readers are transceivers that generate such fields. On-board clocks are not considered realistic for low-cost tags, but crude timers can be based on discharg-

DOI: 10.4018/978-1-4666-1990-6.ch004

ing capacitor (Juels, 2004). Readers implement a radio interface to tags and a high-level interface to the Verifier. The Verifier is a trusted entity that processes private tag data. The channel that links readers to the Verifier is assumed to be secure because hardware constraints are not so tight here and, common security protocols can be used.

In 2004, Ari Juels introduced the security context of a new RFID application, which he called a *yoking-proof* (Juels, 2004), that involves generating evidence of simultaneous presence of two tags in the broadcast range of an RFID reader. There are several practical scenarios where such proofs can substantially expand the capabilities of RFID-based systems. For example:

- In manufacturing: to automatically check that all the components of a kit, or components that are part of a consignment, are accounted for. For example, a component is only shipped if a safety cap is attached.
- Pharmaceutical distribution: to automatically check that drugs are accompanied with information leaflets when they are shipped (or dispensed).
- At airports: for security, to automatically check that passengers are accompanied by their baggage.
- In a battlefield context: for security, weaponry or equipment may have to be linked to specific personnel (who are the only ones that can use or operate it).

Although it is straightforward to design solutions for the case when the Verifier is online, since it is sufficient for individual tags to authenticate themselves to the Verifier, the case when the Verifier is not online is challenging both from a security and an engineering point of view. In particular, offline solutions require that tag interrogations (reader scanning's) are restricted to broadcasting challenges that are valid only for a short time period, and that collecting tag responses to generate a grouping-proof should be completed during this period. Therefore, research on grouping-proofs has focused on the offline case.

In the yoking-proof, an RFID reader first activates all the tags in its range and then interrogates those responding tags that are paired (yoked). The interrogation involves (*i*) establishing a communication channel with each one of the tags that are paired, (*ii*) collecting and relaying tag responses, and finally (*iii*) generating a proof of "simultaneous presence" of paired tags in the broadcast range of the reader. This proof can (later) be verified by the Verifier.

Saito and Sakurai observed (Saito and Sakurai, 2005) that Juels's yoking-proof is subject to an interleaving attack in which the adversary combines flows from different sessions, and proposed the use of time stamps. They also extended this proof to groups withtags, which they called grouping-proofs. Piramuthu replaced the time stamps by random numbers to prevent attacks that collect prior responses and combine these to forge a grouping-proof (Piramuthu, 2006). Peris-Lopez et al. combined the strengths of yoking-proofs with the grouping-proofs to address some of their weaknesses (Peris-Lopez et al., 2007). Burmester et al. extended the grouping-proofs to address anonymity and unlinkability in a strong modular security framework. Several EPCGen2 compliant grouping-proofs have been recently published to enhance impatient medication safety.

The field of grouping-proofs is very active, and many works have been published so far. This chapter will detail the setting for RFID grouping-proofs and discuss the threat model for such applications. General guidelines for designing secure grouping-proofs will be proposed, and examples of grouping-proofs that are provably secure in a strong security framework will be presented. With these goals in mind, the remainder of this chapter is organized as follows. Section 2 describes the RFID setting for grouping-proofs. Section 3 provides a review of related published work. Next,

Section 4 discusses the adversarial threat model, and Section 5 proposes some guidelines for secure RFID applications. Section 6 analyses some recently published EPCGen2 compliant schemes and this is followed by a discussion on security (Section 7). Finally, in Section 8 we discuss the challenges for extending the functionality of RFID grouping-proofs and conclude the chapter.

2. THE SETTING FOR RFID GROUPING-PROOFS

Throughout this chapter, we assume the following regarding the environment that characterizes group scanning applications (Burmester et al., 2008):

- RFID tags are passive, *i.e.,* they have no power of their own, and have very limited computation and communication capabilities. However, we assume that they are able to perform basic cryptographic operations such as generating pseudo-random numbers and evaluating pseudo-random functions. Furthermore, tags cannot communicate with each other directly, only through an RFID reader that can establish wireless channels to link them.

- Most proofs rely on timeout assumptions. In particular, that the protocol will always terminate within a certain interval of time, which, in many cases, is a feature of the protocol itself; and if not, it is not difficult to impose this restriction on the interaction between tags and readers. RFID tags do not maintain clocks or keep time. However, the activity time span of a tag during a single session can be limited using techniques such as measuring the discharge rate of capacitors (Juels, 2004).

- RFID readers are trusted to manage the interrogations of tags. They enable the tags of a group to generate a grouping-proof during an interrogation session, and keep a record of such proofs for each session.

- The Verifier (backend Server) is a trusted entity, which may share some secret information with the tags such as cryptographic keys. The Verifier has a secure channel (private and authenticated) that links it to the (authenticated) RFID readers.

- The design of grouping-proofs generally focuses on security issues at the protocol layer and not on physical or link layer issues, such as the coupling design, the air-RFID interface, and the power-up and collision arbitration processes.

2.1 Offline versus Online Grouping-Proofs

The (trusted) Verifier can be online or offline and different solutions are required in each case. In the offline mode, the RFID reader does not enjoy continuous connectivity with the Verifier, and delayed confirmation may be acceptable. For instance, this may be the case with supply chain applications, due to the increased fragmentation and outsourcing of manufacturing functions. A supplier of partially assembled kits may perform scanning activities that will be verified later when the kits are assembled at a different site. The interaction of the Verifier with the reader is restricted to broadcasting a challenge that is valid for a short time span, collecting responses from the tags (via RFID reader intermediates), and checking for legitimate group interaction –the Verifier in this mode never unicasts messages to particular groups of tags. The Verifier communicates with the readers through authenticated channels. In contrast, in the online mode, the reader has continuous connectivity with the Verifier.

Research on grouping-proofs has focused on situations in which the Verifier is offline while the tags are being scanned, as this is the most challenging case. As observed earlier, solutions for the online case can be easily obtained by having individual tags authenticate themselves to the Verifier, which will then decide which scanned tags belong to a group.

2.2 Static versus Dynamic Groups

RFID grouping-proofs can establish the presence of two types of groups: *static* or *dynamic*. Static grouping-proofs can only establish the presence of pre-defined groups of tags; this excludes "new" groups, or subgroups of established groups (that are not established). Dynamic grouping proofs can generate proofs for any group of authorized tags that is simultaneously scanned.

Generating a *subgrouping-proof* can be seen as a variant of generating a dynamic grouping-proof. Due to the unreliability of the radio interface, the larger is the number of the tags of a group, the higher the probability of interrogation failure. So, if a large group can be divided into several subgroups, each with a small number of tags and each subgroup interrogated separately, then the interrogation process will be completed at an earlier stage if there is an error (Leng et al. 2010). In particular the reader does not need to interrogate again all subgroups, only those for which no proof was generated. Leng et al. in their protocol, propose to use a tree based anti-collision algorithm to facilitate the subgrouping of the RFID tags.

2.3 Sequential versus Concurrent Proofs

A grouping-proof protocol can be executed in two distinct modes: *sequential* and *concurrent*. In the sequential mode, the grouping-proof is usually generated by having each tag authenticate a message sent by the previous tag: this involves *message chaining*. The reader is in charge of forwarding all the messages. This may cause scalability issues, if the number of tags in the group is large. The reason is that a reader needs to relay messages from one tag to another so that each tag in the chain can authenticate the message the previous tag sent.

A variant of this model makes use of a "pallet tag", which is attached to a pallet with tagged products. The pallet tag has more computational resources than ordinary RFID tags and acts as a representative of all tags that are in the pallet. Pallet tags can be re-used several times and for different applications, and therefore their cost can be amortized.

If the protocol is sequential, then the time it takes to identify a group of n tags is linear in n. This may lead to aborted interrogations when the number of tags is large. By contrast, for concurrent proofs, the messages sent and the corresponding confirmations are transmitted independently, and the checking is distributed.

2.4 Reading Order Dependent versus Reading Order Independent

Several grouping-proofs are *reading order* dependent. This means that the tags must be read in specific order for every interrogation. Although this property can be confused with the previous one, they are different. Indeed, although a concurrent protocol must be order independent, a reading order independent protocol can be sequential.

Lien et al. (Lien et al, 2008) point out that order dependent grouping-proofs are inefficient, have higher interrogation failure rates and make anonymity difficult to implement. In the next section we shall describe their reading order independent protocol, that uses the exclusive-or operation to generate a grouping-proof.

2.5 Missing Tags Identifier

Finally, we describe a property that we have called the *"missing tags identifier."* With this property we distinguish those protocols for which the Verifier is able to identify the missing tags when the grouping-proof is not valid. If the missing tags can be identified then there are several actions that the Verifier can take to establish the nature of the fault.

Table 1. A summary of notations

Notation	Description
D	Tag database at the Verifier (or backend Server)
tag_X	Tag X in D
k_x	Secret key of the tag X
c_X	Counter of the tag X
$MAC_{(.)}(\cdot)$	Keyed message authentication code
$f_{(.)}(\cdot)$	Keyed pseudo-random function
$P_{X_1 X_2 \cdots X_n}$	Grouping-proof for the tags $X_1 X_2 \cdots X_n$

In general, when a grouping-proof is invalid, the Verifier cannot find out the reason for the failure. The reason can be missing tags, transmission error, or faulty tags. If the Verifier cannot identify the nature of the failure, then the choice of further actions to coping with a grouping-proof failure is very limited (Leng et al., 2010).

3. BACKGROUND

For the rest of this chapter, we will use the notations summarized in Table 1.

3.1 Yoking-Proofs (Juels)

The yoking-proof proposed by Ari Juels (Juels, 2004) is a proof that a pair of RFID tags (tag_A, tag_B) has been scanned simultaneously (within a short time period) by an RFID reader. This proof is for offline applications. As pointed out by Juels, the notion of two-party *presence* is closely related to two-party *key agreement* (Shoup, 1999).

The protocol is presented in Figure 1. The tags (tag_A, tag_B) have identifiers A, B, secret keys k_A, k_B that are shared with the Verifier but not

the reader, counters c_A, c_B, and use a keyed message authentication code (MAC) and a keyed hash function f to generate a proof of simultaneous presence. The reader scans the tags sequentially, links pairs of responding tags, and is in charge of forwarding messages and generating the proof.

In the protocol the reader transmits to the first tag, tag_A, the message "left proof" and to the second tag, tag_B, the message "right proof" to indicate their roles. Then tag_A and tag_B generate and transmit the challenges,

$$a = \left(A, c_A, r_A = f_{k_A}[c_A]\right) \quad \text{and} \quad b = (B, c_B, m_B = MAC_{k_B}[a, c_B]),$$

respectively, and tag_A authenticates the challenges (a, b). The challenges and the authenticator constitute a proof P_{AB} of simultaneous presence. The Verifier checks that the proof P_{AB} is valid by first re-computing the challenges (a', b'), then computing the authenticator:

$$m'_{AB} = MAC_{k_A}[a', b'],$$

and finally checking that $m'_{AB} = m_{AB}$. On timeout, or incorrect input, any party will abort.

Juels also proposed a minimalist version of this proof that does not use counters, based on a one-time MAC that is a simplified version of the symmetric-key Lamport digital-signature (Lamport, 1979). In this version the challenges to be authenticated have fixed length d and the secret key of each tag consists of d secret pairs $(s_i^{(0)}, s_i^{(1)})$, $i = 1, \ldots, d$, with each $s_i^{(j)}$ having length l (roughly 80 bits). The MAC on the bit string $b_1 b_2 \cdots b_d$ is the string $s_1^{(b1)} s_2^{(b2)} \cdots s_d^{(bd)}$.

A more efficient version is obtained by using an (n, k) error-correcting code with block length $n = 120$, message length $k = 32$, and minimum distance d at least 32. In this version, the MAC is computed on a single number rather than a pair of numbers. Figure 2 describes this protocol, with the parties initialized with one-time random values r_A and r_B. For this version the storage require-

Figure 1. The yoking-proof protocol using standard cryptographic primitives

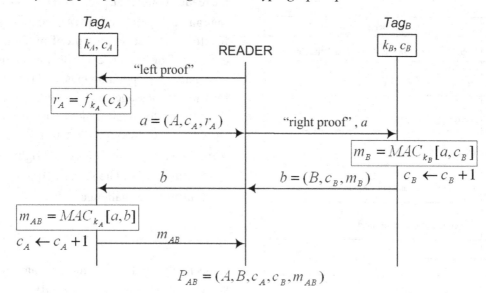

ments for each tag are only 360 bits: 120 for the random number and 240 for the secret pairs (Juels, 2004).

3.2 Grouping-Proof (Saito and Sakurai / Piramuthu)

Saito and Sakurai (Saito and Sakurai, 2005) observed that the minimalist version of the protocol proposed by Juels is vulnerable to a replay attack. Figure 3 describes the attack. This is an active attack and has two stages. In the first stage the adversary challenges tag_A by using the number r' to get $a = (A, r_A)$ and $m'_A = MAC_{k_B}[r']$. In the second, that takes place at a later time, the adversary uses these values to challenge tag_B and get the authenticator m_B of r_A. Now the adversary

Figure 2. A one-time yoking-proof protocol using a minimalist MAC

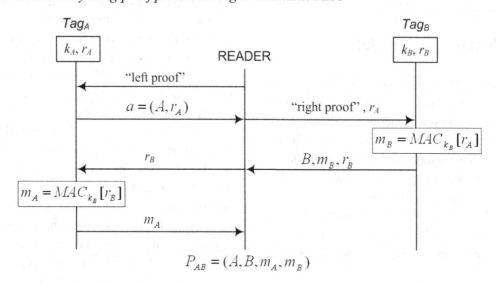

has all the parts of a forged grouping-proof $P_{AB} = (A, B, m_A', m_B)$.

However this proof contains the wrong authenticator m_A', as noted by Piramuthu (Piramuthu, 2006). To overcome this problem, the correct authenticator $m_A = MAC_{k_B}[r_B]$ must use the number r_B. If the numbers r_A, r_B, are shared with the Verifier, then the Verifier can check that the proof is forged.

There is another problem with one-time yoking-proofs, and indeed with any one-time grouping-proof that does not contain temporal information. If the adversary is the first one to interrogate the pair of tags (tag_A, tag_B) then the adversary can use the generated proof to impersonate these tags to an authorized reader at a later time. Since the generated proof will be valid, the Verifier will accept that the pair of tags (tag_A, tag_B) was scanned some time later, by which time the tags may be separated.

To address this issue Saito and Sakurai presented another version of the yoking-proof that relies on time stamps provided by the Verifier (Saito and Sakurai, 2005). This is shown in Figure 4.

Saito and Sakurai extended this proof to groups with an arbitrary number of tags by using a *pallet tag*. They called this proof a *grouping-proof* (we also use this name for proofs of simultaneous presence of $n \geq 2$ tags in preference to yoking-proofs, because yoking refers to pairs of tags). Figure 5 describes such a proof. In the protocol, the reader receives a time stamp *TS* from the Verifier and sends the time stamp to the n tags participating in the protocol. The pallet tag encrypts the MACs by using symmetric key encryption and submits the ciphertext to the reader.

Piramuthu identified a problem with time stamps (Piramuthu, 2006). An adversarial reader can interrogate a tag using a time stamp for some future point in time, and then use this same time stamp later to interrogate another tag, thus violating the simultaneous interrogation requirement. Time stamps can be predicted, allowing the adversary to collect prior responses and combine them to forge proofs. Accordingly, Piramuthu proposed the use of random numbers instead of

Figure 3. A replay attack against a yoking-proof

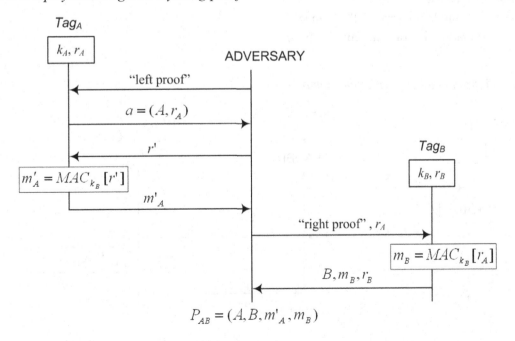

Figure 4. A yoking-proof with time stamp

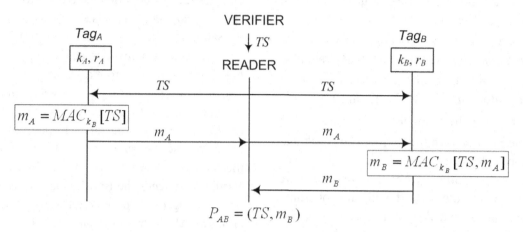

time stamps. Piramuthu's protocol is given in Figure 6.

In the protocol the reader receives a random number r from the Verifier and sends it to both tags. This number is used as a seed to generate the numbers r_A and r_B of the tags. The authenticator m_B of tag_B depends on r and r_A, while the authenticator m_A of tag_A depends on m_B and r_A. Thus the values: r, r_A, r_B, m_A, and m_B are strongly linked. Note that the value r_B is not authenticated –this will lead to the attack by Peris et al. described below.

Finally Piramuthu observed that since no private information is transmitted during the interrogation, and since the data transmitted is refreshed every time the protocol is run, the proposed grouping-proof provides both *user privacy* and *location privacy*.

Peris et al. (Peris-Lopez et al., 2007) noted that in Piramuthu's proof the value r_B is not authenticated, and furthermore there is symmetry in the components of the proof $P_{AB} = (r, r_A, r_B, m_A, m_B)$ that can be exploited in a *multi-session* attack. The attack is shown in Figure 7. In a first session, that is passive, the adversary eavesdrops on an interrogation of tag_A and tag_B to get the values:

Figure 5. A grouping-proof with time stamp

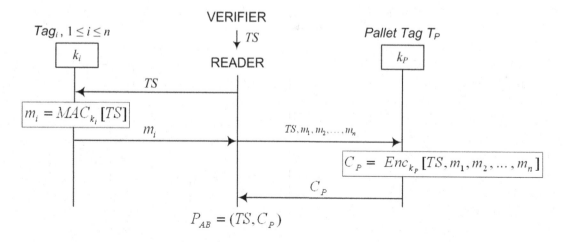

Figure 6. A grouping-proof using random numbers

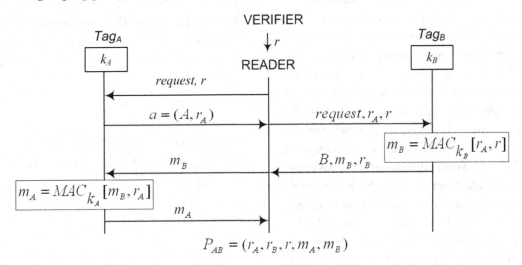

The values m_B, r_A, are used later to interrogate tag_X to get the authenticator $m_X = MAC_{k_X}[m_X, r_A]$. Now it is easy to see that $P_{XB} = (r_X, r_B, r, m_X, m_B)$

$r, \ r_A, \ r_B$ and $m_B = MAC_{k_B}[r_A, r]$ (m_A is not needed).

is a valid proof that tag_X and tag_B were simultaneously scanned, which is false.

Finally, Lin et al. (Lin et al., 2007) pointed out that Piramuthu's proof may suffer from interference when multiple readers are present. A tag can be queried almost simultaneously by two readers,

Figure 7. A multi-session attack

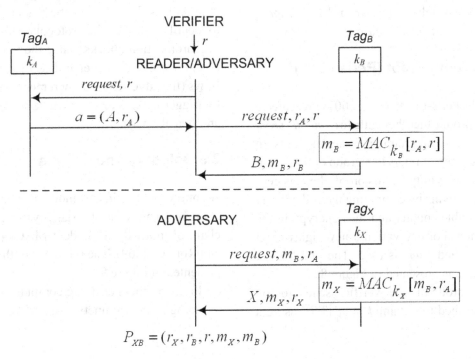

and then, the tag might have problems determining what messages to sign (a *race* condition).

There have been several attempts to extend the yoking-proof to capture anonymity. Piramuthu claimed that the grouping-proof in Figure 6 provides user privacy and location privacy because the transmitted data is refreshed every time the protocol is run. However, in this protocol the tags send *immutable* identifiers so the anonymity offered is very weak. Furthermore, the adversary can use the identifiers to link the interrogation sessions, which is a violation of location privacy.

Bolotnyy and Rose (Bolotnyy and Robins, 2006) proposed an extension of the yoking-proof in which the tags use *mutable* pseudonyms. To compute a pseudonym each tag selects a random number and hashes it with a keyed hash function *f*, using its secret key. It then sends to the reader together with its yoking-proof challenge this random number and its hashed value. For example, if tag_A is interrogated by a reader, it will generate a random number r, compute $a = f_{k_A}(r)$ and send to the reader the pair (r, a) together with its challenge. This can then be used by the Verifier to determine which secret key was used to compute a given r, and thus identify tag_A. This approach only works when the Verifier is online, not for the offline case.

3.4 Clumping Proof (Peris et al.)

Peris et al. (Peris-Lopez et al., 2007) presented a grouping-proof which they called a *clumping proof* (a clump is a small group or cluster of plants. In an RFID context it refers to a group of tags.) that addresses anonymity. In this proof, the tags generate pseudonyms by combining their identifiers with a mutable counter and then encrypting the combination. For encryption a new lightweight function, called *Num*, is used. The proof is for two tags and is presented in Figure 8.

In the protocol, the Verifier sends to the reader a hashed timestamp $t = g_{k_V}[TS]$, using a

keyed hash function (k_V is the secret key of the Verifier). The reader then partitions the hash into two halves $t = (t_{MSB}, t_{LSB})$, and sends t_{MSB} to tag_A. Then tag_A responds by sending the triple

$$Y = (a_1, a_2, counter_A)$$

to the reader. The reader then sends to tag_B the pair (t_{LSB}, a_2), and tag_B responds by sending the triple

$$Z = (b_1, b_2, counter_B) .$$

The proof consists of: Y, Z, an authenticator of tag_A for (a_2, b_2), and the hashed timestamp t.

In this protocol the reader has to find which pairs of tags should be linked (those belong to a group) and this can only be done by using the authenticators a_2, b_2, which are computed using the secret keys of the tags. Since only the Verifier knows the secret keys of the tags, this proof must also be online.

Peris et al. suggest that, to reduce the computational cost of the Verifier, the counters could be transmitted in the clear, together with the other values. However, if the counters are used to identify tags then the protocol is subject to de-synchronization attacks, since the values of the counters are updated regardless of the received flows (the adversary can increment the counters by triggering sessions, via rogue readers, which are then aborted).

3.5 Bolotnyy and Robins

Bolotnyy and Robins extended the yoking-proof to groups of more than two tags by using a circular chain of mutually dependent MACs (Bolotnyy and Robins, 2009). The protocol for this proof is presented in Figure 9.

In the protocol each tag computes a MAC of a message that is a function of a MAC computed

Figure 8. A clumping proof

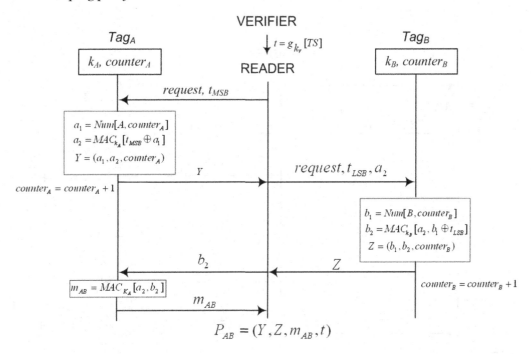

by the preceding tag in the yoking-proof chain. To avoid replay attacks and allow temporal ordering of the proofs, each tag increments its counter immediately after it sends its computations to the reader. To prevent an adversary from replaying old proofs, the authors suggest that the Verifier stores counter values of tags obtained from the latest verified yoking-proof in which the tags of the group participated.

3.6 Burmester et al.

Burmester et al. presented several grouping-proof protocols (Burmester et al, 2008). The main feature of these protocols is that the tags of the group share a *group key*. This makes it possible for the tags of a group to (*a*) recognize each other, and (*b*) be distinguished by RFID readers who can then link them in the offline mode (even though they do not share the group key). In particular, this makes it possible to avoid the generation of corrupted proofs.

RFID tags have extremely limited computational capabilities and cannot verify each other's computations. This implies that in the offline mode *unrelated* tags can take part in a grouping-proof session, and this will only be detected by the Verifier at some later time, not by the reader. While, from an authentication perspective this may not represent a security threat, in many practical applications it is an undesirable waste of resources, and could be characterized as a DoS vulnerability.

To appreciate how accidental pairing may create challenges to real-world applications, consider an application in which a grouping-proof is used to ensure that all the components of a product are contained in a shipment pallet. An offline RFID reader is configured to take temporary measures after a failed grouping-proof –*e.g.,* notify an assembly worker. This capability is denied if a tag (accidentally or maliciously) engages in grouping-proof session with unrelated tags. Accidental occurrences of this type may be quite likely, particularly with anonymous grouping-proofs, and are facilitated by the fact that the scanning

Figure 9. A circular chain yoking-proof

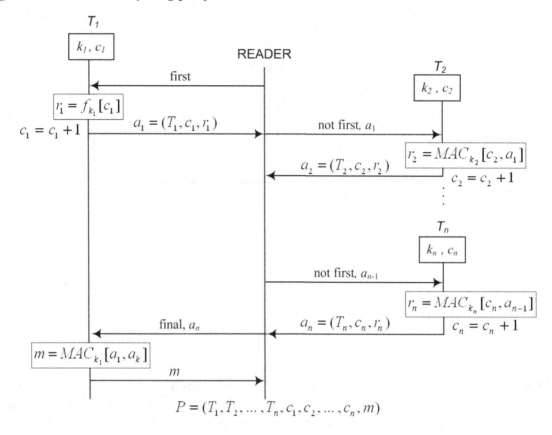

$$P = (T_1, T_2, ..., T_n, c_1, c_2, ..., c_n, m)$$

range of readers may vary according to different environmental conditions (Burmester et al. 2008).

Another feature of the protocols presented by Burmester et al. is that their security is proven in a strong security framework that supports *modularity*. Since RFID systems are typically components of larger network systems, it is preferable to design them so that their security is preserved when the protocols are executed in arbitrary composition with other (secure) protocols.

We shall not describe these protocols in this section, but leave that for Section 6 since we will present new versions of these protocols.

3.7 Lien et al.

Lien describes a reading order independent protocol (Lien et al, 2008), which is based on Piramuthu's protocol and the idea of the "pallet tag".

This protocol is presented in Figure 10. The reader receives a random number r from the Verifier and broadcasts it to every tag T_i, including the pallet tag T_P. Using r as the seed, each tag T_i calculates its random number r_i and sends this along with its identifier A_i. T_P also generates its random number r_P. The reader then sends the triples (A_{T_P}, A_i, r_i) to the pallet tag, with A_{T_P} an identifier for the pallet tag, regardless of their order. After receiving the n pairs, T_P uses its secret key k_P to calculate $m_{pi} = MAC_{k_P} [r_i, r_P]$, and sends this together with the tag identifiers back to the reader, who forwards these to the tags. The tags now compute the authenticators $m_i = MAC_{k_i} [m_{pi}, r_i]$ which are received by the pallet tag through the reader in any order of arrival. After receiving the n pairs, T_P computes the authenticator

Figure 10. The reading order independent protocol of Lien et al.

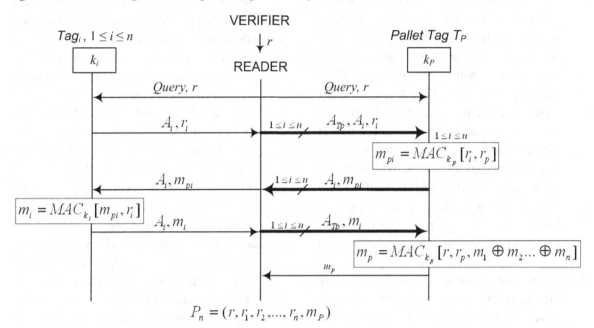

$$P_n = (r, r_1, r_2, ..., r_n, m_P)$$

$$m_P = MAC_{k_p}\left[r, r_P, m_1 m_2 \cdots m_n\right],$$

and sends this to the reader. The reader generates the proof $P_n = (r, r_1, \ldots, r_n, m_P)$. The exclusive-or operation is commutative and is used to make the grouping-proof order independent.

4. ADVERSARIAL MODEL

4.1 Attacks on RFID System

Several types of attacks have been described in the literature. For RFID systems, these are exacerbated by the characteristics, restrictions, and mobility of tags, allowing them to be manipulated at a distance by covert readers. Below we list some of the more important attacks.

Denial-of-Service (DoS) attacks: The adversary causes tags to assume a state from which they can no longer function properly. Tags become either temporarily or permanently incapacitated.

Sniffing attacks: These are confidentiality attacks on the air interface, in which the adversary eavesdrops opportunistically on the information exchanged between the legitimate tag and the reader. RFID technology is wireless; communication between parties is made in the open air medium and not only does it offer many opportunities for an adversary to eavesdrop, but also to block, inject, jam or intercept and modify the transmitted messages.

Skimming Attacks: These are confidentiality attacks on tags, in which an unauthorized reader –the adversary, interacts with a tag to obtain data stored in its memory. Meanwhile, the owner of the tag remains unaware. This attack is feasible because most RFID tags are *promiscuous:* they are designed to respond to reader scans automatically, without requiring any kind of authentication.

Tracking/Traceability attacks: These are privacy attacks on the air interface and/or the tag. The adversary combines sniffing and skimming attacks to track a tag over a period of time.

Tracking is feasible if the adversary is able to decide with probability better than negligible if messages from different (eavesdropped) sessions involve the same tag or not. Unlike other point-to-point technologies, RFID is typically intended to keep track of highly mobile items. Thus, an individual tag can be traced by tracking it, or tracking its group. It is important to note that it is not necessary to know the identity of a tag to track it. It is sufficient for the adversary to identify certain patterns of transmitted messages: *e.g.,* the level of signals, the time of response, being part of a group or using the same encrypted response.

Replay/Spoofing attacks: These are integrity attacks on tags or the reader, in which the adversary impersonates a tag or the reader without knowing the tags secret key(s). The adversary simply uses messages previously sent by the reader or the tag to impersonate them in the current session. These messages can be gathered via sniffing or skimming. Tags typically respond with the same static identifier. In order to avoid a replay attack, the authenticators must change dynamically from session to session. Interleaving attacks, in which the adversary constructs valid messages by combining flows from different sessions, can be considered as particular case of replay attacks.

(Crypto) Analysis attacks: These are confidentiality and integrity attacks, performed on the implemented security mechanisms (of the tags, the reader or both). The adversary collects and analyzes exchanged messages to infer sensitive information. Exchanged messages leak information that can be used by the adversary to reduce (or eliminate) the security of a cryptographic primitive; *e.g.,* by breaking the cryptographic algorithm or recovering the secret keys. Economic pressures keep the price of the tags as low as possible, and therefore the available resources for protecting them are very constrained. Security primitives must be very simple.

Impersonation attacks: These are integrity attacks on tags or the reader, in which the adversary uses leaked sensitive information to mimic the behavior of one of the parties to another. These attacks are usually preceded by an analysis attack to extract sensitive information. A replay attack is a particular case of an impersonation attack for which the previous analysis is not necessary because all the needed information is obtained directly from the parties involved.

Denial-of-Service attacks (DoS): These are availability attacks, performed on tags, the reader and/or the air interface, in which the adversary prevents the system from operating as intended, temporarily or permanently. DoS attacks can be carried out in different ways. For example, the adversary can cause a tag to assume a state from which it can no longer function, by desynchronizing the reader and tag, by disabling the tag logically (*kill* command) or physically (*high power* RF field), or simply by appropriating the tag. Intentional jamming can also be used to prevent normal communication.

Off-line Man-In-The-Middle (MITM) relay attacks: These are integrity attacks performed on the tag and reader, in which the adversary inserts or modifies messages exchanged by the genuine parties. This kind of attack can be seen as a combination of replay and analyses attacks. It is basically a replay attack in which the messages relayed are adapted (or created) following an analysis attack.

On-line Man-In-The-Middle relay attacks: These are integrity attacks performed on the tag and reader, in which the adversary interposes between the parties and exchanges their messages. They are similar to the off-line attacks described above, with the exception that the adversary relays the (possibly modified) messages on the fly (on-line).

Side-channel attacks: These are integrity and confidentiality attacks performed on the tag. The adversary exploits information leaked by the physical implementation of protocols (Mangard et al., 2007). The physical implementation of cryptographic algorithms can leak information about secret data through side channels such as the fluctuation in power consumption, timing information, electromagnetic radiation, fault induction, etc. RFID tags do not have batteries and are passively (electromagnetically) powered, which makes them particularly vulnerable to this kind of attack.

Physical attacks: These are integrity attacks performed on the tag. They are hardware layer attacks in which the adversary is able to disassemble the chip and access its memory/resources directly (*i.e.* physically). Due to economic reasons, proper tamper protection is reserved for high-cost RFID devices (*e.g.* smart cards). Hence, most RFID tags can be physically attacked and controlled by the adversary. Sometimes side-channel attacks are considered also as physical attacks. Here, we have preferred to distinguish between them as a way to discriminate between non-invasive and invasive attacks. Side-channel attacks are usually non-invasive. Normally, they are reversible and do not leave evidence of the attack. By contrast, invasive attacks are usually destructive, and thus irreversible and leave evidence of the attack. We could also mention semi-invasive attacks where only some external parts of the chip are removed. Many of these are reversible.

Tag cloning: These are integrity attacks performed on the tag, in which the adversary creates fake clones of genuine tags after capturing their most relevant information. The adversary can capture this information via either invasive (physical) or non-invasive attacks. The counterfeit tag can be expected to behave identically to the original (*e.g.* for

access control), but it may also behave in a malicious way, *e.g.* to prevent a grouping-proof from being generated.

Virus attacks: These are integrity attacks, performed on the tag – but affect the middleware, in which the adversary inserts a virus in the memory of a tag in order to infect the back-end database (Rieback et al., 2006). These attacks take advantage of the fact that they are usually dismissed by system developers and thus the information sent by the tags is implicitly trusted. Multiple vulnerabilities can be exploited: injection of SQL code, causing buffer overflows in C or C++ codes or the inclusion of any kind of malicious script. Once a database is infected, the virus can spread to other tags and later to other middleware when such tags are scanned.

Note that sometimes it is not very easy to distinguish the boundary between one attack and another, where one starts and another finishes, and in many cases an adversary will combine more than one of these attacks to achieve his goals.

4.2 Threat Model

RFID tags are a challenging platform from an information assurance standpoint. Their extremely limited computational capability implies that traditional multi-party computation techniques for securing communication protocols are not feasible, and that instead lightweight approaches must be considered. Yet the robustness and security requirements for RFID applications can be quite significant. Ultimately, security solutions for RFID applications must take as rigorous a view of security as other types of applications (Burmester et al., 2009).

Accordingly, we assume a *Byzantine adversary*. In this model all legitimate entities (the Verifier, the RFID readers, the tags) and the adversary A have polynomially bounded resources. The adversary A controls the delivery schedule of the communication channels, and may eaves-drop into, or modify, their contents and may also initiate new communication channels and directly interact with authorized parties. In particular, the adversary can attempt to perform impersonation, reflection, man-in-the-middle, and any other passive or active attacks that involve reader-to-tag communication. The adversary A may also instantiate new channels, and directly interact with authorized parties. For grouping-proofs, adversarial attacks target three main objectives:

- RFID tags T : A can impersonate a legitimate RFID reader R, and collect responses in an attempt to forge a grouping-proof.
- An RFID reader R : A can impersonate a tag T in an attempt to get the reader R to generate a grouping-proof, or to get information to forge a grouping-proof.
- The communication channel between tag(s) T and the reader R (the *wireless medium*): A can eavesdrop on the messages exchanged, try to modify or block them.

A Formal Model for Strong Privacy

Juels and Weiss (Juels and Weiss, 2006) proposed a model for RFID systems that addresses a variety security related weaknesses. This is based on an earlier simulation model for authenticated key exchange by Bellare et al. (Bellare et al., 2000). Here we describe a modified version of the Juels and Weiss model proposed by Ouafi and Phan (Ouafi et al., 2008). In this model, in each interrogation session i, the tags T and the reader R interact as specified by a RFID protocol. At the end of a session i the parties involved (tag/ reader) output Accept if their view of the execution is correct. The adversary A can interact with the parties during the execution of the protocol by using one or more of the following three queries:

- Execute (R, T, i): This models passive attacks, where the adversary A gets access to an honest execution of the protocol by eavesdropping.
- Send (U_1, U_2, i, m): This models active attacks by allowing the adversary A to impersonate any instance of party U_1 (reader or tag) and send a message m of its choice to some instance of party U_2 (tag or reader).
- Corrupt (T, K, K'): This allows the adversary A to learn the stored secret key K of the tag T, and replace it by the key K'. It captures the notion of *forward security* and the extent of the damage caused by compromising the keys of tags.

The case where all the tags and the reader are compromised is not considered, as in this case the adversary can always forge grouping-proofs.

Finally, in the design of RFID grouping-proof, another interesting property to achieve is *untraceability*, in which a tag cannot be traced by an adversary A who eavesdrops on tag interrogations. This property is captured by the oracle query:

- Test$_{\text{Upriv}}$ (U, i): This query does not correspond to any of the abilities of the adversary A or any real-world event. It is used to define *untraceability* in terms of *indistinguishability*. If a party U has accepted and the oracle is asked this query, then depending on a randomly chosen bit b the adversary A is given T_b from the set $\{ T_0, T_1 \}$ —consisting of two earlier tag interrogation flows, with exactly one linked to party U. The adversary must guess the bit b.

Untraceability is defined in terms of a game G that involves the adversary A, the reader R, the tags T and an oracle, in which:

- Phase 1 (Learning): A can use any of the queries Execute, Send and Corrupt at will, to acquire a history of reader-tag interrogation flows.
- Phase 2 (Challenge):

At some point during G the adversary A selects a *fresh* session (Bellare et al., 2000) to be tested, and sends the query Test$_{\text{Upriv}}$ to the oracle. A is then given by the oracle the challenge T_b selected randomly from the set $\{T_0, T_1\}$.

A (b)A continues making the queries Execute, Send and Corrupt at will.

Phase 3 (Guess): Finally the adversary A outputs a bit b', which is its guess of the value of b.

The success of the adversary in winning the game G and thus breaking the notion of privacy is quantified in terms of the adversary's advantage in distinguishing whether it can trace the tag U. For more details the reader is referred to Ouafi and Phan (Ouafi et al., 2008).

The Universal Composability Framework

RFID systems are often used as components of larger and more complex applications and therefore one should consider protocols that are secure under composability with arbitrary applications. The Universal Composability (UC) framework (Canetti, 1995, 2000, 2001) specifies a particular approach to security formalization, which guarantees that a UC-based security proof for a protocol remains valid if is composed with other protocols (modularity), or if is executed in arbitrarily concurrent settings (including interleaving executions of).

The UC framework defines a *real-world* simulation, an *ideal-world* simulation, an *emulation* that translates protocol runs of from the real-world to the ideal-world, and an interactive *environment* that captures whatever is external to the current protocol execution (see Figure 11). In particular, the components of UC security formalization are:

- A mathematical model of real executions of the protocol. In this model, honest parties are represented by probabilistic, polynomial-time (PPT) Turing machines that correctly execute as specified, and adversarial parties that can deviate from in an arbitrary manner. The adversarial parties are controlled by a single PPT adversary that:
  a. Has full knowledge of the state of the adversarial parties;
  b. Can arbitrarily schedule the communication channels and activation periods of all parties, both honest and adversarial, and
  c. Interacts with the environment in arbitrary ways, in particular can eavesdrop on all communications.

In the UC framework there is also an additional PPT entity, called the *environment* that generates the initial inputs of all parties, reads their final outputs and interacts in an arbitrary way with the adversary during the execution of protocol.

- An idealized model of executions of , where the security properties do not depend on the correct use of cryptography, but instead on the behavior of an *ideal functionality* F_ρ, a trusted party that all parties may invoke to guarantee correct execution of particular protocol steps. The ideal-world adversary A' is controlled by F_ρ, to mimic the behavior of the real-world adversary A. The functionality F_ρ can be thought of as the *formal specifications* of protocol ρ.

- A proof that, for each PPT adversary A for ρ, there is a PPT simulator S that translates real-world runs of ρ in the presence of A into ideal-world protocol runs of ρ in the presence of a simulated adversary

Figure 11. Architecture of security proofs in the Universal Composability Framework

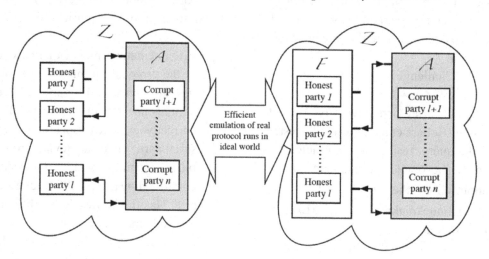

$A' = S(A)$ such that, no environment Z can distinguish (with better than negligible probability) whether it is communicating with an instance of ρ and A in the real-world or with F_ρ and A' in the ideal-world. Formally,

$$\forall A \exists S : View_Z^{real}(A, \rho) \approx View_Z^{ideal}(A', F_\rho).$$

The main feature of the UC framework is that the UC security of a composite system can be derived from the UC security of its components without need for holistic reassessment of its robustness.

In this chapter, we are mainly concerned with security issues at the protocol layer and not with physical or link layer issues, such as the coupling design, the power-up and collision arbitration processes and the air-RFID interface.

5. GUIDELINES FOR SECURE RFID APPLICATIONS

Here we present effective strategies that can be used to thwart the attacks described in the previous section.

- To prevent the adversary from causing tags to assume an unsafe state, tags share with the Verifier secret keys (assuming here that the cost of public-key cryptography in tags is too high for the application in question, a likely scenario), which only are updated after mutual authentication. When a tag is challenged by an RFID reader it will generate a response using this key.

- To prevent cloning attacks the Verifier should be able to check a tag's response, but the adversary should not be able to access a tag's identifying data. The response must therefore corroborate (but not reveal!) knowledge of the tag's private data. Of course it should be hard for the adversary to extract private data from the tag's response. This can be assured by using cryptographic one-way functions.

- To prevent unauthorized tracing the adversary should not be able to link tag responses to particular tags. This can be guaranteed by randomizing the values of the tags' responses. Since all entities of an RFID system are assumed to have polynomially bounded resources, it is sufficient for these values to be cryptographically pseudorandom.

- To prevent the adversary from constructing valid transcripts by combining flows from different sessions, the flows of any particular session should be strongly linked. This can be assured by binding all messages in a session to the secret key and to fresh (pseudo-) random values. Although the inclusion of time stamps could be suggested to achieve this, its use in clear, without any encryption, is not recommended because future proofs could be generated in advance as they can be predicted. So, simple random fresh numbers or encrypted versions of timestamps are recommended, which should also have an explicit or an implicit time of validity. The Verifier will not accept proofs that were collected with expired timestamps. This temporal validity is crucial for RFID grouping-proofs. This defines the time resolution for the Verifier's scanning: simultaneity of a group interrogation cannot be determined to a finer time interval.

- Modularity and Reusability. Security protocols are often deployed in a variety of contexts with similar security requirements, that is, where the adversary is subject to similar computational and communication bounds. This practice can introduce vulnerabilities. For instance a protocol is often analysed under the implicit assumption of operating in isolation, and therefore may fail in unexpected ways when combined with other instances of its own or other protocols –see *e.g.* (Burmester and Medeiros, 2009). To guarantee security against such attacks it is necessary to model security in a modular, concurrency-aware model. The Universal Composability framework discussed earlier, in addition to capturing threats arising from concurrency, allows for secure protocol re-use –*e.g.,* as a building block for more complex applications.

- Computing capabilities and performance (Peris-Lopez et al., 2010). RFID tags are devices with severely restricted memory and computing power, and the RFID communication channel is especially unreliable. The design of a grouping-proof should take this into account. The number of computations as well as the messages transmitted on the channel by tags should be minimized (Burmester and Munilla, 2011). According to memory requirements, protocol designers should limit the use of non-volatile tags' memory in which identifiers and secret keys are stored. On the other hand, parallel scanning (or the use of subgroups) is preferred. Sequential RFID scanning can be slow, particularly if the group is large, and can lead to aborted sessions.

- Separation of duties. The Verifier (a trusted entity) manages trust associations and the security of the RFID system, whereas RFID readers (potentially untrusted) support reliability. Separation of duties requires that tags share private data with the Verifier –at the security layer, which is not shared with RFID readers –at the tag-identification layer. We have static (ephemeral data is shared), and dynamic (a function generating data is shared) separation. To address non-cryptographic DoS attacks at the tag-identification layer in the offline case, RFID readers should share with the tags private data supporting reliability associations. Indeed, the RFID protocol should make it possible to filter out unwanted traffic as early as possible, so that the parties are not overwhelmed by unnecessary data.

- Forward security. An attacker may get access to the secret key(s) stored on tag's memory as these devices are typically not tamper proof and therefore susceptible to physical manipulation. So, privacy of past communications should be guaranteed even if a tag is compromised some time later. However, this property is not easy to achieve. The Verifier and tags should update their shared cryptographic material but they have to keep synchronized. For grouping-proofs, in contrast to normal RFID authentication protocols where only two parties involved, several tags and an (untrusted) reader participate in the generation of the proof with the Verifier taking part some time later (in offline-mode). Furthermore, updating of cryptographic material is difficult because only a subgroup of tags in the group may be involved in a session.

- Secure tag location. Most of RFID protocols are subject to location-based attacks, including so-called mafia fraud attacks or (on-line) relay attacks. Indeed, a relay attack is always feasible from a theoretical point of view. Similarly, grouping-proofs will be also subject to attacks which relay messages faster than the time span defined by the Verifier's time stamps. If protection against this kind of attack is required, the protocol should include (mount) some distance bounding mechanism to estimate the distance (Munilla et al. 2009); a common method is to use the *round-trip time* of messages exchanged.

6. RECENTLY PROPOSED SCHEMES

6.1 The EPC Global UHF Class 1 Gen 2 Standard

The EPC Global UHF Class 1 Gen 2 standard, commonly known as EPCGen2, was ratified by ISO in 2006 as an amendment to the 18000-6 standard (EPCGlobal). EPCGen2 defines the physical and logical requirements for a passive-

backscatter RFID system operating in the 860 MHz - 960 MHz frequency range.

The standard specifies two protocol layers, the *physical* and the *tag identification* layer. These define the procedures and commands for identifying tags in a multiple-tag environment. The hardware requirements for EPCGen2 are:

i. a 16-bit Pseudo-Random Number Generator (PRNG) and,

ii. a 16-bit Cyclic Redundancy Code (CRC).

6.2 EPCGen2 Compliant Grouping-Proofs for Medical Services

Huang and Ku (Huang and Ku, 2009) proposed an EPCGen2 compliant grouping-proof designed to enhance inpatient medication safety. Thus, it uses a 16-bit PRNG to authenticate protocol flows and a 16-bit CRC to hash strings to 16-bits. Each tag has a 16-bit private key PIN and a 32-bit identifier EPC, which it shares with the Verifier. The tag uses its PRNG to generate session authenticators C, with initial value

$$C = PRNG(PIN).$$

The protocol is given in Figure 12.

The RFID reader starts the interrogation by broadcasting a timestamp TS. The tags then respond by sending their identifiers EPC and a message authenticator M, computed as follows:

$$M = CRC(EPC,C) \text{ C R, where}$$
$$R = PRNG(TS).$$

This grouping-proof has several weaknesses which we list below:

1. This proof does not guarantee privacy: the tags can easily be traced (linked) from the identifiers EPC. Consequently it is not ap-

Figure 12. Huang-Ku's EPCGen2 compliant proof for medical services

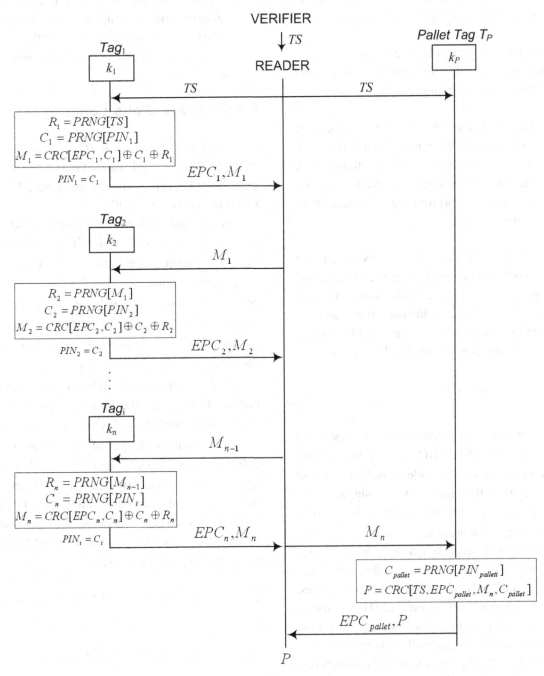

propriated for medical services that require privacy.

2. In the protocol RFID readers are identified with the Verifier (backend Server): either they share the private keys of the tags —a

violation of the guidelines of the previous section, or they have online access to the Verifier.

3. When the Verifier is offline, tags do not (and cannot) check that the RFID reader is

authorized. Thus an adversarial reader can de-synchronize tags –a DoS attack.

The adversary can obtain a proof for any selected timestamp TS, in particular for a future TS. This leads to a replay attack.

4. The tags use a PRNG to generate a message authentication code, by replacing its current state with the message to be authenticated. PRNGs are not intended to be used in such a way, and may not provide security guarantees if this is done.

From the definition of the protocol, it is not clear how the RFID reader can identify the tags of a group. For example, which tags, and from which groups, should be the ones to respond to a reader timestamp call TS? A reasonable assumption is that the group is arranged in a cycle

$$\left(tag_1, \ldots, tag_n, Pallet\, tag\right),$$

and the protocol is executed sequentially, starting from tag_1. Since EPCGen2 tags cannot eavesdrop on communication, the reader must first establish channels with the tags of each group, and then send to each tag_i the authenticator M_{i-1} that it received from the previous tag in the cycle, using an allocated channel. When the number of tags in the group is large this can slow down the interrogation and lead to aborted sessions and failures.

Chien-Yang-Wu (Chien et al., 2011) proposed a variant of this grouping-proof that addresses some of its weaknesses. To deal with de-synchronization attacks, the time-stamp is encrypted with a private key that the Verifier and the reader shares. However there are still unresolved issues:

-   It does not address privacy (unlinkability).
-   It is subject to related key attacks (Burmester and Medeiros, 2008).

-   An adversary can participate in authorized interrogations causing proofs to be generated that will only be discarded later by the Verifier, leading to DoS attacks (memory overflow).

6.3 Kazahaya (Peris-Lopez et al.)

Peris-Lopez et al. present in (Peris-Lopez et al., 2010) an RFID grouping-proof that is EPCGen2 compliant. Tags are limited to invoking a PRNG function and bitwise XOR.

The tags are divided into groups, which are identified by a group identifier ID_{group} to prevent the participation of unrelated tags in the proof. Tags have a unique identifier ID_i and store two private keys: a secret group key k_{group} that is used to prove membership in the group, and a key k_i for authentication. For each tag, the Verifier stores the tuple $\{ ID_i, ID_{group}, k_i, k_{group} \}$.

The protocol is described by Figure 13. In the initialization phase, the verifier computes encrypted timestamps $t_n = f_{k_v}(Timestamp)$, where k_v represents the secret key of the Verifier, which will be valid within a limited time window. For tag_A and tag_B, with identifiers ID_A and ID_B respectively and belonging to the same group, the protocol starts by having the reader send the timestamp t_n to the tag_A. Then, tag_A generates two random numbers $\{r1_A, r2_A\}$, computes:

$$MA_{group} = PRNG\left(ID_{group} r1_A\, PRNG(k_{group}) PRNG(t_n)\right)$$

$$M_A = PRNG\left(ID_A r2_A\, PRNG(k_A) PRNG(t_n + 1)\right)$$

and sends $\{r1_A, r2_A, MA_{group}, M_A\}$ to the reader. The tag_B receives $\{r1_A, t_n, MA_{group}, M_A\}$, checks if MA_{group} is correct and if so, then it generates two random numbers $\{r1_B, r2_B\}$, computes:

Figure 13. Kazahaya protocol

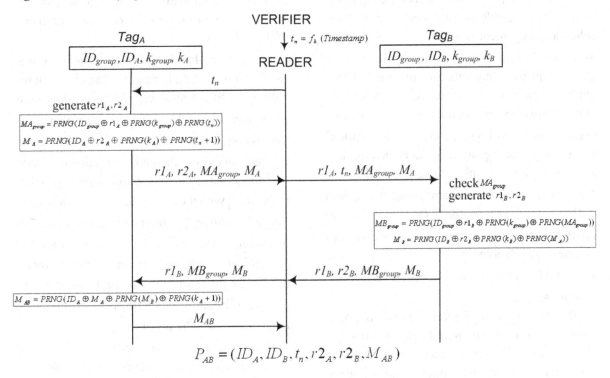

$$MB_{group} = PRNG\left(ID_{group}\, r1_B\, PRNG(k_{group})\, PRNG(MA_{group})\right)$$

$$M_B = PRNG\left(ID_B\, r2_B\, PRNG(k_B)\, PRNG(M_A)\right)$$

and sends $\{r1_B, r2_B, MB_{group}, M_B\}$ to the reader. The reader submits $\{r1_B, MB_{group}, M_B\}$ to the tag_A which checks if MB_{group} is correct. If it is correct, then tag_A computes the final message and sends the result to the reader:

$$M_{AB} = PRNG\left(ID_A\, M_A\, PRNG(M_B)\, PRNG(k_A + 1)\right)$$

Finally, the reader generates the proof

$$P_{AB} = \{ID_A, ID_B, t_n, r2_A, r2_B, M_{AB}\}.$$

6.4 Two Provably Secure Grouping-Proofs

We present here two RFID grouping-proofs (Burmester et al, 2008). The first proof does not provide anonymity (there are many applications for which anonymity is not a concern. For these, simpler alternatives are preferred because of their lower complexity and smaller circuit footprint). For this proof the messages that the tags send to the RFID reader include a group identifier ID_{group}. The second proof adds support to anonymity. For this proof no identifier is passed to the reader. Mutable pseudonyms are used whose values depend on the group identifier, a group key, and the Verifier's challenge, but the dependency is known only to the Verifier. Thus, only the Verifier is able to match the proof with a given group of tags: this guarantees unlinkability and anonymity.

In these grouping-proofs, a specific tag in the group plays the role of "initiator", transmitting either a group identifier ID_{group}, or a random password PS –if there are privacy concerns a scheme with a rotating initiator with a randomized schedule can be implemented. All the tags of the group share a secret group key k_{group} that is used to prove membership in the group, and additionally the initiator tag stores an identification key k_{tag_1} that is used to authenticate protocol flows. The Verifier stores for each group the values (ID_{group}, k_{tag_1}) in a database D. This is used to verify the correctness of grouping-proofs. Tags instances are denoted as tag_i, $i = 1, 2,...$, and the secret key for instance tag_i is written in shorthand as k_i.

To authenticate sessions and to confirm group membership a (keyed) pseudorandom function $f(.;.)$ is used. We can construct such functions with variable input and output length by using lightweight LFSR pseudorandom number generators as a building block (Goldreich, 2001). This allows for full optimization of security/efficiency tradeoffs (Burmester et al., 2009).

The grouping-proof protocols start with an RFID reader broadcasting a random challenge r_{sys} which is obtained from the trusted Verifier at regular intervals. This challenge defines the scanning period: the period should be long enough to make it possible for group interrogations to be completed, but groups should be scanned at most once during this period. Both grouping-proofs are static: they cannot generate proofs for new groups, or ad hoc (sub)groups constituted *on-the-fly*.

A Basic Grouping-Proof

The basic grouping-proof is presented in Figure 14, with initiator tag_1. For simplicity there is only one other tag, tag_2. The case when there are $n > 2$ tags will be discussed later. The initiator tag stores four numbers: $(ID_{group}, k_{group}, k_1, sn)$. The last

number is a session number, used to identify sessions. The other tag(s) store only two numbers: (ID_{group}, k_{group}).

The protocol has three phases. In the first phase the RFID reader challenges the tags in its range with r_{sys} and the tags respond by sending their group identifier ID_{group}. The initiator also sends the session number sn_1. In the second phase the tags get linked by channels through the reader. The third phase has two parts. First tag_1 challenges tag_2 with the authenticator aut_1, and tag_2 responds with aut_2', obtained by using the pseudorandom function f. Then tag_1 verifies the authenticator aut_2' and (if correct) proves that it has verified the presence of tag_2 by sending the group confirmation cnf_1 to the reader. Finally the reader generates the proof P_{group}.

Each phase can be executed concurrently with all the tags in the group (tags can evaluate aut immediately after being linked), except that the third phase is initiated by tag_1. The various phases cannot be consolidated without loss of some security feature, or worse, of determinate outcome. In fact, if we removed the first phase (r_{sys}) the protocol would be subject to a full-replay attack. If we removed the second phase (the exchange of group identifiers ID_{group}), the reader would not be able to match the tags, so that the group itself would be undefined.

Phase three consists of three rounds of communication, and each is crucial to provide the data for the grouping-proof. If we were to suppress the exchange of aut_1 and aut_2', or did not implement timeout, then replay attacks would be successful. Also, the implementation of the third round enables an authorized reader to detect certain protocol failures immediately, namely those that lead the initiator tag to timeout. The update of the number sn immediately after it is sent by tag_1 allows the state of the interrogation to be updated even if the protocol round should

Figure 14. A basic grouping-proof

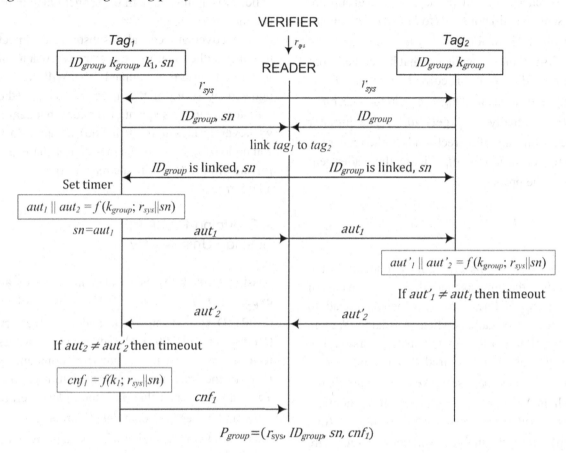

$$P_{group} = (r_{sys}, ID_{group}, sn, cnf_1)$$

be interrupted. This, along with timers, prevents replay attacks.

It should be pointed out that since authorized readers cannot verify grouping-proofs, a determined adversary can cause an RFID reader to generate a proof P_{group} in which only r_{sys} and ID_{group} have correct values. Such a proof will of course be rejected by the Verifier. Furthermore, it is important to note that P_{group} is an actual proof of simultaneous scanning of the tags of a group by a reader only when all the tags of the group are not compromised; *i.e.* a compromised tag may impersonate several tags to the reader, although any such proof would not be accepted by the Verifier.

There are several ways to extend the basic grouping-proof to groups of more than two tags.

One of them is to arrange the tags of a group in a *logical ring*:

$$(tag_1, tag_2, \ldots, tag_n),$$

with indices taken in a cycle $mod\, n$, so $tag_{n+1} = tag_1$. In the ring grouping-proof each tag_i will compute two authenticators: aut_{i-1} and aut_i. The first is used to authenticate the preceding tag_{i-1} in the ring, while the second is sent as an authenticator to its successor tag_{i+1}. The authenticators are obtained by evaluating

$$f\left(k_{group}; r_{sys} \| j\right), \quad j = i-1, i\, .$$

To check that all the tags in the group are present, the initiator sends to tag_2 the authenticator aut_1. Then, for $0 < i \leq n$, each tag_i first checks that the authenticator that it received from the preceding tag_{i-1} is correct, and then (if correct) sends to the next tag tag_{i+1} in the ring aut_i; until eventually tag_1 gets aut_n. The initiator checks this, and if correct sends to the reader the group confirmation cnf_1. Finally the reader generates the proof:

$$P_{group} = \left(r_{sys}, \ ID_{group}, sn, cnf_1 \right).$$

Phase 3 of this protocol is sequential, and therefore the time taken to identify all the tags of the group is linear in n, which can lead to aborted interrogations when n is large. We can modify Phase 3 to get a concurrent version. For this purpose it is required that the tags tag_i, $1 < i \leq n$, also have secret keys k_i that are shared with the Verifier. For this variant, the authenticators (i, aut_i) are broadcast independently (one step); similarly the confirmations (i, cnf_i) are broadcast independently. In this case the proof generated by the reader is:

$$P_{group} = \left(r_{sys}, ID_{group}, sn, cnf_1, \ldots, cnf_n \right).$$

Remarks

Although the use of a group key makes it possible for the tags to assure each other presence, it does not constitute part of the proof and an adversary can still fool the RFID reader. The adversary may substitute the confirmations of initiator tags with bogus confirmations (this may not be feasible in a confined wireless medium). Since these are computed using secret keys shared with the Verifier, they cannot be checked by the reader or the other tags. This is a DoS threat: the reader will generate and store a proof, which of course is not valid. This applies to both the sequential and the concurrent versions.

The adversary can also substitute or inject missing authenticators, for example when an incomplete group is scanned. This will be recognized by some of the interrogated tags (the initiator tag in the sequential version), but again, whatever the tags do (or do not do), the adversary can undo (do), because the tags do not share any private information with the reader. Again this is a DoS threat.

A Grouping-Proof with Session Unlinkability

Session unlikability is defined in terms of an experiment Exp_{Adv}^b, $b = \{0, 1\}$, that involves the RFID system, an oracle and the adversary (Burmester et al, 2008). The adversary has access to a history of earlier tag interrogations and is given by the oracle two specific tag interrogations int_1 and int_2, such that int_1 took place before int_2, and either int_1 completed normally, or an intermediate interrogation with the tags involved in int_1 completed normally.

Exp_{Adv}^0 corresponds to the event that the same tag was involved in int_1, int_2, while Exp_{Adv}^1 corresponds to the event that different tags were involved. The *advantage* of the adversary is:

$$Adv_{Adv} = \left| Prob\left[Exp_{Adv}^0 = 1 \right] - Prob\left[Exp_{Adv}^1 = 1 \right] \right|.$$

We have *session unlinkability* if, the advantage of the (any) adversary is negligible (in terms of the security parameter of the protocol, *e.g.*, the length of the secret keys). That is, it is hard to decide whether the same tag was involved in both. This captures a strong form of anonymity, since if tags can be distinguished then they can also be linked.

To get session unlinkability the group identifiers ID_{group} are replaced by randomized group pseudonyms PS. Again we first consider the case when there are two tags in the group, the initiator

tag_1 and the responder tag_2. To prevent de-synchronization failures or attacks, the responder maintains both a "current" and an "earlier" version of the state of its group pseudonym. For this purpose it stores in non-volatile memory two values of a pseudorandom number: r^{old} and r^{cur}. The initiator maintains only the current state and stores the current value r. The pseudorandom numbers are initialized with the same random value. When the communication is not disrupted (naturally, or intentionally by the adversary), r^{cur} has the same value as r. Otherwise one of r^{old}, r^{cur} always has the value of r.

The protocol is presented in Figure 15, and has three phases. In the first phase the RFID reader challenges the tags in its range with r_{sys}. Then the initiator computes the group pseudonym PS by evaluating $f(k_{group}; r_{sys} \| r)$ and sends this to the reader. Here k_{group} is the group key. The responder computes the pseudonyms PS^{old}, PS^{cur} by evaluating

$$f(k_{group}; r_{sys} \| r^j), j \in \{old, cur\},$$

and sends these to the reader.

In the second phase the reader identifies groups by selecting those sets of tags that have at least one pseudonym in common with PS. The reader informs the tags of its selection and links them. In the third phase the initiator challenges tag_2 with the authenticator aut_1 and tag_2 responds with aut_2', obtained by using the pseudorandom function f and the group key k_{group}. Finally tag_1 proves that it has verified the presence of tag_2 by sending the group confirmation cnf_1 to the reader. This is computed using the pseudorandom function f and the initiator's key k_1. Then the reader generates the proof

$$P_{group} = \left(r_{sys}, PS, sn_1, cnf_1 \right),$$

in which the pseudonym PS is linked to the challenge r_{sys} (via the group key k_{group} and the pseudorandom number r) and the challenge r_{sys} is linked to the confirmation cnf_1 (via the secret key k_1 and the session number sn_1).

The Verifier keeps in a database D the values: $(r_{sys}, PS, k_{group}, k_1, r)$ that link the secret key k_1, the group key k_{group} and the group pseudonym PS for the interrogation session r_{sys}. D is doubly indexed by PS and k_1. The pseudonyms are updated with each successful execution of the protocol (the group key k_{group} and the numbers r are used for this purpose). To prevent replay attacks, each tag_i in the group uses its own session number sn_i in the computation of its authenticator aut_i.

The grouping-proof with session unlinkability can be extended to groups of n tags by using an approach similar to that of the basic proof.

7. DISCUSSION ON SECURITY

The formal security specifications for both of the above described grouping-proofs capture the requirements for a proof of simultaneous presence of the tags of a group. The functionality of the basic grouping proof F_{group} comprises the expected behaviour of a grouping-proof. That is, to generate the challenges of authorised readers, respond on behalf of non-corrupted tags (the functionality F of an application does not address the behavior of adversarial entities: this is not part of its specifications. If a protocol UC-realizes F then it is resilient against adversarial behavior), generate grouping-proofs, and implement time-outs.

The functionality of the grouping proof with session unlinkability F_{su_group} comprises the expected behaviour of a grouping-proof that provides *session unlinkability*. This is similar to F_{group}

Figure 15. A grouping-proof with session unlinkability

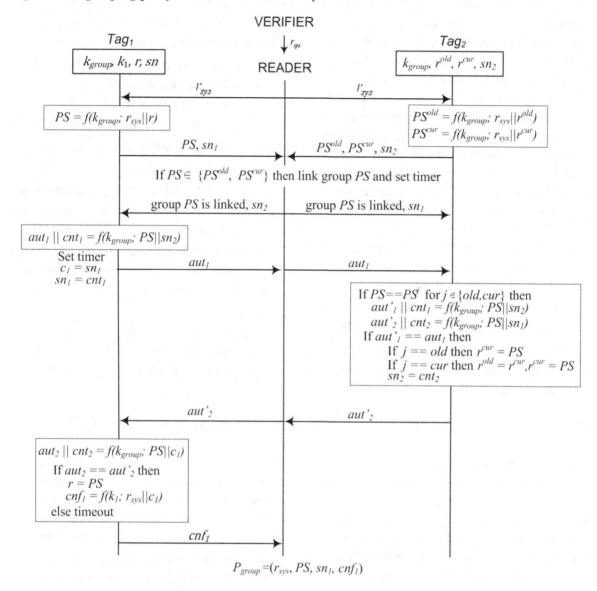

$$P_{group} = (r_{sys}, PS, sn_1, cnf_1)$$

except that authorized readers decide which tags are to be grouped based on group pseudonyms.

It is shown in (Burmester and Medeiros, 2008) that:

The basic grouping-proof UC-realizes the functionality F_{group}.

The grouping-proof with session unlinkability UC-realizes F_{su_group}.

For more details, we refer the reader to (Burmester and Medeiros, 2008).

8. CONCLUSION

In this chapter we have reviewed the literature on RFID grouping-proofs that originated with Ari Juel's yoking-proof. We discussed the strengths and weaknesses of the proposed grouping-proofs and considered low cost implementations that are EPCGen2 compliant, in particular some implementations for medical services.

We discussed the security aspects of grouping-proofs. RFID systems are typically components of large networks (the Internet of Things) so their security should not be analyzed in isolation, but in a framework that supports modularity and composability. We reviewed such a framework and proposed two grouping-proofs that support modularity and composability. The functionality of the first one is the expected behavior of a grouping-proof, while the second one supports in addition anonymity.

There are several major challenges for extending the functionality of grouping-proofs. The protocols so far presented are for static groups. Designing dynamic grouping-proofs for ad-hoc groups in offline applications is a major challenge, even if we restrict the design to *subgrouping-proofs*. There are many applications where such proofs may be helpful. For example a subgrouping-proof that a specific item is missing from a kit is perhaps as important as a grouping-proof that the kit is complete.

Another extension involves privacy. The last protocol presented supports privacy (session unlinkability) but this does not extend to *forward secrecy*. If the adversary captures a tag, then its state (long term secret key(s)) can be used to identify that tag in its earlier interrogations. Although there are several RFID "single tag" protocols that support forward security, extending these to grouping-proofs for practical settings is elusive (the solution suggested in (Burmester et al, 2008) is not practical), and a major challenge.

ACKNOWLEDGMENT

This work has been partially supported by Ministerio de Ciencia e Innovación (Spain) and the European FEDER Fund under project TIN2011-25452.

REFERENCES

Bellare, M., Pointcheval, D., & Rogaway, P. (2000). Authenticated key exchange secure against dictionary attacks. *Eurocrypt '00. LNCS, 1807*, 139–155.

Bolotnyy, L., & Robins, G. (2006). *Generalized yoking-proof, for a group of radio frequency identification tags*. International Conference on Mobile and Ubiquitous Systems (MOBIQUITOUS). San Jose, CA.

Bolotnyy, L., & Robins, G. (2009). Generalized "yoking-proofs" and inter-tag communication. In Turcu, C. (Ed.), *Development and implementation of RFID technology* (pp. 447–462). doi:10.5772/6538

Burmester, M., Tri van Le, de Medeiros, B., & Tsudik, G. (2009). Provably secure ubiquitous systems: Universally composable RFID authentication protocols. *ACM Transactions on Information and System Security (TISSEC), 12*(4).

Burmester, M., & de Medeiros, B. (2008). The security of EPC Gen2 compliant RFID protocols. In Bellovin, S. M., Gennaro, R., Keromytis, A. D., & Yung, M. (Eds.), *LNCS 5037* (pp. 490–506). Springer. doi:10.1007/978-3-540-68914-0_30

Burmester, M., de Medeiros, B., & Mota, R. (2008). Provably secure grouping-proofs for RFID tags. In Grimaud, G., & Standaert, F. X. (Eds.), *CARDIS* (*Vol. 5189*, pp. 176–190). Lecture Notes in Computer Science Springer.

Burmester, M., & Medeiros, B. (2009). On the security of route discovery in MANETs. *IEEE Transactions on Mobile Computing, 8*(9), 1180–1188. doi:10.1109/TMC.2009.13

Burmester, M., & Munilla, J. (2011). Lightweight RFID authentication with forward and backward security. *ACM Transactions on Information and System Security, 14*(1). doi:10.1145/1952982.1952993

Canetti, R. (1995). *Studies in secure multiparty computation and application.* Ph.D. thesis, Weizmann Institute of Science.

Canetti, R. (2000). Security and composition of multiparty cryptographic protocols. *Journal of Cryptology, 13*(1), 143–202. doi:10.1007/s001459910006

Canetti, R. (2001).Universally composable security: A new paradigm for cryptographic protocols. In *Proceedings of the IEEE Symposium on Foundations of Computer Science (FOCS'01)* (pp. 136—145).

Chien, H. Y., Yang, C. C., Wu, T. C., & Lee, C. F. (2011). Two RFID-based solutions to enhance inpatient medication safety. *Journal of Medical Systems, 35*(3), 369–375. doi:10.1007/s10916-009-9373-7

Global, E. P. C. (n.d.). *EPC tag data standards, vs. 1.3.* Retrieved from http://www.epcglobalinc.org/standards/EPCglobal_TagData Standard TDS Version 1.3.pdf.

Goldreich, O. (2001). *The foundations of cryptography.* Cambridge, UK: Cambridge University Press. doi:10.1017/CBO9780511546891

Huang, H. H., & Ku, C. Y. (2009). A RFID grouping-proof protocol for medication safety of inpatient. *Journal of Medical Systems, 35*(3), 369–375.

Juels, A. (2004). Yoking-proofs for RFID tags. *Proceedings of the First International Workshop on Pervasive Computing and Communication Security* (pp. 138-143).

Juels, A., & Weiss, S. A. (2007). Defining strong privacy for RFID. *Proceedings of PerCom, 07,* 342–347.

Lamport, L. (1979). *Constructing digital signatures from a one way function.* Technical Report CSL-98, SRI International, October 1979.

Leng, X., Lien, Y., Mayes, K., & Markantonakis, K. (2010). An RFID grouping proof protocol exploiting anti-collision algorithm for subgroup dividing. In *International Journal of Security and Networks (IJSN) Special Issue on "Security and Privacy in RFID Systems,* 2010.

Lien, Y., Leng, X., Mayes, K., & Chiu, J.-H. (2008). *Reading order independent grouping proof for RFID tags.* In 2008 IEEE International Conference on Intelligence and Security Informatics, June 17-20, 2008, Taipei, Taiwan.

Lin, C.-C., Lai, Y.-C., Tygar, J. D., Yang, C.-K., & Chiang, C.-L. (2007). Coexistence proof using chain of timestamps for multiple RFID tags. In *The Proceeding of APWeb/WAIM International Workshop, LNCS 5189,* (pp. 634-643). Springer-Verlag.

Mangard, S., Oswald, M. E., & Popp, T. (2007). *Power analysis attacks- Revealing the secrets of smart cards.* Springer 2007.

Munilla, J., & Peinado, A. (2009). Distance bounding protocols for RFID. In Kitsos, P. (Ed.), *Security in RFID and sensor networks* (pp. 151–169). CRC Press. doi:10.1201/9781420068405.ch7

Ouafi, K., & Phan, R. C.-W. (2008). Privacy of recent RFID authentication protocols. *Proceedings of ISPEC, LNCS, 4991,* 263–277.

Peris-Lopez, P., Hernandez-Castro, J. C., Estevez-Tapiador, J. M., & Ribagorda, A. (2007). Solving the simultaneous scanning problem anonymously: Clumpling proofs for RFID tags. In *Third International Workshop on Security, Privacy and Trust in Pervasive and Ubiquitous Computing.*

Peris-Lopez, P., Orfila, A., Hernandez-Castro, J. C., & van der Lubbe, J. C. A. (2011). Flaws on RFID grouping-proofs. Guidelines for future sound protocols. *Journal of Network and Computer Applications, 34*(3). doi:10.1016/j.jnca.2010.04.008

Piramuthu, S. (2006). *On existence proofs for multiple RFID tags*. In IEEE International Conference on Pervasive Services, Workshop on Security, Privacy and Trust in Pervasive and Ubiquitous Computing.

Rieback, M., Crispo, B., & Tanembaum, A. (2006). *Is your cat infected with a computer virus?* In Pervasive Computing and Communications, Pisa, Italy, March 2006. IEEE Computer Society Press.

Saito, J., & Sakurai, K. (2005). Grouping-proof for RFID tags. *19th International Conference on Advanced Information Networking and Applications* (pp. 621-624).

Shoup, V. (1999). *On formal models for secure key exchange* (version 4), 15 November 1999. Revision of IBM Research Report RZ 3120 (April 1999).

Chapter 5
Elliptic Curve Cryptography on WISPs

Michael Hutter
Institute for Applied Information Processing and Communications (IAIK), Graz University of Technology, Austria

Erich Wenger
Institute for Applied Information Processing and Communications (IAIK), Graz University of Technology, Austria

Markus Pelnar
Institute for Applied Information Processing and Communications (IAIK), Graz University of Technology, Austria

Christian Pendl
Institute for Applied Information Processing and Communications (IAIK), Graz University of Technology, Austria

ABSTRACT

In this chapter, the authors explore the feasibility of Elliptic Curve Cryptography (ECC) on Wireless Identification and Sensing Platforms (WISPs). ECC is a public-key based cryptographic primitive that has been widely adopted in embedded systems and Wireless Sensor Networks (WSNs). In order to demonstrate the practicability of ECC on such platforms, the authors make use of the passively powered WISP4.1DL UHF tag from Intel Research Seattle. They implemented ECC over 192-bit prime fields and over 191-bit binary extension fields and performed a Montgomery ladder scalar multiplication on WISPs with and without a dedicated hardware multiplier. The investigations show that when running at a frequency of 6.7 MHz, WISP tags that do not support a hardware multiplier need 8.3 seconds and only 1.6 seconds when a hardware multiplier is supported. The binary-field implementation needs about 2 seconds without support of a hardware multiplier. For the WISP, ECC over prime fields provides best performance when a hardware multiplier is available; binary-field based implementations are recommended otherwise. The use of ECC on WISPs allows the realization of different public-key based protocols in order to provide various cryptographic services such as confidentiality, data integrity, non-repudiation, and authentication.

DOI: 10.4018/978-1-4666-1990-6.ch005

INTRODUCTION

Wireless Identification and Sensing Platforms (WISPs) are based on Radio Frequency Identification (RFID) technology. They provide the same capabilities as RFID tags but feature additional functionalities like sensing of data from the near proximity. Typical data that is collected by WISPs are the environmental temperature, light, or 3D acceleration of the device or of a targeted object. Among the most interesting features of WISPs is the possibility to perform customized operations using assembled microcontrollers. This allows individual handling of data collection, processing, and monitoring of sensed information. However, for most of the applications, several security and privacy issues arise such as protecting sensitive data that have been collected by WISPs and that are transmitted over the air interface. This chapter addresses the implementation of Elliptic Curve Cryptography (ECC) on such platforms in order to overcome these concerns.

ECC is a public-key technique that has gained much importance especially in environments which provide only low resources. It features a high level of security while needing only small key sizes compared to other existing public-key techniques like RSA. There exist many ECC-based cryptographic protocols that have been standardized and evaluated over many years which encourage its use in security-related applications. The main operation in ECC is the scalar multiplication $Q = k \cdot P$ of a point P on an elliptic curve. This operation involves the group operations of addition and doubling of curve points which again are based on finite-field arithmetic. One of the most resource-consuming finite-field operations is the scalar multiplication which constitutes about 75% of the total runtime. Therefore, the performance of scalar multiplication largely determines the efficiency of ECC on WISPs. If the microcontroller which is assembled on the WISP provides a dedicated hardware multiplier, a word-size mul-

tiplication can be performed in a single clock cycle. If no hardware multiplier is supported, multiplication has to be implemented with the help of addition and shift operations which significantly reduce the overall performance.

In order to evaluate ECC on WISPs, we make use of the WISP4.1DL UHF tag developed by Intel Research Seattle. The WISP consists of a tiny low-resource microcontroller (the MSP430F2132) that is attached to a dipole antenna. Next to the microcontroller, the tag features several sensors such as temperature, light, and 3D accelerometer which allows a broad range of RFID and sensor-node applications. There already exist many publications that use the WISP as a demonstrator platform, e.g.,

(Yeager, Sample, & Smith, 2008), (Saxena & Voris, 2009), (Yeager, Holleman, Prasad, Smith, & Otis, 2009), (Smith, Fishkin, Jiang, Mamishev, Philipose, Rea, Roy, Sundara-Rajan, 2005). Only a few publications presented implementations of cryptographic algorithms on the WISP. (Chae, Yeager, Smith, & Fu, 2007), for example, implemented the block cipher RC5 and demonstrated the feasibility of symmetric cryptography on that platform.

In this chapter, we evaluate the feasibility of asymmetric cryptography on WISPs by implementing ECC. We make use of the Montgomery ladder scalar multiplication over the smallest recommended NIST elliptic curve over prime fields, i.e., P-192. Furthermore, we implemented ECC over binary fields using a comparable elliptic curve standardized by ANSI X9.62 using 191 bits. In order to meet the low-resource constraints of WISPs, we applied several optimization techniques. First, we applied different field-multiplication methods that reduce the memory and computational requirements to a minimum. Second, we make use of state of the art ECC formulae to provide efficient computation on the algorithmic level. As a result, we show that a scalar multiplication can be performed in 8.3 seconds at 6.7 MHz on the WISP4.1DL device (featuring no hardware multiplier) and only 1.6 seconds on the WISP when a hardware multiplier is supported. Our binary-field based implementation needs about 2 seconds without needing a hardware multiplier. ECC over prime fields is therefore recommended on WISPs which feature multiplication support, ECC over binary fields is recommended otherwise.

After the introduction, we will give a short overview on ECC. We will describe the basic principles of the public-key technique and explain the different parameters of ECC. Afterwards, we will present the WISP platform we used for demonstration. First, the hardware is described in detail by focusing mainly on the features of the MSP430 microcontroller. Second, the firmware and programming tools are presented. The following sections give details about the implementation of ECC on the WISP. We start by describing the basic prime-field arithmetic of modular addition, subtraction, multiplication, squaring, and inversion. After that, we will describe the used formulae for the Montgomery ladder scalar multiplication. The same description is then given for our binary-field implementation. Next, the results are presented and the performance over both type of fields are described. Future research directions and conclusions are given finally.

ELLIPTIC CURVE CRYPTOGRAPHY

An elliptic curve E is a mathematical structure that is defined over a finite field (or often referred to as Galois field) GF(q). It can be represented by the long Weierstrass equation, i.e.:

$$E: \quad y^2 + a_1 xy + a_3 y = x^3 + a_2 x^2 + a_4 x + a_6,$$

where a_1, a_2, a_3, a_4 and a_6 are constants in GF(q) which need to satisfy the prerequisite that the discriminant of E is not zero. $P = (x, y)$ denotes a point on the elliptic curve E. The set of all points on the elliptic curve together with the point at infinity O is denoted by $E(\text{GF}(q))$. It forms an Abelian group that allows addition and doubling of elliptic-curve points by following the chord-and-tangent rule. Using these two operations, a scalar multiplication can be performed which constitutes the main operation in ECC.

A scalar multiplication involves two points P and Q and performs $Q = k \cdot P$, where k is a scalar such that $0 \leq k < Ord(P)$ and $Ord(P)$ is the order of E which equals to the distinct number of points on the elliptic curve E. The mathematical hard problem to determine k from P and Q is known as the Elliptic Curve Discrete Logarithm Problem (ECDLP). The ECDLP can be solved in exponential runtime which is a major advantage to other existing problems that are based for example on the Integer-factorization problem or the discrete-logarithm problem which can be solved in sub-exponential runtime.

An elliptic curve can be defined over any finite field GF(q). If the characteristic q is neither 2 nor 3, the Weierstrass equation can be represented over prime fields and it simplifies to the form

$$y^2 = x^3 + a_4 x + a_6.$$

If the characteristic of the field is 2 (e.g. binary-extension field GF(2^m)), the Weierstrass equation can be transformed to the isomorphic elliptic curve

$$y^2 + xy = x^3 + a_2 x^2 + a_6.$$

Depending on the underlying finite field, different arithmetic (prime or polynomial) has to be applied to calculate a point addition or point doubling.

Among the most determining factor in terms of scalar-multiplication performance is the representation of elliptic-curve points. Elliptic-curve points can be represented in different coordinate systems. *Affine* representation makes use of two coordinates (x, y) to represent a point on the elliptic curve. For the group operations of point addition and point doubling, they require inversions in GF(q) which are by far the most expensive field operations. By using *projective* coordinates, it is possible to avoid such inversions at the cost of an additional coordinate. The choice of coordinate representation depends on the implementation and architecture of the executing processor. If inversion is a very slow operation it is basically recommended to use projective coordinates otherwise it would be better to use affine coordinates to provide best performance.

Scalar Multiplication Methods

There exist several methods to perform a scalar multiplication on elliptic curves. One of the most common ways is the binary method which corresponds to the additive version of the classical square and multiply method. It can be implemented in a for-loop that iterates through the binary representation of the scalar from left-to-right or from right-to-left. In each iteration, a point doubling operation is performed; a point addition is only performed when the bit of the scalar is one. Thus, the method is fast but provides a runtime that depends on the secret scalar. This fact allows implementation attacks to reveal the scalar by observing side-channel information such as the power consumption or the execution time (Kocher, Jaffe, & Jun, 1999), (Mangard, Oswald, & Popp, 2007).

Next to the binary method, there exist other scalar multiplication methods that apply techniques to further improve the performance. A typical way to achieve this is to make use of a recoding representation of the scalar or to spend a certain amount of memory to pre-compute intermediate values of the scalar multiplication. Among the most prominent methods in view of recoding is the binary method in Non-Adjacent Form (NAF). There, the representation of the scalar is modified such that instead of using (0, 1) as basis, the triplet of (-1, 0, 1) is used. Using this representation, it is possible to maximize the number of zeros within the scalar multiplication. As a result, this minimizes the number of required point additions and therefore the total runtime of the algorithm.

An extension of the binary NAF is the windowed NAF method (often also referred as width-ω NAF or NAF$_\omega$). The idea is to slice the representation of the scalar into several pieces and to process ω digits at a time. Therefore, multiples of the base point are pre-computed and used within the scalar multiplication. Due to the pre-computation, additional memory is needed. However, using this method the number of addition operations can be significantly decreased since the average density of non-zero digits among all width-ω NAFs is approximately $1/(\omega+1)$.

If the elliptic-curve point is known and fixed during the multiplication, further (off-line) pre-computation can be performed to improve the efficiency. With a window of width ω, it is necessary to compute and store 2^ω points. The runtime complexity is then reduced by a factor of about ω. Typical methods are the fixed-base windowing method (or often referred as BGMW according to the names of their inventors Brickell, Gordon, McCurley, and Wilson) or fixed-base comb methods first presented by Lim and Lee (Lim & Lee, 1994).

All the described methods above have two disadvantages with regards to WISPs. First, they require a certain amount of additional memory which is often not available on such platforms. In fact, memory is one of the most area consuming components in hardware designs so that the resources especially on RFID-based devices are very limited. Second, most of the described methods do not provide resistance against implementation attacks such as timing attacks or Simple Power Analysis (SPA), cf. (Kocher, Jaffe, & Jun, 1999), (Mangard, Oswald, & Popp, 2007).

A method that does not require additional memory and which provides implicit resistance against such attacks is the Montgomery ladder (Montgomery, 1987), (Joye & Yen, 2003). The Montgomery ladder performs a point doubling and a point addition in each loop iteration and provides therefore a constant runtime for the scalar multiplication. Note that it does not involve any dummy operations, which is commonly used to prevent side-channel attacks (e.g. double-and-add-always), but provides a very regular structure instead. This allows combining the doubling and addition formulae in order to improve speed and to reduce the memory requirements. Another advantage of this method is that it allows the computation of addition and doubling without needing the y-coordinates. The entire scalar multiplication can be performed with x-coordinate only operations, thus needing less intermediate variables. This is an important fact that is especially interesting for devices which have only limited resources available such as RFID-tags or WISPs. Due to these reasons, we based our ECC implementation on the Montgomery ladder scalar multiplication. Furthermore, we make use of projective coordinates in order to avoid costly field inversions.

In the following, we evaluate the implementation of both prime and binary-field elliptic curves. We make use of the recommended NIST curve P-192 and the recommended ANSI X9.62 curve c2tnb191v1. Note that both types of elliptic curves provide nearly the same key size so that their implementation can be fairly compared. The algorithms used are described in detail after the following introduction into WISPs.

WIRELESS IDENTIFICATION AND SENSING PLATFORMS

Wireless Identification and Sensing Platforms (WISPs) are similar to Wireless Sensor Nodes (WSNs). Both types of platforms are able to monitor the physical condition of the environment by measuring the temperature, light, or the acceleration of an object. This allows the integration of WISPs into many applications such as industrial monitoring of machine health, motion detection,

air-pollution recognition, temperature control in greenhouses, or in agriculture to monitor the water level of tanks, for instance.

Among the most significant difference between WISPs and WSNs, however, is the type of power supply. While WSNs typically get powered by a battery, WISPs draw the power from the electromagnetic field of a reader. This feature makes WISPs more interesting and applicable to certain applications where WSNs cannot be easily employed. In the following, we describe the WISP4.1DL in a more detail.

The WISP4.1DL UHF Tag

The WISP4.1DL has been developed by Intel Research Seattle in order to provide a development platform for new RFID and sensing applications. A picture of a WISP4.1DL tag is shown in Figure 1. It is similar to a passively powered RFID tag operating in the UHF frequency range of about 900 MHz featuring an ultra-low power general-purpose microcontroller from Texas Instruments (the MSP430F2132). The used microcontroller is a Reduced Instruction Set Computer (RISC) processor providing a 16-bit architecture, 8 KB of flash memory, 512 byte of RAM, and 16 working registers where only 12 can be used for general purpose.

In particular, the MSP430 family has been especially designed for low-resource applications. It provides various operating modes that can be used to personalize the microcontroller in terms of high speed or low power. Such operating modes range from an active mode (AM) to a low-power mode (LPM4). These different modes basically differ in the number of sub-modules being disabled to reduce the power consumption of peripherals. In combination with the supported voltage supervisor of the WISP tag, this feature can be used to significantly extend the uptime and reading range of the tag. The WISP tag includes several sensors such as a temperature sensor, a light-level detec-

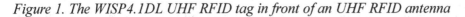

Figure 1. The WISP4.1DL UHF RFID tag in front of an UHF RFID antenna

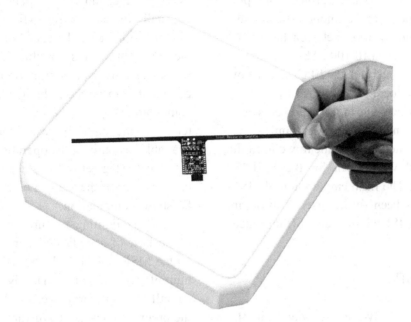

tor, and a 3D accelerometer. These sensors allow the realization of a broad range of applications. (Sample, Yeager, & Smith, 2008) have been the first who reported a security-enabled application by implementing the symmetric block cipher RC5 on the WISP. (Saxena & Voris, 2009) extended the use of the accelerometer in order to generate random numbers which are important in the field of cryptography. As another example, (Yeager, Holleman, Prasad, Smith, & Otis, 2009) extended the RFID antenna in order to serve as a capacitive touch sensor. Because of the combination of both computation and sensing capabilities, WISPs are perfectly suited for human-activity detection as stated by (Smith, Fishkin, Jiang, Mamishev, Philipose, Rea, Roy, & Sundara-Rajan, 2005) since they can deliver motion-detection capabilities of active sensor beacons in the same battery-free form factor as RFID tags.

The Firmware

The communication between WISP tags and interrogators (readers) is established over the EPC Class-1 Generation-2 UHF RFID protocol (EPCglobal, 2008). It has been standardized in ISO/IEC 18000-6C, which defines the physical and logical requirements for a passive backscatter, interrogator-talks-first (ITF) RFID system.

The latest stable release of the firmware—currently the r65 (including the version HW4.1 SW6.0)—provided by Intel Research Seattle implements main parts of the EPC Class-1 Gen-2 protocol. The firmware allows the configuration of the peripherals such as the temperature sensor or the accelerometer. The sensed data can be transmitted to the reader by either implementing custom commands of the protocol or by using simple EPC read/write commands. Additionally, data encoding can be switched between Miller-2 and Miller-4 modulation with the latter as default setting.

Reader Setup and Programming Tools

We used the Speedway Revolution R220 (Impinj, 2012) UHF reader in our experiments. It supports the EPC Class 1 Generation 2 protocol and provides two high performance monostatic antenna ports. The communication between the reader and a PC is done over a 10/100BASE-T network. Based on this connectivity, the EPC-global Low Level Reader Protocol (LLRP) v1.0.1 is used as application interface. The transmit power is up to 32.5 dBm using an external power supply.

As a development environment, we used the IAR Embedded Workbench for the MSP430 microcontroller. The flash emulation tool MSP-FET430UIF has been further used to program and debug the WISP tag over an USB interface.

ECC ON WISPS

In the following, we give details about the implementation of an ECC framework that runs on the WISP4.1DL RFID tag. Since only low resources are available, several optimizations are necessary to allow the execution of ECC-based protocols on that platform. We start with a description of the requirements of ECC on WISPs. After that, we will describe the ECC framework. Next, we describe the finite-field and ECC group arithmetic over prime fields. Finally, we describe the binary-field arithmetic implementation and the used ECC formulae.

Requirements

One of the most limiting resources of the WISP4.1DL platform is the memory. In fact, the MSP430F2132 is shipped with 8 kB of □ ash memory where 3.2 kB are needed only for the EPC Class-1 Generation-2 protocol. Thus, implementations are limited to only 4.8 kB. In addition, the MSP430F2132 provides 512 bytes of volatile RAM. The RFID-protocol implementation needs about 200 bytes so that only 312 bytes are available for ECC. Considering these restrictions, it is important to carefully balance between speed and memory consumption. Optimizations such

as unrolling of individual operations (such as it is usually the case in finite-field multiplication routines to increase the speed of execution) are therefore not always possible. Since the tag is powered passively, it is furthermore necessary to pay attention to the energy budget. Time-extensive computations need to be separated into parts after which the tag has to test if enough energy is available to continue the operation. If this is not the case, the tag gets into a sleeping mode where the capacitor of the tag can be charged again to finish the computation.

Another limiting resource in view of ECC is the lack of a dedicated hardware multiplier on the MSP430F2132. In fact, the speed of the multiplication operation largely determines the overall ECC performance, as already stated in the previous sections. Hardware multipliers can perform a multiplication within only a few clock cycles whereas a few hundred clock cycles are needed if no dedicated multiplier is available. Note that next to the MSP430F2132, many other microcontrollers of the MSP430x2xx family feature a hardware multiplier. For example, (Gouvêa & Lopez, 2009) made use of such a multiplier (multiply-accumulate operation) to speed up the computation of Pairing-based cryptography on a MSP430 microcontroller.

In the following, we therefore consider two different implementations for WISP tags. The first implementation provides ECC oper $p \equiv 2^{192}$-2^{64}-1 ations without a multiplier (this is the case for the WISP4.1DL). The second implementation considers a WISP which assembles a microcontroller that features a hardware multiplier. For the latter case, we simulated the performance of ECC and compare the results with the results obtained for the WISP4.1DL.

Elliptic Curve Cryptography Framework

We implemented a general framework that provides the basic functionalities to implement ECC-based protocols on WISPs. Considering security and performance, we decided to base our implementation on a recommended NIST elliptic curve (NIST, 2009). Due to the limiting resources, we applied the smallest recommended NIST curve over prime fields which is over $GF(p_{192})$ where the used prime is a Mersenne-like prime defined as .

The entire framework has been implemented in Assembly and C language. Most algorithms have been implemented in Assembler language to speed-up the computation. The framework consists of five main modules: a random number generator, the SHA-1 hash-function, a support function, a finite-field arithmetic module, and an elliptic-curve arithmetic module.

As a random number generator (RNG), we implemented the algorithm proposed by (Marsaglia, 2003), (Brent, 2004). The algorithm provides a good trade-off between cryptographic security and computational complexity. It mainly consists of simple shift and XOR operations and can be therefore efficiently implemented on the MSP430 microcontroller. For the initialization of the RNG, we used the accelerometer as also proposed by (Saxena & Voris, 2009 to generate a random seed. In order to guarantee random initialization, the device is locked until stochastic movement of the WISP tag is detected. In general, the Marsaglia algorithm takes three parameters a, b, and c; we have chosen $a=1$, $b=3$, and $c=10$ in our implementation. For generating a 32-bit random number this means that one left shift, 3 right shift, and 10 left-shift operations have to be performed, i.e., 14 shift operations in total. Since we have to generate a 192-bit number, 6 rounds are calculated where a 32-bit number is generated in each round. Finally, we check if the generated number is smaller than the order of the elliptic curve in order to fulfill the requirements to perform a valid scalar multiplication. The RNG implementation needs 42 lines of Assembler code and 36 lines of C code.

We also implemented the SHA-1 hash function which is typically used in many cryptographic signature schemes, e.g. ECDSA. Hash functions are basic building blocks of cryptographic protocols. They are mainly used for authentication, integrity, and non-repudiation services. Due to memory reasons, we assume that the data which has to be hashed is shorter than the block size of the algorithm, which is 512 bits. This limits the code and memory requirements of our implementation. We implemented the SHA-1 hash function using three Assembler routines: one routine performs a left rotation of a word, one routine calculates the needed addresses to access the words of the current message block, and one routine which generates the message digest. First, constants such as H0,...,H4 or K0,...,K3 that are stored in flash memory and loaded into RAM. Second, the State A,B,C,D,E is initialized with the constants H0,...,H4. After that, the message digest is calculated by performing 80 rounds according to the SHA-1 specification. In the individual rounds, logical operations are performed on the State such as AND, OR, or XOR. These operations can be implemented very efficiently using the MSP430. The constants K0,...,K3 are added and the State is finally shifted one position to the right. After the 80 rounds, the State A,B,C,D,E is accumulated to H0,...,H4 which then holds the final message digest. Most of the variables could be hold in available registers, the rest is maintained in RAM. The entire SHA-1 implementation needs 279 lines of Assembler code.

The support function module is used to provide basic operations for Integer arithmetic such as comparisons, operand copies, or initialization of array elements. These operations are needed mainly for the finite-field arithmetic and the elliptic-curve arithmetic implementation.

The finite-field arithmetic module and the elliptic-curve arithmetic module contain the main operations used to support ECC on the WISP tag. They are described in the following subsections.

$$a + b = c$$

Finite-Field Arithmetic over Prime Fields

Both ECC group operations of point addition and doubling are based on underlying finite-field operations. In order to obtain an efficient performance for scalar multiplication, it is important to optimize these operations as much as possible. In fact, finite-field operations are heavily used in loops so that thousands of clock cycles can be saved for scalar multiplication by reducing only one single clock cycle in the underlying finite-field arithmetic.

The MSP430F2132 microcontroller provides a 16-bit architecture. This means that all prime-field operations are based on 16-bit operands, i.e., a 192 bit field element is stored in a 12-word array structure. In particular, we decided to represent each field element in little-endian representation. Thus, indirect auto-increment addressing can be used that is supported by the MSP430 microcontroller. This special addressing mode provides base-address increment after the fetch operation without any additional overhead. This allows efficient array processing and avoids additional clock cycles.

Addition and Subtraction

The addition of field elements is implemented via a loop that iterates through the twelve words of the operands starting with the least significant word. The operand words are loaded from memory and added using the *ADD* and *ADDC* instruction of the MSP430. In order to speed up the addition operation, we unrolled the loop. In ad-

dition to an out-of-place version, we also implemented an in-place version of the prime-field addition where the result overwrites one input operand, i.e., $a \leftarrow a + b$. This has the advantage that special addressing instructions can be used to save execution cycles. In fact, 34% of the total number of clock cycles can be saved while only increasing the overall code size insignificantly. After addition, the result has to be reduced modulo the prime p. This can be done by simply subtracting the prime if the result is larger than the modulus p. The prime field subtraction has been also implemented by unrolling the instructions. In contrast to modular addition, the modulus p has to be added if the result is smaller than zero.

Multiplication and Squaring

Prime-field multiplication and squaring operations consume most of the runtime of a scalar multiplication. They have therefore a significant impact on the overall performance of ECC. Consequently, it is crucial to put optimization efforts into these routines. As for modular addition and subtraction, prime-field multiplication and squaring in $GF(p)$ consist of a multiplication step (resulting in a double-precision number) and a followed reduction step modulo a prime p.

Two basic multi-precision multiplication algorithms are common: the row-wise standard schoolbook method (also called operand-scanning form) and the column-wise Comba method (also called product-scanning form) (Comba, 1990). Both algorithms process the words in two loops (an outer loop and an inner loop). In 2004, (Gura, Patel, Eberle, & Shantz, 2004) introduced a hybrid method that combines the row-wise and column-wise techniques. The idea behind this method is to make advantage of the two basic multi-precision multiplication algorithms to increase the performance. The Comba method is therefore used in the outer loop of the algorithm and the schoolbook method is used in the inner loop. Still, the number of required registers and needed memory accesses strongly depend on the choice of the algorithm parameter d. Note that there exist also a newer multiplication technique (called operand-caching multiplication) (Hutter & Wenger, 2011). However, it showed that operand caching does not provide any speed improvements compared to the classical Comba or hybrid method because of the underlying hardware architecture and the given instruction set of the MSP430.

If no hardware multiplier is available on the MSP430, the 16-bit operand-multiplication routine needs four of the twelve registers available. In addition to that, two registers are needed for counter variables as loops have to be used to keep the code size small. Another three registers are necessary to hold the addresses of the operands. With only three remaining registers, it is not possible to implement the hybrid method with parameter $d = 2$ as it would require seven available registers. Thus, for the WISP tag without hardware multiplier, the hybrid method has to be implemented with $d = 1$ which actually corresponds to the standard Comba method. An MSP430 with dedicated hardware multiplier can implement the hybrid method with $d = 2$. This is possible as no additional registers are needed for the 16-bit multiplication. In addition to make the hybrid method with $d = 2$ feasible, we had to use loop unrolling which makes two registers available previously needed for counter variables. As preferable side effect, loop unrolling comes with a significant performance improvement to the cost of substantial code size increase.

For squaring of a field element $c = a^2$, we decided to implement a dedicated squaring operation instead of reusing the multiplication operation. This needs additional code memory but increases the efficiency of scalar multiplication since squaring can be computed faster than multiplication. This is due to the fact that only 78 of the 144 partial products actually have to be computed in case of 192-bit operands and 16-bit word

size. The remaining partial products can be substituted through cheap shift operations. Additionally, the number of memory accesses and thus required clock cycles can be reduced as only one operand is present. In case of no hardware multiplier, the reduction of partial-product calculations, which are extremely expensive, clearly outweighs the little overhead required by the squaring routine. Note that we used the squaring operation only for the implementation of the MSP430 without a hardware multiplier. For the MSP430 with a hardware multiplier, no squaring operation has been implemented. This is due to the fact that the overhead becomes more decisive as the multiplication operation of the dedicated multiplier is much faster than the software multiplication. Thus, a less significant speed improvement is obtained by a squaring operation compared to the MSP430 implementation without a hardware multiplier.

In order to fit the final result in the underlying field GF(p), the 384-bit product or square has to be reduced modulo the prime p. As the used NIST prime p is a generalized-Mersenne prime, the reduced result can be computed by simple additions, i.e., a fast NIST reduction, c.f (Hankerson, Menezes, & Vanstone, 2003). The reduction operation can be therefore performed very efficiently.

Elliptic-Curve Arithmetic over GF(p)

We implemented the Montgomery ladder to perform the scalar multiplication $Q = k \cdot P$. There are several reasons for this decision. First, the Montgomery ladder does not need any precomputations and does not need extra memory on the WISP as compared to other scalar multiplication methods. This is important to keep the needed resources as low as possible. Second, the method provides resistance against several implementation attacks, as already stated in the previous sections. Timing attacks, Simple Power

Algorithm 1. Montgomery ladder for elliptic curves over GF(p)

Input: $P=(x,y)$ and $k=(k_{n-1},...,k_0)$ with $k_{n-1}\neq0$ Output: $Q=kP$
1: $(X_0,Z_0)\leftarrow P$; $(X_1,Z_1)\leftarrow 2P$;
2: $X_0\leftarrow X_0\times Z_1$; $X_1\leftarrow X_1\times Z_0$; $Z\leftarrow Z_0\times Z_1$;
3: **for** $i=n-2$ downto 0 **do**
4: $R_2\leftarrow Z^2$, $R_3\leftarrow R_2+R_2$, $R_3\leftarrow R_3+R_2$, $R_1\leftarrow Z\times R_2$, $R_2\leftarrow 4b\times R_1$,
5: $R_1\leftarrow X_{1-k[i]}^{~2}$, $R_5\leftarrow R_1+R_3$, $R_4\leftarrow R_3^2$, $R_1\leftarrow R_1-R_3$, $R_3\leftarrow X_{1-k[i]}\times R_1$,
6: $R_5\leftarrow R_5+R_3$, $R_5\leftarrow R_5+R_5$, $R_3\leftarrow R_5+R_2$, $R_1\leftarrow R_1-R_3$, $R_3\leftarrow X_{k[i]}^{~2}$,
7: $R_1\leftarrow R_1+R_3$, $X_{k[i]}\leftarrow X_{k[i]}-X_{1-k[i]}$, $X_{1-k[i]}\leftarrow X_{1-k[i]}+X_{1-k[i]}$, $R_3\leftarrow X_{1-k[i]}\times R_2$,
8: $R_4\leftarrow R_4-R_3$, $R_3\leftarrow X_{k[i]}^{~2}$, $R_1\leftarrow R_1-R_3$, $X_{k[i]}\leftarrow X_{k[i]}+X_{1-k[i]}$,
9: $X_{1-k[i]}\leftarrow X_{k[i]}\times R_1$, $X_{1-k[i]}\leftarrow X_{1-k[i]}+R_2$, $R_2\leftarrow Z\times R_3$, $Z\leftarrow x_P\times R_2$,
10: $X_{1-k[i]}\leftarrow X_{1-k[i]}-Z$, $X_{k[i]}\leftarrow R_5\times X_{1-k[i]}$, $X_{1-k[i]}\leftarrow R_3\times R_4$, $Z\leftarrow R_2\times R_5$,
11: *test_power_supply()*;
12: **end for**
13: **Return** $Q=(X_0,Z)$.

Analysis (SPA) attacks as well as many fault attacks (e.g. safe-error attacks) are not feasible on this kind of scalar multiplication method. Third, it allows further reducing the memory requirements by using *x*-coordinate only formulae that do not involve the computation of *y*-coordinates. In addition to that, we applied a common-*Z* technique (often referred as Co-*Z* in literature) where the involved elliptic-curve points share a common coordinate (Meloni, 2006), (Lee & Verbauwhede, 2007), (Goundar, Joye, & Miyaji, 2010), (Hutter, Joye, & Sierra, 2011). Hence, only three coordinates (X_1,X_2,Z) have to be maintained in memory instead of four (X_1,Z_1,X_2,Z_2).

The point addition and point doubling operations have been performed using the formulae of (Hutter, Joye, & Sierra, 2011). The proposed formulae have been especially designed for low-resource devices and perform all operations *out-of-place* which reduces the memory requirements by one temporary register. We applied Algorithm 5, cf. (Hutter, Joye, & Sierra, 2011), needing ten multiplications, five squarings, and sixteen additions (including subtractions) to perform a (differential) addition and doubling operation. We allocated eight intermediate variables of twelve words each (thus needing 192 bytes of RAM only for the scalar multiplication). In order to keep the memory requirements to a minimum, we decided

to reuse four of these intermediate variables also for other operations of the framework.

Algorithm 1 shows the implemented Montgomery ladder in Co-Z representation, where $n=192$.

Due to the high energy costs of a scalar multiplication, we decided to monitor the energy budget during this operation to increase the reading range. Therefore, after each loop iteration within the scalar multiplication, a check of the power supply is performed. Thus, the available energy is tested 190 times for one Montgomery-ladder execution. If needed the device is put into sleeping mode to recover energy.

Finite-Field Arithmetic over Binary Fields

We also implemented an ECC framework for the WISP4.1DL that is based on binary-extension fields. In general, operations over binary fields involve arithmetic with polynomials. One major advantage of binary-field arithmetic is the carry-less processing of operands. This reduces the complexity and increases the speed of an ECC implementation significantly. Moreover, since all operations are done using polynomials, mainly XOR, Shift, and ADD operations have to be performed which can be realized very efficiently on common microcontrollers. In the following subsections, we describe the individual modular-arithmetic operations.

Addition and Subtraction

An addition of two finite-field elements over $GF(2^m)$ can be realized by a simple XOR operation instead of an arithmetic addition over $GF(p)$. The significant difference between prime field and binary field addition is therefore that no carry is propagated by XORing the operands. Without producing a carry bit, no modular reduction is needed to limit the size of the result. Moreover, since addition and subtraction are the same operation over $GF(2^m)$, no extra code is required to support also subtraction.

Multiplication and Squaring

Multiplication over $GF(2^m)$ has a similar complexity as multiplication over $GF(p)$. However, microcontrollers typically lack the support of a binary-field multiplier so that the multiplication has to be implemented using other available instructions. While one hardware multiplication would costs one or a few clock cycles, software implementations would need several hundred clock cycles. Thus, prime-field based implementations of ECC are often faster on conventional microcontrollers since they can simply make use of the hardware multiplier even though the arithmetic over $GF(2^m)$ is much easier and can be performed more efficiently than over $GF(p)$.

If no hardware multiplier for binary fields is available on the WISP, multiplication has to be implemented in software. There exist several ways to realize efficient multiplication in this case. For our experiments, we decided to implement the *right-to-left comb method* for polynomial multiplication (Hankerson, Menezes, & Vanstone, 2003). Compared to other polynomial multiplication methods, no pre-computation is required and the runtime is still acceptable. This is done by minimizing the number of costly shift operations. Starting with the rightmost (least significant) bit of the first polynomial, the second polynomial is added to the intermediate result. This addition is only performed if this bit is set (i.e. 1).

By analyzing a polynomial squaring operation, the advantage of binary-extension fields becomes clear. During a prime-field multi-precision squaring, only halve of the intermediate products need to be calculated. Thus, squaring over $GF(p)$ still scales quadratically with the size of the Integer. For $GF(2^m)$, in contrast, the binary-field squaring operation scales linearly with the size of the polynomial. A binary squaring of a random $a(z)$ is obtained by inserting a 0 bit between consecutive bits. But instead of performing shift operations, a pre-computed 8-bit table with 16 entries has been used to calculate the square of a polynomial.

Elliptic-Curve Arithmetic over $GF(2^m)$

Similar to the arithmetic over $GF(p)$, we applied the Montgomery ladder as scalar multiplication method over binary fields. For efficient addition and doubling of elliptic-curve points, we applied the projective-version of the formulae proposed by López and Dahab (López & Dahab, 1999). The formulae need only 6 multiplications and 5 squarings for a combined addition and doubling. Note that these formulae are nearly twice as fast as the formulae over prime fields. Additionally, this formula only needs 6 registers, whereas 7 registers are needed for the prime-field based Montgomery ladder.

Algorithm 2 shows the Montgomery ladder for $GF(2^{191})$.

RESULTS

In the following, we present experimental results of our ECC framework running on the WISP4.1DL. All implementations have been compiled using the IAR Assembler v5.10.4 and the IAR C/C++

Algorithm 2. Montgomery ladder for elliptic curves over GF(2^m)

Input: $P=(x,y)$ and $k=(k_{n-1},\ldots,k_0)$ with $k_{n-1}\neq 0$ and $c=b^{2^{(m-1)}}$ **Output:** $Q=kP$
1: $(x,y)\leftarrow P$;   2: $X_0\leftarrow x$; $Z_0\leftarrow 1$; $Z_1\leftarrow X_0{}^2$; $X_1\leftarrow X_1\times Z_1$; $X_1\leftarrow X_1+b$;   3: **for** $i=n-2$ downto 0 **do**   4: $X_{1-k[i]}\leftarrow X_{1-k[i]}\times Z_{k[i]}$, $Z_{1-k[i]}\leftarrow Z_{1-k[i]}\times X_{k[i]}$, $T_1\leftarrow X_{1-k[i]}\times Z_{1-k[i]}$,   5: $Z_{1-k[i]}\leftarrow Z_{1-k[i]}+X_{1-k[i]}$, $Z_{1-k[i]}\leftarrow Z_{1-k[i]}{}^2$, $X_{1-k[i]}\leftarrow Z_{1-k[i]}\times x$,   6: $X_{1-k[i]}\leftarrow X_{1-k[i]}+T_1$, $X_{k[i]}\leftarrow X_{k[i]}{}^2$, $Z_{k[i]}\leftarrow Z_{k[i]}{}^2$, $T_1\leftarrow Z_{k[i]}\times c$,   7: $Z_{k[i]}\leftarrow Z_{k[i]}\times X_{k[i]}$, $T_1\leftarrow T_1{}^2$, $X_{k[i]}\leftarrow X_{k[i]}{}^2$, $X_{k[i]}\leftarrow X_{k[i]}+T_1$ 8: *test_power_supply();*   9: **end for**   10: **Return** $Q=(X_0,Z_0)$.

Compiler v5.10.6 (Kickstart LMS). First, we provide results of two different ECC implementations over prime fields. One that has been optimized for WISP tags that feature an MSP430 with no hardware multiplier and one implementation for WISP tags with an MSP430 that supports a hardware multiplier. For a fair comparison of the two implementations, we omitted the EPC Gen-2 protocol implementation for all simulations. On the one hand we used the MSP430F233 microcontroller as a device that features a hardware multiplier, and on the other hand we used the MSP430F2132 that is assembled on the WISP4.1DL tag. In order to obtain practical results for the WISP4.1DL, the existing firmware has been flattened to allow a running system containing the EPC class-1 Generation-2 protocol as well as the ECC framework. Second, we provide results of our binary-field implementation and discuss the performance on the WISP. a^2

Performance over GF(p)

Most effort in optimizations has been put into the finite-field multiplication and squaring operation. We decided to implement the hybrid-multiplication method. Table 1 shows the amount of required clock cycles for multi-precision multiplication and squaring as a function of the different system settings. As it was expected, the worst case is the setting without hardware multiplier and the parameter $d=1$ of the hybrid-multiplication method. The performance can be improved significantly by using a device with hardware multiplier (about 90% improvement). We have to mention that the implementation of the hybrid-multiplication method with $d=1$ has not been optimized for usage with hardware multiplier. So there is still a lot of room for optimizations by specially fitting the hybrid-multiplication method to devices that feature a hardware multiplier. As it can be seen in Table 1, the amount of required clock cycles for a multi-precision multiplication can roughly be halved by applying loop unrolling and using $d=2$. Nevertheless, this optimization is not always

Table 1. Performance of 192-bit multi-precision multiplication and squaring

System setting	Hybrid multiplication [Cycles]	Squaring [Cycles]
WISP without HWM ($d=1$)	25,350	14,361
WISP with HWM ($d=1$)	5,046	3,363
WISP with HWM ($d=2$)	2,581	-

Table 2. Flash-memory requirement of the multi-precision multiplication with and without a dedicated squarer

System setting	Code size without dedicated squarer [Bytes]	Code size with dedicated squarer [Bytes]
WISP without HWM ($d = 1$)	1,076	1,572
WISP with HWM ($d = 1$)	1,020	1,376
WISP with HWM ($d = 2$)	4,236	-

Table 3. Performance of a 192-bit scalar multiplication using the Montgomery ladder running on different WISP-tag settings

System setting	Without squaring [Cycles]	With squaring [Cycles]
WISP without HWM ($d = 1$)	63,257,925	54,630,581
WISP with HWM ($d = 1$)	17,376,758	15,761,884
WISP with HWM ($d = 2$)	10,289,883	-

practicable as the code size increases significantly as shown in Table 2. The increase in code size is caused by the loop unrolling when applying $d = 2$.

Another improvement can be achieved by introducing squaring () instead of multiplication ($a \cdot a$). If no hardware multiplier is available, the number of clock cycles can be reduced roughly by the factor 1.7 to the cost of an increased code size. This improvement compared to the implementation of (Cohen, Miyaji, & Ono, 1998) can be explained by the high costs of partial-product calculations. For the system settings with a hardware multiplier, the improvement is less significant as the major improvement through the use of the hardware multiplier itself. So if a computation of the partial products for a multi-precision multiplication is relatively expensive, usage of squaring is recommended because the advantage of clock-cycle reduction outweighs the drawback of an increased code size. Since squaring does not provide a major improvement in the runtime compared to the hybrid-multiplication method with $d = 2$, squaring has not been implemented for this system setting.

The optimizations described before have been applied on the lowest implementation level. As expected and shown in Table 3, the support of a

hardware multiplier provides best performance for the scalar multiplication. If no hardware multiplier is available, a squaring implementation is recommended on the WISP tag as it reduces the amount of clock cycles from 63,257,925 to 54,630,581, i.e., about 13.6%. Thus, up to 10^7 clock cycles can be saved. Furthermore, it shows that the best performance has been obtained by using a hardware multiplier in combination with the hybrid-multiplication method with parameter $d = 2$. This setting is about six times faster than the fastest method without hardware multiplier.

We simulated the entire 192-bit scalar multiplication using the Montgomery powering ladder. It needs about $5.5 \cdot 10^7$ clock cycles on the WISP4.1DL platform. This corresponds to 8.2 seconds on the WISP running at a clock frequency of 6.7 MHz. Running at this clock frequency—which actually is the highest frequency feasible for succeeding the scalar multiplication with our hardware setup—the maximum distance to the reader is about 2 centimeters. Although the power supply is tested after each Montgomery-ladder round, the scalar multiplication will not finish at distances further than the 2 centimeters as one round exceeds the number of cycles which can be processed with the power available at this distance. If reading ranges larger than the 2 centimeters are necessary, either the clock frequency can be reduced which leads to a longer computation time or an additional capacitor can be assembled on board. Reducing the clock frequency from 6.7 MHz to 6.1 MHz results in a computation time of 9.0 seconds and a maximum distance to the reader is extended to about 10 centimeters. Correspondingly, a frequency of 3.98 MHz leads to 13.8 seconds and a frequency of 0.98 MHz leads to 56.1 seconds at a maximum distance of 40 centimeters.

In order to increase the reading distance, we recommend soldering a large electrolytic, tantalum or super capacitor into the power supply of the WISP4.1DL device (Intel Research Seattle, 2010), (WISP Wiki, 2012). For this, we tried different supercaps between 0.047 and 0.1 µF and achieved reading distances up to 2 meters. The

large capacitor allows the WISP tag to provide enough energy for cryptographic computations even at larger distances. However, it needs some time to charge the capacitor to work properly at these distances.

In addition, we have to note that replacing the MSP430F2132 with the MSP430F233, the total number of clock cycles can be significantly reduced by the factor of 5 since the MSP430F233 features a dedicated hardware multiplier. This results in an estimated computation time of about 1.5 seconds at 6.7 MHz and plays a major role in the energy and power consumption of the WISP tag.

Table 4 presents an overview of the required code size for the different system settings. All configurations except the configuration with the hardware multiplier and $d = 2$ support squaring. The code size of the additional framework modules is listed in Table 5.

Table 4. Memory requirements of the WISP tag for different framework-level implementations

WISP without HWM	Code [Bytes]	Const [Bytes]	Data [Bytes]
Multi-precision arithmetic	1,572	0	0
Finite-field arithmetic	2,532	0	96
ECC arithmetic	3,944	210	176
WISP with HWM ($d = 1$)	**Code [Bytes]**	**Const [Bytes]**	**Data [Bytes]**
Multi-precision arithmetic	1,376	0	0
Finite-field arithmetic	2,346	0	96
ECC arithmetic	3,758	206	326
WISP with HWM ($d = 2$)	**Code [Bytes]**	**Const [Bytes]**	**Data [Bytes]**
Multi-precision arithmetic	4,236	0	0
Finite-field arithmetic	5,206	0	96
ECC arithmetic	6,618	206	326

Table 5. Memory requirements of the WISP tag for additional framework modules

Module	Code [Bytes]	Const [Bytes]	Data [Bytes]
SHA-1 hash function	1,012	30	10
Random Number Generation (RNG)	218	0	4
Modified EPC Class 1 Gen 2 protocol	3,260	66	181

Table 6. Memory requirements of the WISP tag for the binary-field ECC implementation

WISP without HWM	Code [Bytes]	Const [Bytes]	Data [Bytes]
Scalar multiplication	2,518	142	234

Performance over GF(2^m)

Table 6 shows the required resources and the amount of required clock cycles for a point multiplication over binary fields. The applied elliptic curve *c2tnb191v1* is standardized in X9.62. Utilizing its Mersenne-like irreducible polynomial, a field reduction is possible within just 493 clock cycles. For that operation, only XOR and shift instructions are necessary. Consequently, a field multiplication can be done on average within 10,532 clock cycles, a field squaring within 1,522 clock cycles, and a field addition within 139 clock cycles. Compared to the NIST P-192 implementation (not using the hardware multiplier), the multiplication over *c2tnb191v1* is 2.4 times faster and the squaring is more than 9.4 times faster. As a result, the scalar multiplication for the binary field is 3.8 times faster. It needs only 14,281,000 clock cycles to perform a 191-bit scalar multiplication using the Montgomery ladder. This corresponds to about 2.1 seconds on the WISP4.1DL running at 6.7 MHz.

Using the hardware multiplier, the NIST P-192 field multiplication is 4 times faster than the field multiplication over the curve *c2tnb191v1*. The NIST P-192 field squaring still is 1.7 times slower. Consequently, the performance of the point multiplication differs by only 28%.

Note that the binary-field implementation has been implemented in C. Only the bit-serial multiplication and addition has been implemented and optimized in Assembly. Thus, further improvements are possible by implementing also the squaring operation and even the scalar multiplication in Assembly.

Table 6 shows the memory requirements for a 191-bit scalar multiplication. Compared to the ECC implementation over prime fields, it needs only 2,518 bytes of code. It needs 142 bytes for storing constants and 234 bytes to maintain all variables in RAM.

COMPARISON WITH RELATED WORK

There exist many publications that provide results of ECC implementations on the MSP430 platform. One of the first publications of ECC on the TI MSP430x33x family of microcontrollers is due to Guajardo et al. (Guajardo, Blümel, Krieger, & Paar, 2001). They made use of an elliptic curve over the finite field GF(p) with 128 bits, cf. SECG128. Their implementation needs about 34 million clock cycles for a scalar multiplication, i.e., 3.4 seconds at an operating frequency of 1 MHz.

In 2007, Scott et al. improved the performance of Guajardo et al. by presenting multi-precision calculation results for the MSP430F1611 (Scott & Szczechowiak, 2007). They implemented the hybrid multiplication with d=2 and compared the performance with the classical Comba multiplication. For a 160-bit multiplication, they reported 1,746 clock cycles for the hybrid and 2,065 clock cycles for the Comba multiplication. In 2008, Liu and Ning presented a configurable library, i.e., *TinyECC*, which has been evaluated on different platforms such as the ATmega128, ARM, and the MSP430 (Liu & Ning, 2008). They made use of the Tmote Sky sensor node which integrate an MSP430 microcontroller and compared the performances with other results obtained from other architectures. Also note that a similar framework has been presented by (Szczechowiak, 2008). The authors presented *NanoECC* which is based on the Multiprecision Integer and Rational Arithmetic C/C++ Library (MIRACL).

In 2009, Gouvêa et al. further improved the results of Scott by reducing the number of needed clock cycles for a 160-bit multi-precision multiplication from 1,746 to only 1,586 (Gouvêa & López, 2009). They showed that the classical Comba method is more efficient on that platform than the hybrid-multiplication method. Similar results have been also obtained by (Wenger and Werner, 2011). They reported 1,570 clock cycles for an unrolled 160-bit Comba multiplication on the same platform.

In view of WISPs, one of the first who demonstrated the feasibility of cryptography on the Wireless Identification and Sensing Platform (WISP) is due to Chae et al. (Chae, Yeager, Smith, & Fu, 2007). They presented results of RC5 that is a symmetric block cipher with variable block size (32, 64, or 128 bits), the number of rounds (0 to 255), and the key size (0 to 2,040 bits). They implemented RC5-32/12/16 which corresponds to a block size of 32 bits, 12 rounds, and 16 bytes of secret key. For the key setup, they reported an execution time of 7.93 milliseconds at a reader distance of 30 centimeters. Encryption and decryption with 64-bit messages can be performed in about 1.4 milliseconds. Furthermore, they observed that at a distance of 50 centimeters the power consumption of the WISP tag significantly decreases when flash-memory write instructions are used within the design of the cryptographic implementation. In fact, at a distance of 75 centimeters, no flash-memory write instructions can be performed anymore using the WISP4.1DL.

Next to the implementation of RC5, there exists also a low-resource implementation of the Advanced Encryption Standard (AES) on the WISP4.1DL platform (Szekely, Höfler, Stögbuchner, & Aigner, 2013). The authors implemented AES-128 and optimized it for the MSP430 plat-form. Their implementation allows encrypting of data within 5,432 clock cycles. The code size of the AES implementation is about 2,540 bytes; the total size of the security-enhanced WISP firmware is 5,574 bytes. Furthermore, the AES requires 83 bytes of RAM. As a demonstration application, they used the AES to encrypt measured sensor data (temperature, accelerometer, or light) of the WISP4.1DL. In view of the reading range, the authors stated that they did not notice a difference in the maximum reading distance between a WISP that supports AES encryption/decryption or not. They only identified differences in the throughput that depends on the distance to the reader. In fact, the throughput is reduced by 20% when AES is supported by the WISP. As a proof of concept, they performed measurements at reading distances of 50, 100, and 200 centimeters. At a distance of 50 centimeters, a reader rate of 30.62 has been obtained for the WISP which performs an AES encryption. Without AES encryption the read rate is 37.45. At 100 centimeters, the read rate is only 16.95 for the AES-enabled WISP, and 21.24 for a WISP without AES support. Moreover, they reported a read rate of 7.85 and 9.80 at a distance of 200 centimeters, respectively.

Table 7 gives a comparison between RC5, AES, and ECC in terms of security level, execution time, and reading distance. In order to make a fair comparison, we evaluated the implementations for the same WISP4.1DL platform running at the same frequency of 3 MHz. Note that RC5 and AES are both symmetric primitives and ECC is an asymmetric primitive which explains the large differences in the execution time and reading distance. Furthermore, for ECC over prime fields, we give results for the WISP without a hardware multiplier (WISP4.1DL) and also for the WISP which is shipped with a different microcontroller that features a hardware multiplier (e.g. the MSP430F233 instead of the MSP430F2132). It shows that ECC over prime fields provides best performance needing 3.4 seconds at 3 MHz (when a hardware multiplier is available). Also

Table 7. Implementation comparison of RC5, AES, and ECC on the WISP4.1DL running at 3 MHz

Cryptographic Primitive	Security Level [bits]	Execution Time [ms]	Reading Distance [centimeters]
RC5-32/12/16	128	1.4	75
AES-128	128	1.8	up to 200
ECC-P192 (without HWM)	96	18,200	25
ECC-P192 (with HWM)	96	3,400	N/A
ECC-B191	96	4,700	30

note that the execution time varies depending on the actual reading distance. The closer the WISP tag to the reader, the faster will be the execution time since less power-down sleeping actions have to be performed. In addition, we only give values of reading distances that have been reported by the authors. The reading distance obviously varies depending on the used reader and the radiated power level.

SECURITY CONSIDERATIONS

When implementing ECC or other cryptographic primitives on WISPs, one has to consider several security concerns. In 1996, Kocher introduced an implementation attack where the execution time of a Diffie-Hellman, RSA, or DSS implementation is exploited to extract secret key material (Kocher, 1996). This method is very generic so that it can be also applied to many other cryptographic implementations that do not consider constant runtimes such as ECC. Three years later, Kocher et al. presented an even more powerful attack that allows extraction of the secret key by simply observing power-consumption traces of a given smart-card implementation (Kocher, Jaffe, & Jun, 1999). They showed how to extract the secret key of a Data Encryption Standard (DES) implementation using 1,000 measured power traces. For this attack, they made use of statistical methods (such as the Pearson correlation coefficient) to compare the measured power traces with hypothetical power-consumption values that have been generated for any possible sub-byte of the secret key. For the correct key guess, a significant correlation peak is observed which succeeds the attack. This method is called Differential (or Correlation-based) Power Analysis (DPA).

In a Simple Power Analysis (SPA) attack, only one or a few power traces are needed. By simply observing the power-consumption characteristics, an attack can identify patterns of the algorithm execution that reveal bits of the secret key. This is, for example, the case in a standard square and multiply exponentiation or in the case of the binary method in ECC scalar multiplication. In this scenario, an attacker can simply identify if a point addition is performed within a loop or not (since the power-consumption pattern of point addition and point doubling can be clearly distinguished from each other).

There have been many publications of SPA and DPA attacks on ECC in the last decade. Most of them performed attacks on a double-and-add scalar multiplication such as reported by (Gebotys & Gebotys, 2003) or (Örs, Batina, & Prenel, 2003). Gebotys et al. performed a successful SPA attack on a 192-bit ECC implementation that was running on a DSP VLIW processor. Örs et al. have been able to reveal the secret scalar of a 160-bit scalar multiplication running on an FPGA platform. In 2007, Medwed et al. presented an attack on the ephemeral key of an ECDSA implementation (Medwed & Oswald, 2008). For this attack, they generated power-consumption templates that can be used to extract the value of the ephemeral key using only one single power shot, cf. (Chari, Rao, Rohatgi, 2002). The target implementation was running on an ARM7 microprocessor. They created templates for several intermediate values and performed a matching phase which reveals the ephemeral key bit by bit having a success probability of almost 1.

Next to the demonstration of successful power-analysis attacks, large effort has been made by the cryptographic community to prevent such attacks on different platforms. (Coron, 1999), for example, proposed to randomize the projective coordinates during an ECC scalar multiplication to prevent DPA attacks (Coron's third counter-measure). This countermeasure is very simple to implement and can be easily integrated into WISP implementations without having much overhead in code or RAM size. Note that Coron's first and second countermeasure has been shown to provide weaknesses as demonstrated by (Okeya & Sakurai, 2000). The proposed randomization of the secret scalar provides a certain bias that can be exploited in an attack and the blinding of the point P is weak in randomizing the expression of computed/computing objects.

Side-Channel Attacks on WISPs

Szekely et al. presented a DPA and Electromagnetic Power Analysis (DEMA) attack on the WISP4.1DL platform (Szekely, Höfler, Stögbuchner, & Aigner, 2013). The target of their attack has been an AES-128 implementation. In particular, they targeted the first S-box output of the first round of AES. For the power-consumption measurements, they integrated a 33 Ohm resistor in the power-supply path of the WISP and measured the voltage drop using a differential probe and a digital storage oscilloscope. Note that the authors removed two capacitors (*C14* and *C20*) on the WISP in order to avoid cancellation of side-channel information through smoothing the power consumption during encryption of data.

Furthermore, they used an electromagnetic near-field probe to gather electromagnetic emanations of the tag and amplified the signals using a 30 dB amplifier. For the electromagnetic attack, they measured the emanation of 200 different encryptions and performed a Pearson-Correlation based attack that reveals each byte of the 128-bit secret key of AES.

There exist also other publications that demonstrate the susceptibility of passively powered UHF tags to side-channel attacks. Plos, for example, performed a successful attack on commercially available UHF tags and showed how to extract secret information from tags at a distance of up to 1 meter (Plos, 2008). The author used a self-made dipole antenna to measure the data-dependent backscatter signal in the electromagnetic far field. He measured 10,000 traces and obtained a significant correlation coefficient of about 0.08 which succeeds the attack. Similar results have been also previously shown by Oren & Shamir in 2007 (Oren & Shamir, 2007). They observed that the backscattered signal of passively powered UHF tags contain data-dependent information. By measuring these signals in the far field, they could extract the kill password of commercially available tags.

FUTURE RESEARCH DIRECTIONS

The next generation of WISPs considers the integration of Field Programmable Gate Arrays (FPGA) or Complex Programmable Logic Devices (CPLD). The WISP Version 5 of Intel Research Seattle, for example, will feature a CPLD next to the MSP430 microcontroller. The CPLD can be used to outsource time and memory expensive operations such as the protocol handling or the computation of cryptographic operations. In view of ECC, the finite-field multiplication could be implemented in configurable hardware. Due to the possibility to parallelize complex algorithms and to customize the own design of a multiplier, a significant speed-up can be achieved that would lower the total runtime of scalar multiplication to only a few tens of milliseconds.

Another improvement would be the integration of microcontrollers on WISPs which provide dedicated binary hardware multipliers. In fact, most of the microcontrollers in use today integrate only prime-field based multipliers which encourage the use of prime-field based ECC formulae. However, with the use of binary-multiplier enabled microcontrollers, the performance of ECC could be highly increased through the calculation of a single cycle binary-word multiplication.

Emerging research trends with regards to WISPs go towards low-power ECC designs. In contrast to WSNs, which get typically powered by a battery and which have to meet low-energy requirements, WISPs get powered passively by the field of a reader. Thus, implementations have to meet low-power consumption requirements in order to operate even at a larger reading distance. The mean current consumption of the implementation must not exceed the available energy in the capacitor of the WISP device. Special consideration has to be taken on the energy consumption per cycle rather than energy consumption per operation.

CONCLUSION

In this chapter, we evaluated the feasibility of Elliptic Curve Cryptography (ECC) on the Wireless Identification and Sensing Platform (WISP). We implemented ECC over a 192-bit prime field and over a 191-bit binary field. Several optimizations have been necessary in order to meet the low-resource requirements of passively powered WISPs. We provided results for WISP4.1DL which features no dedicated hardware multiplier and give also results for WISP tags which feature a hardware multiplier, e.g., the MSP430F233 microcontroller. Our prime-field ECC implementation demonstrated that a scalar multiplication using

the Montgomery ladder can be performed within 8.3 seconds on the WISP4.1DL tag running at 6.7 MHz. The same operation can be performed within 1.6 seconds on the MSP430F233 which is an increase of about 80%. The binary-field ECC implementation needs about 2 seconds on the WISP4.1DL tag. Furthermore, it showed that a dedicated squaring implementation on the WISP4.1DL improves the performance by about 14%. With these results, we successfully demonstrated the feasibility of ECC on the WISP. ECC-enabled WISPs can be used to protect the sending of sensed data or to provide authentication services for WISPs and reader devices using asymmetric cryptography. Future research will go towards a more efficient calculation of finite-field multiplications using configurable hardware or application specific ICs.

REFERENCES

Brent, R. P. (2004). Note on Marsaglia's Xorshift random number generators. *Journal of Statistical Software, 11*(4), 1–5.

Chae, M.-J., Yeager, D. J., Smith, J. R., & Fu, K. (2007). *Maximalist cryptography and computation on the WISP UHF RFID tag*. Paper presented at the Conference on RFID Security, Malaga, Spain.

Chari, S., Rao, J. R., & Rohatgi, P. (2002). Template attacks. In B. S. Kaliski Jr., Ç.-K. Koç, & C. Paar (Eds.), *4th International Workshop on Cryptographic Hardware and Embedded Systems - CHES 2002* (pp. 13-28). Berlin, Germany: Springer.

Cohen, H., Miyaji, A., & Ono, T. (1998). Efficient elliptic curve exponentiation using mixed coordinates . In Ohta, K., & Pei, D. (Eds.), *Advances in Cryptogoly – Asiacrypt'99* (pp. 51–65). Berlin, Germany: Springer. doi:10.1007/3-540-49649-1_6

Comba, P. (1990). Exponentiation cryptosystems on the IBM PC. *IBM Systems Journal, 29*(4), 526–538. doi:10.1147/sj.294.0526

Coron, S. (1999). Resistance against differential power analysis for elliptic curve cryptosystems. In Ç.-K. Koç, & C. Paar (Eds.), *First International Workshop on Cryptographic Hardware and Embedded Systems - CHES 1999* (pp. 292-302). Berlin, Germany: Springer.

EPCglobal. (2008). *EPCTM radio-frequency identity protocols class-1 generation-2 UHF RFID protocol for communications at 860MHz-960MHz*, Version 1.2.0. Retrieved October, 2008, from http://www.epcglobalinc.org

Gebotys, C., & Gebotys, R. (2003). Secure elliptic curve implementations: An analysis of resistance to power-attacks in a DSP processor. In B. S. Kaliski Jr., Ç.-K. Koç, & C. Paar (Eds.), *4th International Workshop on Cryptographic Hardware and Embedded Systems - CHES 2002* (pp. 114-128). Berlin, Germany: Springer.

Goundar, R., Joye, M., & Miyaji, A. (2010). Co-Z addition formulae and binary ladders on elliptic curves. In Mangard, S., & Standaert, F.-X. (Eds.), *Cryptographic Hardware and Embedded Systems -CHES 2010* (pp. 65–79). Berlin, Germany: Springer. doi:10.1007/978-3-642-15031-9_5

Gouvêa, C., & Lopez, J. (2009). Software implementation of pairing-based cryptography on sensor networks using the MSP430 microcontroller. In Roy, B., & Sendrier, N. (Eds.), *Progress in Cryptology - INDOCRYPT 2009, LNCS 5922* (pp. 248–262). Berlin, Germany: Springer. doi:10.1007/978-3-642-10628-6_17

Guajardo, J., Blümel, R., Krieger, U., & Paar, C. (2001). Efficient implementation of elliptic curve cryptosystems on the TI MSP430x33x family of microcontrollers. In K. Kim (Ed.), *4th International Workshop on Practice and Theory in Public Key Cryptosystems – PKC 2001, LNCS 1992* (pp. 365-382). Berlin, Germany: Springer.

Hankerson, D., Menezes, A. J., & Vanstone, S. (2003). *Guide to elliptic curve cryptography*. New York, NY: Springer.

Hutter, M., Joye, M., & Sierra, Y. (2011). Memory-constrained implementations of elliptic curve cryptography in Co-Z coordinate representation. In A. Nitaj & D. Pointcheval (Eds.), *Progress in Cryptology - AFRICACRYPT 2011 Fourth International Conference on Cryptology*, LNCS 6737, (pp. 170-187). Berlin, Germany: Springer.

Hutter, M., Medwed, M., Hein, D., & Wolkerstorfer, J. (2009). Attacking ECDSA-enabled RFID devices. In M. Abdalla, D. Pointcheval, P.-A. Fouque, & D. Vergnaud (Eds.), *7th International Conference on Applied Cryptography and Network Security - ACNS 2009* (pp. 519-534). Berlin, Germany: Springer.

Hutter, M., & Wenger, E. (2011). Fast multi-precision multiplication for public-key cryptography on embedded microprocessors. In B. Preneel & T. Takagi (Eds.), *13th International Workshop on Cryptographic Hardware and Embedded Systems - CHES 2011*, (pp. 459-474). Berlin, Germany: Springer.

Impinj. (2012). *Speedway revolution - Superior performance made easy.* Speedway Revolution Reader Specifications. Retrieved January 2012, from www.impinj.com

Intel Research Seattle. (2010). *WISP: Wireless identification and sensing platform.* Retrieved January 2012, from http://seattle.intel-research.net/wisp

Joye, M., & Yen, S.-M. (2003). The Montgomery powering ladder. In G. Goos, J. Hartmanis, & J. van Leeuwen (Eds.), *4th International Workshop on Cryptographic Hardware and Embedded Systems* (pp. 291-302). Berlin, Germany: Springer.

Kocher, P. (1996). Timing attacks on implementations of Diffie-Hellman, RSA, DSS, and other systems. In N. Koblitz (Ed.), *Proceedings of the 16th Annual International Cryptology Conference on Advances in Cryptology, CRYPTO 1996* (pp. 104-113). Berlin, Germany: Springer.

Kocher, P., Jaffe, J., & Jun, B. (1999). Differential power analysis. In M. Wiener (Ed.), *Proceedings of the 19th Annual International Cryptology Conference on Advances in Cryptology, CRYPTO 1999* (pp. 388-397). Berlin, Germany: Springer.

Lee, Y. K., & Verbauwhede, I. (2007). A compact architecture for Montgomery elliptic curve scalar multiplication processor. In S. Kim, M. Yung, & H.-W. Lee (Eds.), *8th International Workshop on Information Security Applications - WISA 2007* (pp. 115-127). Berlin, Germany: Springer.

Lim, C., & Lee, P. (1994). More flexible exponentiation with precomputation. In Y. Desmedt (Ed.), *Proceedings of the 14th Annual International Cryptology Conference on Advances in Cryptology - CRYPTO 1994*, (pp. 95-107). Berlin, Germany: Springer.

Liu, A., & Ning, P. (2008). TinyECC: A configurable library for elliptic curve cryptography in wireless sensor networks. *7th International Conference on Information Processing in Sensor Networks - IPSN 2008* (pp. 245-256).

López, J., & Dahab, R. (1999). Fast multiplication on elliptic curves over GF(2^m) without precomputation . In Koç, C. K., & Paar, C. (Eds.), *Cryptographic Hardware and Embedded Systems – CHES 1999* (pp. 316–327). Berlin, Germany: Springer. doi:10.1007/3-540-48059-5_27

Mangard, S., Oswald, M. E., & Popp, T. (2007). *Power analysis attacks - Revealing the secrets of smart cards*. Berlin, Germany: Springer.

Marsaglia, G. (2003). Xorshift RNGs. *Journal of Statistical Software, 8*(14), 1–6.

Medwed, M., & Oswald, E. (2008). Template attacks on ECDSA. In K. Chung, M. Yung, & K. Sohn (Eds.), *9th International Workshop on Information Security Applications - WISA 2008*, (pp. 14-27), Berlin, Germany: Springer.

Meloni, N. (2006). *Fast and secure elliptic curve scalar multiplication over prime fields using special addition chains*. Cryptology ePrint Archive, Report 2006/216.

Montgomery, P. L. (1987). Speeding the Pollard and elliptic curve methods of factorization. *Mathematics of Computation, 48*(177), 243–264. doi:10.1090/S0025-5718-1987-0866113-7

National Institute of Standards and Technology (NIST). (2009). *FIPS-186-3: Digital signature standard* (DSS). Retrieved June, 2011, from http://www.itl.nist.gov/fipspubs/

Okeya, K., & Sakurai, K. (2000). Power analysis breaks elliptic curve cryptosystems even secure against the timing attack. In B. K. Roy & E. Okamoto (Eds.), *First International Conference in Cryptology in India Progress in Cryptology - INDOCRYPT 2000*, (pp. 178-190). Springer.

Oren, Y., & Shamir, A. (2007). Remote password extraction from RFID tags. *IEEE Transactions on Computers, 56*(9), 1292–1296. doi:10.1109/TC.2007.1050

Örs, B., Batina, L., & Prenel, B. (2003). Hardware implementation of elliptic curve processor over GF(p). *International Journal of Embedded Systems, 3*(4), 229–240. doi:10.1504/IJES.2008.022394

Plos, T. (2008). Susceptibility of UHF RFID tags to electromagnetic analysis. In T. Malkin (Ed.), *The Cryptographers' Track at the RSA Conference - CT-RSA 2008*, (pp. 288-300). Springer.

Saxena, N., & Voris, J. (2009, November). *Accelerometer based random number generation on RFID tags*. Paper presented at the 1st Workshop on Wirelessly Powered Sensor Networks and Computational RFID, Berkely, California.

Scott, M., & Szczechowiak, P. (2007). *Optimizing multiprecision multiplication for public key cryptography*. Cryptology ePrint Archive, Report 2007/299. Retrieved January, 2012, from http://eprint.iacr.org

Smith, J. R., Fishkin, K. P., Jiang, B., Mamishev, A., Philipose, M., & Rea, A. D. (2005). RFID-based techniques for human-activity detection. *Communications of the ACM*, (48): 39–44. doi:10.1145/1081992.1082018

Szczechowiak, P., Oliveira, L., Scott, M., Collier, M., & Dahab, R. (2008). NanoECC: Testing the limits of elliptic curve cryptography in sensor networks. In R. Verdone (Ed.), *5th European Conference on Wireless Sensor Networks - EWSN 2008*, (pp. 305-320). Berlin, Germany: Springer.

Szekely, A., Höfler, M., Stögbuchner, R., & Aigner, M. (2013). Security enhanced WISPs: Implementation challenges. In Smith, J. R. (Ed.), *Wirelessly Powered Sensor Networks and Computational RFID, Springer (to appear in 2013)*.

Wiki, W. I. S. P. (n.d.). Retrieved January, 2012, from http://wisp.wikispaces.com

Yeager, D., Holleman, J., Prasad, R., Smith, J., & Otis, B. (2009). NeuralWISP: A wirelessly powered neural interface with 1-m Range. *IEEE Transactions on Biomedical Circuits and Systems*, 3(6), 379–387. doi:10.1109/TBCAS.2009.2031628

Yeager, D. J., Sample, A. P., & Smith, J. R. (2008). WISP: A passively powered UHF RFID tag with sensing and computation . In Ahson, S. A., & Ilya, M. (Eds.), *RFID handbook: Applications, technology, security, and privacy* (pp. 261–278). CRC Press.

KEY TERMS AND DEFINITIONS

Asymmetric Primitive (n): In cryptography, where two distinct keys are used to establish a cryptosystem

Advanced Encryption Standard (AES): a symmetric encryption algorithm which has been standardized by NIST in 2000

Computer Science (n): Study of transferring, processing, and storing of information

Differential Power Analysis (DPA): a method to extract secret information out of physical side channels

Elliptic Curve Cryptography (ECC): A public-key cryptography primitive that operates on elliptic curves

Elliptic Curve Digital Signature Algorithm (ECDSA): A standardized digital-signature generation and digital-signature verification scheme based on elliptic curves

Finite Field (n): An algebraic structure that contains a finite number of elements

Loop Unrolling (v): Unfolding all instructions of a loop without applying a loop instruction

Mersenne Prime (n): A prime number that is a power of two minus one

Montgomery Ladder (n): An efficient scalar multiplication that allows omitting the y-coordinate of elliptic curve points

MSP430: Is a by Texas Instruments designed 16-bit mixed-signal microprocessor

Radio Frequency Identification (RFID): Wireless communication technology that allows identification of objects in the near proximity of a reader

Reduced Instruction Set Computing (RISC): Is a design strategy that tries to simplify the within a CPU used instructions

Rivest Cipher 5 (RC5): A stream cipher designed by Ron Rivest

RSA (n): A public-key cryptographic system which stands for the inventors Rivest, Shamir, and Adleman

Scalar Multiplication (n): Mathematical operation that defines a vector space in linear algebra

Secure Hash Algorithm-1 (SHA-1): Is a by NIST standardized hash algorithm

Side-Channel Attacks (n): A kind of implementation attacks that exploits physical characteristics of devices to extract secret information

Simple Power Analysis (SPA): Makes use of one or a few power-consumption traces to reveal secret-key information by simple observation

Wireless Identification and Sensing Platform (WISP): Is a passive RFID tag designed by Intel Research with an externally programmable interface

Chapter 6
Low Complexity Minimal Instruction Set Computer Design using Anubis Cipher for Wireless Identification and Sensing Platform

J. H. Kong
The University of Nottingham Malaysia, Malaysia

L.-M. Ang
Edith Cowan University, Australia

K. P. Seng
The Sunway University, Malaysia

ABSTRACT

This chapter presents a low complexity processor design for efficient and compact hardware implementation for WISP system security using the involution cipher Anubis algorithm. WISP has scarce resources in terms of hardware and memory, and it is reported that it has 32K of program and 8K of data storage, thus providing sufficient memory for design implementation. The chapter describes Minimal Instruction Set Computer (MISC) processor designs with a flexible architecture and simple hardware components for WISPs. The MISC is able to make use of a small area of the FPGA and provides security programs and features for WISPs. In this chapter, an example application, which is Anubis involution cipher algorithm, is used and proposed to be implemented onto MISC. The proposed MISC hardware architecture for Anubis can be designed and verified using the Handel-C hardware description language and implemented on a Xilinx Spartan-3 FPGA.

DOI: 10.4018/978-1-4666-1990-6.ch006

INTRODUCTION

In a typical RFID system, assets, products and objects are usually labeled with a tag. Each tag contains a very small microchip with a limited computational, storage capabilities and a coupling element. Such devices are usually being classified according to their memory type and input power source (either active or passive). In (Peris-Lopez, Hernandez-Castro, Tapiador, & Ribagorda, 2009), the author proposed a method for tag classification that bases on which were the operations supported on-chip. High-cost tags are divided into two class types: simple and full fledged. The full-fledged tags are able to support conventional cryptography like symmetric encryption, cryptographic one-way functions and even the resource-consuming public key cryptography. Simple tags can support random number generators and one-way hash functions. Likewise, there are also two classes divided for low-cost tags: lightweight and ultra-lightweight. Lightweight tags usually support random number generation but not cryptographic hash function. Moreover, ultra-lightweight tags can only compute simple bitwise operations such as: XOR, AND, OR, etc. SO, ultra-lightweight tags pose great design challenges as they are expected to be widely used.

On the other hand, the Intel Research Seattle had come up with an improved version of RFID tags called the WISP (Wireless Identification and Sensing Platform). WISP is a wireless, battery-free platform for sensing and computation that is powered and read by a standards-compliant Ultra-High Frequency (UHF) RFID reader. The notable features of WISP are battery free sensing, UHF communication, and a fully programmable 16-bit flash microcontroller with analog to digital converter. The conventional security mechanism for lightweight devices is cryptographic solutions. Since there are limited resources on WISP, some security trade off, between the security strength, on-chip-resources and low-power; has to be made

to abide to the WISP standard. In (Sample, Yeager, Powledge, Mamishev, & Smith, 2008), the author has implemented the RC5 algorithm on WISP, which is the first strong cryptographic algorithm to be implemented on a UHF RFID tag. WISP has set a new standard for RFID applications. When we observe popular approaches taken by researchers to meet the area constraints in RFID tags are to implement lightweight encryption algorithms, so that the processing requirement are less computationally demanding and result in smaller hardware implementations. The Tiny Encryption Algorithm (TEA), proposed by P. Israsena from the National Electronics and Computer Technology Centre (NECTEC) is an example of a lighter encryption algorithm. The TEA is a Feistal cipher that incorporates only XOR, ADD and SHIFT operations (Israsena, 2006). Other lightweight encryption hardware implementations which have been proposed can be found in (Engels, Fan, Gong, Hu, & Smith, 2010) and (Hell, Johansson, & Meier, 2007).

However, the use of lightweight encryption algorithms poses greater security risk as compared to stronger encryption algorithms such as the Advance Encryption Standard (AES, also known as Rijndael) (Technology, 2001). Several researchers have proposed low complexity hardware architectures for the AES algorithm (Good & Benaissa, 2006; Rouvroy, Standaert, Quisquater, & Legat, 2004). In (Good & Benaissa, 2006), a very small FPGA Processor for AES has been proposed by Tim Good and Mohammed Benaissa. Their low complexity ASIP structure and compact hardware area gives greater advantage of implementing a stronger cipher on a smaller hardware area. One of designer for AES, Vincent Rijnmen, together with Paulo S. L. Barreto has come up with a new cryptographic primitive design called the Anubis. Anubis is a Rijndael variant that uses involutions for the various operations. The involution nature of Anubis allows low-cost hardware and compact software implementations to use the same operations for both encryption and decryption. Both the Substitution box and the mix columns operations are involutions. Anubis is expected to behave as good as can be expected from a block cipher with the given block and key length. Cipher strength-wise; this implies that the most efficient key-recovery attack for Anubis is exhaustive key search. Otherwise, this cipher is considered secure.

After further analysis, the fundamental operation used in Anubis is XOR, which is very similar to AES. Base on these justifications, we are inspired to build a processing engine, with custom-designed ALUs to accommodate the Anubis' operations. It is reported the Anubis is secured and a variant of AES. Given the nature of Anubis with its involution structure, it is a very good candidate to be considered in-line with the lightweight ciphers for small foot-print implementations. The involution structure makes the Anubis a self-inversing cipher, where no decryption circuit or a separate decryption program is needed. This benefits the design's point as the program size and the circuitry is predicted to be significantly smaller using Anubis. With the added advantage of reports stating there is no known attacks on Anubis other than exhaustive key search, which is brute force attack; this cipher is assumed to be on par with the famous AES in term of security strength since it's a variant of AES. Increasing key size would further increase its strength, so this is a reasonable option as a lightweight cipher for RFID and WISP applications.

In this chapter, we present a very small-area footprint computer architecture and its hardware implementation for WISP using the Anubis block cipher algorithm. To meet the cost and area constraint requirements in WISP, the proposed implementation makes use of a minimal instruction set computer (MISC) processor architecture with custom-designed hardware modules to perform the Anubis encryption. The MISC architecture is an extension of the One Instruction Set Computer (OISC) and the Ultimate Reduced Instruction

Set Computer (URISC) architectures proposed in (Frenger, 2000; Gilreath & Laplante, 2003; Parhami, 1987). The proposed MISC Anubis architecture uses only a single block RAM and a very small amount of chip area, making it suitable for implementation in RFID tags and WISP. Firstly in section 1, we will discuss about some background on the Anubis cipher, then some knowledge on computer architectures and moving to a brief history and details on WISP. In section 2, we will present the MISC Anubis as a solution for the versatile WISP on security applications and the last section is the conclusion.

BACKGROUND AND LITERATURE REVIEW

Review of Light-Weight Ciphers for RFID and WISP

Low-end devices such as RFID and WISP tags are used in many environmental applications. This affects the steep requirement to provide security (and privacy) on to the system. There most common problem for providing secure primitives in these devices is the extremely constrained environment. The cryptographic primitive has to be in a small footprint with reduced power consumption and with sufficient processing speed. In order to fulfill these stringent requirements, secure block ciphers have to be developed.

There are a lot of lightweight ciphers proposals in the recent years. For example: the KTANTAN cipher proposed in (Canni, re, Dunkelman, Kne, & evi, 2009). The authors proposed a new family of very hardware-efficient and oriented block ciphers, which shares an 80-bit key. The cipher, KATAN, is composed of three different block sizes, with 32, 48, or 64-bit block size and the second cipher, KTANTAN, contains the other three ciphers with the same block sizes. From the article, the author stated that the secret keys in 'hard-coded' or 'burnt' into the device, which means that the device is having their own non-renewable master key and highly dependent on the secrecy of this key. As for implementation results, the KTANTAN cipher has been implemented in 462 GE (Gate Equivalent) while achieving encryption speed of 12.5 KBit/sec (at 100 KHz). And the KTANTAN48, which is the version that is recommend for implementation on RFID tags uses 588 GE. The KATAN64, which is the largest cipher among the candidates of the family, uses 1054 GE with a recorded throughput of 25.1 Kbit/sec (at 100 KHz).

The structure of the KATAN and the KTAN-TAN ciphers is simple. The plaintext is first loaded into two registers with the size of the cipher block size. In each round, several bits are taken from the registers and enter two nonlinear Boolean functions. The output of the Boolean functions is loaded to the least significant bits of the registers, after they were shifted. To ensure sufficient mixing is executed, 254 rounds of the cipher are executed. The authors have devised several mechanisms used to ensure the security of the cipher, while maintaining a small design foot print. The first cipher they presented is the use of an LFSR instead of a counter for counting the rounds and to stop the encryption after 254 rounds. As there are only 254 rounds, the authors proposed an 8-bit LFSR with a sparse polynomial feedback can be used. The LFSR is initialized with some state, and the cipher has to stop running the moment the LFSR arrives to some predetermined state. In (Canni, et al., 2009), the author has implemented the 8-bit LFSR counter with a result of gate equivalent of 60 gates while using an 8-bit counter that took 80 gate equivalents. It is expected that the speed and critical path for the LFSR is shorter than the one for the 8-bit counter and another advantage for using LFSR is the that when one of the bits taken from it, the sequence is expected to remain random with a uniform number of 1's and 0's in alternative way.

Like any other block ciphers, the security of the KATAN and KTANTAN depends on the secrecy of the encryption key. One of the problems that may arise in such a simple cipher construction such as KATAN is related to self-similarity attacks such as the slide attacks. For example, in (Andrey Bogdanov, 2008) where the analysis of a linear slide attack done on Keeloq concluded that it is because of the key is used repetitively, making it susceptible to several slide attacks. From the first glimpse, it shows that the KTANTAN proposal is weak against such attack due to the hard-coded keys. The authors in (Canni, et al., 2009) designed the cipher in a way to have the key loaded into an LFSR with a primitive feedback polynomial. This solution helps the KATAN family to achieve security against the slide attack. And of course the design nature of the KTANTAN of having the hardcoded keys is making it less favorable. The only solution is thru key generation. Thus, the only means to prevent a slide attack is by generating a simple, non-repetitive sequence of bits from the master key. The author has same up with an irregular key generation method using the "round counter" LFSR, which produces easily, computed bits, which at the same time follow a non-repetitive sequence, together with a third customized block which is used to prevent the self-similarity attacks and increases the diffusion-confusion of the cipher. In overview, the ciphers actually have two distinct round functions. The choice of the round function is made according to the most significant bit of the round-counting LFSR. This irregular update also increases the diffusion of the cipher, as the nonlinear update affects both the differential and the linear properties of the cipher. In the end, the only difference between a KATAN cipher and KTANTAN is the way the key is stored and the sub keys are derived, making it possible to design a very compact circuit to support all six variants with various key lengths.

On the other hand, we have the ultra light weight cipher, PRESENT (A. Bogdanov, et al.,

2007) However the author stated, despite the research on implementation advances, the AES is not suitable for extremely constrained environments such as RFID tags and sensor networks. The author presented an ultra-lightweight block cipher namely, PRESENT. It is reported that the hardware requirements for PRESENT are competitive with today's leading compact stream ciphers, with the design of the cipher and at 1570 GE. Another more dedicated implementation of PRESENT in 0.35 µm CMOS technology reported to utilize 1000 GE (Rolfes, Poschmann, Leander, & Paar, 2008) with the same design in 0.25 µm and 0.18 µm CMOS technology consumes 1169 and 1075 GE, respectively. PRESENT is an example of an SP-network and consists of 31 rounds. The block length is 64 bits and two key lengths of 80 and 128 bits are supported. The 80-bit key PRESENT is more than adequate security for the low-security applications typically required in tag-based deployments. In each of the 31 PRESENT rounds consists of an XOR operation to introduce a round key K_i for $1 \leq i \leq 32$, where K_{32} is used for post-whitening, a linear bitwise permutation and a non-linear substitution layer. The non-linear layer uses a single 4-bit S-box S which is applied 16 times in parallel in each round.

In (Xiao, Hu, & Kumar, 2007), a sensor based telecardiology system has been proposed by F. Hu, S. Kumar, and Y. Xiao. They have implemented a security scheme using the Skipjack Algorithm. Comparison has been made between Triple-DES, RC5 and Skipjack. The Triple-DES is too slow for software implementation in sensors and RC5 requires the key schedule to be pre-computed, which uses 104 extra bytes of RAM per key. In (Woo Kwon, Hwaseong, Yong Ho, & Dong Hoon, 2008), HIGHT, RC5 and Skipjack were evaluated for which has the greater advantage of implementing a suitable cipher on a smaller hardware area. On the other hand, in (Law, Doumen, & Hartel, 2006), comparisons were made to KASUMI, Camelia, Twofish, RC5, RC6,

Rijndael and MISTY1. Skipjack is the winner of all the categories compared. Skipjack produces the shortest expanded key, uses the least code and data memory, appeared to be slightly less energy-efficient as compared to Rijndael when size-optimized, and it is the most energy-efficient cipher when speed-optimized. With this information, we understand that Skipjack is a considerable candidate for lightweight ciphers because of its area of publications shows that it falls into the category of sensor network implementations, which points back to the small-footprint ciphers area of research.

Skipjack is a block cipher developed by the U.S. National Security Agency (NSA). It was first proposed as the encryption algorithm in a U.S. government-sponsored scheme of key escrow and used only for encryption. The design was initially secret but was later declassified on June 23rd, 1998. Skipjack is a 64-bit cipher that utilizes an 80-bit crypto variable. It encrypts and decrypts 4-words (8-bytes) of data blocks, alternating between two sets of stepping rules. A Skipjack with full 32 rounds proceed with applying 8 rounds of Rule A and 8 rounds of Rule B and a repetition of both rules (Rule A and Rule B in precedence) to the plaintext and accumulating a total of 32 rounds. In detail, the algorithm takes in the input w_{ik}, $1 \leq i \leq 4$, (i.e., $k = 0$ for the beginning step). The counter starts at 1 and step according to Rule A for 8 steps and switch to Rule B for 8 steps. The algorithm returns to Rule A for 8 steps and finally proceed to Rule B for 8 steps. The counter increments by one after each step are executed. The final output is $wi32$, $1 \leq i \leq 4$.

In some related previous work, the author from (Chae, 2007) used a RC5 cipher for providing a back-end encryption on the WISP tag. In (Oren & Feldhofer, 2009), the author presents a full-fledge public-key identification with 4682 gate equivalents. This author's work focuses on providing an encryption for the tag to secure its payload ID to an authorized reader. Whereas in

(Tae Youn, Ji Young, & Dong Hoon, 2008) the author describes the RFID tag search system in detail and the author suggested using AES (Feldhofer & Wolkerstorfer, 2007) to protect the ID by comparing with other algorithms such as the ECC and SHA-256. On the other hand, the author in (Vaudenay, 2006) uses the XOR cryptography and an LFSR (Linear Feedback Shift Register) random generator. However the author did not provide the implementation results.

From the findings above, we do understand that there are a lot of choices out there for small foot print design. The cipher strength has to be put into consideration when it concerns the overall security and protection under the wing of the chosen cipher. Secondly, the cipher complexity when coupled with its hardware designs. Memory components are always scarce and this drives the design trend. System throughput is an issue in applications that requires real-time high speed processing, so the design decision revolves between hardware occupancy and system throughput.

A Brief Review of AES (Advanced Encryption Standard)

The AES is a symmetric block cipher that uses data blocks of 128 bits as input and using cipher keys with lengths of 128, 192 or 256 bits. The basic unit for processing in the AES algorithm is a byte and the input, output and cipher key bit sequences are processed as 'array of bytes', which is a state array storing the input bytes in the AES algorithm. The state array is a two-dimensional array of bytes. It consists of 4 rows of bytes. Each row contains N_b bytes where N_b is the block length (128 bits) divided by 32. The input plain text is stored in the state array at the start of the encryption process and it is executed over a number of rounds determined by the key size. The round function can be broken down into four different byte-oriented transformations: Sub Bytes, Shift Rows, Mix Columns and Add Round Key (Technology, 2001).

For a complete AES encryption, a total of 10 rounds (Nr = 10) of substitution and permutation are required. All four byte-oriented transformation together with a 128 cipher key are applied onto the input plain text. For each round, another unique cipher key will be generated by the Key Expansion algorithm. In every round, the Round Key will be added to the plain text before going through rounds of S-Box Substitution (Sub Bytes), Mix Columns, and Shift Rows.

Shift Rows Transformation

In the Shift Rows transformation, it involves a cyclic shifting of the bytes in the last three rows with a different number of byte offset. The first row is not shifted. The bytes in the second row (r=1) are cyclically shifted left by one byte and the bytes of the subsequent rows are shifted by 2 and 3 bytes. For $0 \leq r < 4$ and $0 \leq c < Nb$, the Shift Rows Transformation can be expressed as: S'r,c = Sr,c + shift (r, Nb) mod Nb, where the shift value shift(r, Nb) depends on the row number r.

Mix Columns Transformation

The Mix Columns transformation operates on the each columns of the State Array. By treating each column as a four-term polynomial, they are considered as polynomials over GF(28) and multiplied modulo $x4 + 1$ with a fixed polynomial a(x), given by a(x) = {03}x3 + {01}x2 + {01}x + {02}.

Sub Bytes Transformation

The Sub Bytes transformation implements a non-linear cipher. Each byte, S r,c of the state is substituted with another byte, S'r,c independently, using a substitution table called the S-Box.

Add Round Key Transformation

During each round of the encryption process, a unique key has to be added to the current state. Each Round Key consists of Nb words from the key schedule of the current round, generated by the Key Expansion algorithm. Those Nb words are each added into the columns of the state, such that, [S0,c, S1,c, S2,c, S3,c] = [S'0,c, S'1,c, S'2,c, S'3,c] XOR [wround * Nb+c] for $0 \leq c < Nb$. The first Round Key addition occurs when round = 0. The subsequent rounds are added with Round Keys generated by the Key Expansion algorithm.

Key Expansion Algorithm

The Key Expansion algorithm generates unique cipher keys for each round of the Add Round Key Transformation. The initial round key is filled by an initial cipher key. The key expansion generates a total of Nb (Nr + 1) words and the resulting key schedule consists of a linear array of 4-byte words, denoted [wi], with i in the range $0 \leq i <$ Nb(Nr + 1).

The RotWord() takes in a word of four-bytes as input, performs a permutation, and returns the cyclic shifted word, similarly to a Shift Row transformation. The SubWord() function that, which is a Sub Bytes transformation is applied to each of the bytes to produce a substituted output word. The Round Constant, Rcon[i], holds the values given by [xi-1, {00}, {00}, {00}], with xi-1 being powers of x (x is denoted as {02}) in the field GF (2^8). Note that 'i' starts at 1, not 0. It can be seen from Figure 10, that the first Nk words of the expanded key are filled with the cipher key. Every following word, w[i], equals to the XOR of the previous word, w[i-1].

For words in positions that are a multiple of Nk, a transformation is applied to w[i-1] prior to the XOR, followed by an XOR with a round constant, Rcon[i]. This transformation consists of a Shift Row of the bytes in a word (RotWord()), followed by the application of a byte substitution onto all four bytes of the word (SubWord()).

Anubis (Variant of AES)

As we have briefly introduced AES in the previous section, there will be a lot of structural similarities observed as we go on presenting Anubis in much detail. An 'involution' is an operation whose inverse is the same as the forward operation. In other words, when an involution is run twice, it is the same as performing a 'no operation'. This allows low-cost hardware and compact software implementations to use the same operations for both encryption and decryption. The Anubis (Surhone, Tennoe, & Henssonow, 2010) is a Rijndael variant that uses 'involutions' for the various operations. It is a block cipher designed by Vincent Rijmen and Paulo S. L. M. Barreto as a candidate for the the the NESSIE project.

The Anubis is able to operate on data blocks of 128 bits, accepting various keys of length 32N-bits (N varies from 4 to 10). The number of rounds is dependent on the key size. Typically, 12 rounds for the 128-bit keys and plus one extra round for each additional 32 key bits. Each round consists of sixteen parallel substitution lookup tables, a linear transformation (matrix transposition followed by multiplication by a constant MDS diffusion matrix) and the round key addition. The Anubis is not Feistal cipher (Barreto, 2001a, 2001b; Biryukov, 2003; Raddum, 2002). It is designed in a way that all the round transformation components to be involutions. The S-box and the diffusion matrix of Anubis were chosen in a way which guarantees that encryption and decryption are the same operation except in the round sub-keys, which has to be inverted for the decryption round. This involution property allows the reduction of the chip area in hardware implementation and the code size for programs.

There are two versions of the Anubis cipher; the original implementation uses a pseudo-random S-box and subsequently, the S-box was modified to be more efficient to implement in hardware. The newer version of Anubis is called the "tweaked" version of Anubis. The cipher is not patented and has been released by the designers for free public use. The Anubis belongs to the same family of block ciphers as the ANUBIS winner algorithm, RIJNDAEL and their respective algorithmic differences are being sketched in the Table 1.

The Anubis was designed according to the Wide Trail Strategy (Barreto, 2001a; Surhone, et al., 2010). The round transformation in Anubis is composed of various invertible transformations with their own requirements and functionality. The cipher state is internally (similar to AES) viewed as a matrix in $M_{4x4}[GF(2^8)]$. The input block can be viewed as an array of bytes. The 128-bit data blocks and the 32N-bit keys must be mapped to the internal matrix format with the function μ, $GF(2^8)^{4N} \rightarrow M_{Nx4}[GF(2^8)]$ and it's inverse.

$$\mu(a) = b \Leftrightarrow b_{ij} = a_{4i+j}, 0 \le i \le N-1, 0 \le j \le 3 \quad (1)$$

In the following sections, we will define the constants and component mappings that build up the Anubis primitives.

The Transposition (τ)

The mapping τ is a matrix transposition operation and it is also an involution whether it is performed during the forward or inverse operations. This transformation is simply to transpose its argument in the state array:

$$\tau(a) = b \Leftrightarrow b_{ij} = a^t, b_{ij} = a_{ij}, 0 \le i, j \le 3 \quad (2)$$

The Linear Diffusion Layer (θ)

The diffusion mapping layer, θ is a linear mapping based on the (Barreto, 2001a; Chae, 2007; Daemen, Knudsen, & Rijmen, 1997) MDS code where H is symmetric and unitary.

$$\theta(a) = b \Leftrightarrow b = a \cdot H \quad (3)$$

Where,

$$H = \begin{bmatrix} 01 & 02 & 04 & 06 \\ 02 & 01 & 06 & 04 \\ 04 & 06 & 01 & 02 \\ 06 & 04 & 02 & 01 \end{bmatrix} \quad (4)$$

Table 1. The difference between AES and Anubis

	AES (Rijndael)	Anubis
GF(2^8) reduction polynomial	$x^8 + x^4 + x^3 + x + 1$ (0x11B)	$x^8 + x^4 + x^3 + x^2 + 1$ (0x11D)
Block size (bits)	128, 192, or 256	128
Key size (bits)	128, 192, or 256	128, 160, 192, 224, 256, 288, or 320
No. of rounds	10, 12, or 14	12, 13, 14, 15, 16, 17, or 18
Key schedule	dedicated *a priori* algorithm	key evolution (variant of the round function), plus key selection (projection)
Origin of the S-box	mapping $u \rightarrow u^{-1}$ over GF(2^8), plus affine transform	pseudo-random involution
Origin of the round constants	polynomials x^i over GF(2^8)	successive entries of the S-box

Figure 1. The 'tweaked' version of S-Box for Anubis with recursive structure

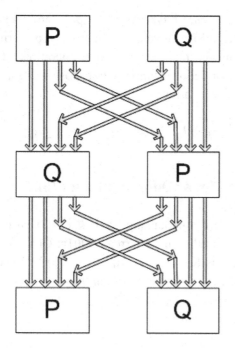

The Nonlinear Layer (γ)

The substitution box used in Anubis is a self-inverse S-box. In the first version of Anubis, the S-Box was generated in a pseudo-random manner and had no structure but this hinders the hardware implementation. In the 'tweaked' version of Anubis, an alternative S-Box is proposed in (Barreto, 2001b; Brumley, 2010) for efficient hardware implementation. The P and Q tables are pseudo-randomly generated involutions with similar design criteria as the S-Box. The recursive structure of Anubis S-Box with P & Q boxes are shown in Figure 1 and the lookup tables for the mini P and Q boxes are shown in Table 2.

$$\gamma(a) = b \Leftrightarrow b_{ij} = S[a_{ij}], 0 \le i \le N-1, 0 \le j \le 3 \tag{5}$$

Table 3 shows the complete Anubis boxes in Boolean form, which results to a hardware implementation of 108 gates (with 18 logic gates each for P & Q boxes) (Barreto, 2001b). For Figure 2 and 3, the schematic diagram of the P and Q boxes are shown respectively. The substitution box S is applied to all the bytes individually where:

The Key Addition (σ[k])

This transformation is a bitwise addition (XOR) of a key matrix $k \in M_{Nx4}\left[GF\left(2^8\right)\right]$ to the state array where:

$$\sigma[k](a) = b \Leftrightarrow b_{ij} = S[a_{ij}], 0 \le i \le N-1, 0 \le j \le 3 \tag{6}$$

Round Constants (c′)

The r-th round constant for r > 0 is a matrix c^r where:

$$c_{0j}^r = S[4(r-1) + j], 0 \le j \le 3,$$

$$c_{ij}^r = 0, 1 \le i \le N, 0 \le j \le 3, \tag{7}$$

Table 2. Mini-boxes P and Q

u	0	1	2	3	4	5	6	7	8	9	A	B	C	D	E	F
$P[u]$	3	F	E	0	5	4	B	C	D	A	9	6	7	8	2	1
$Q[u]$	9	E	5	6	A	2	3	C	F	0	4	D	7	B	1	8

Table 3. The Boolean description of the mini-P and Q boxes

$z = P[u]$	$z = Q[u]$
$t0 = u0 \oplus u1$	$t0 =\sim u0$
$t1 = u0 \oplus u3$	$t1 = u1 \oplus u2$
$t2 = u2 \ \& t1$	$t2 = u2 \ \& t0$
$t3 = u3 \ \& t1$	$t3 = u3 \oplus t2$
$t4 = t0 \ | t3$	$t4 = t1 \ \& t3$
$z3 = t2 \oplus t4$	$z0 = t0 \oplus t4$
$t5 =\sim t1$	$t5 = u0 \oplus u1$
$t6 = u1 \ \& u2$	$t6 = t1 \oplus t2$
$t7 = u3 \ | z3$	$t7 = t5 \oplus t3$
$t8 = t5 \oplus t6$	$t8 = t7 \ | t6$
$z0 = t7 \oplus t8$	$z2 = u2 \oplus t8$
$t9 = u2 \ \& t8$	$t9 = t8 \ \& u0$
$t10 = t6 \oplus u3$	$t10 = u3 \ \& t7$
$t11 = t10 \ | t9$	$t11 = t10 \oplus t2$
$z2 = t0 \oplus t11$	$z1 = t9 \oplus t11$
$t12 = t3 \oplus t9$	$t12 = u2 \ | z0$
$t13 = t8 \ | z3$	$t13 = t7 \oplus t11$
$z1 = t12 \oplus t13$	$z3 = t12 \oplus t13$

a. The Key Schedule

The Anubis key Schedule is a form of key expansion transformation and it is much more complex as compared to the 'Key Expansion' in AES. The cipher key $K \in GF(2^8)^{4n}, 4 \leq n \leq 10$ is expanded onto a sequence or round keys = $K^0, K^1, K^2, \ldots K^R$, with $K^r \in M_{4x4}[GF(2^8)]$.

The preparation of the key schedule is defined as the following:

Firstly, the mapping of the master key to the internal matrix format:

$$k^0 = \sigma[K] \qquad (8)$$

Secondly the expansion of the keys, also known as the r-th round key evolution:

$$k^r = \left(\sigma[c^r] \circ \theta \circ \gamma \circ \pi\right)\left(k^{r-1}\right), \ r > 0, \qquad (9)$$

Lastly, the after the key is expanded, the key selection function is applied:

$$K^r = \left(\tau \circ \omega \circ \gamma \circ\right)\left(k^r\right), \ 0 \leq r \leq R; \qquad (10)$$

The sub operations of the key selection are shown below:

The Cyclical Permutation (π)

The cyclical permutation shifts each column of its argument downwards by j positions.

$$\dot{A}(a) = b \Leftrightarrow b_{ij} = a_{(i-j) mod N, j}, 0 \leq i \leq N-1, 0 \leq j \leq 3 \qquad (11)$$

The Key Extraction (ω)

The key extraction function E is a linear mapping with the following matrix:

$$V = \begin{bmatrix} 1 & 1 & 1^2 & 1^3 \\ 1 & 2 & 2^2 & 2^3 \\ 1 & 6 & 6^2 & 6^3 \\ 1 & 8 & 8^2 & 8^3 \end{bmatrix} \qquad (12)$$

Given that:

$$\dot{E}(a) = b \Leftrightarrow b = V \cdot a , \ (\text{with } N = 4 - 1) \ (13)$$

Figure 2. The logical expression of P-box

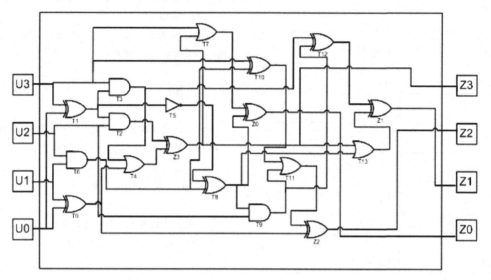

Figure 3. The logical expression of Q-box

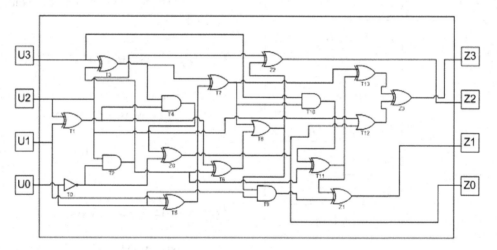

Ultimate Reduced Instruction Set Computer (URISC)

The effect of electronic growth rate in the 21th century has been triple-folding the computing capability available to users. For many applications, the highest-performance micro-processors of today is far more superior than the supercomputer of less than 10 years ago. At the abstract level, basic computer architecture usually consists of the following sub-systems: a set of input-output devices (I/O), central memory storage and a central processing unit (CPU). Usually the CPU has many specialized registers and an internal micro-memory. Computer architecture is usually a detailed specification of the computational, communication, and data storage elements (in terms of hardware) of a computer system. The architecture of a computer organization usually determines which computation operations can be performed and what forms of data and program organization will perform efficiently and optimally.

In the 1970s and 1980s, computer architects had put more focused on the instruction set. During those times, statisticians often had to be skilled FORTRAN programmers for their jobs. In other words, many of them were very familiar with the assembly language for a particular computer that they wrote sub-programs directly using the computer's basic instruction set. Old machines like the Intel 80x86 families of processors and the Motorola 680x0 processors all had multiple addressing modes, variable-length instructions, and large instruction sets.

As these computer technology advances by the middle of the 1980s, such machines were described as complex instruction set computers (also known as CISC). One of the greater advantages of such computer architectures is that each of their instruction performs specific task efficiently for specific computing requirements. This has made computer performance to be fine-tuned on large tasks with various characteristics. These architectures did have the advantage that each instruction/addressing-mode combination performed its special task efficiently, making it possible to fine-tune performance on large tasks with very different characteristics and computing requirements.

When we use the term 'instruction set', it is a set of commands that constitutes the language that describes a computer's functionality and operations. Instruction sets in a processor are akin to the functions in procedural programming language, in that both take parameters and return a result. Most instruction sets make references to memory locations, registers or pointers to a memory location. The referenced memory locations will eventually contain the processed data, which will be used again to produce new data. Hence, any computer processor can be viewed as a machine for taking data and undergo transformation and operations on it to become new data, through instructions.

In 1980, David Patterson and Ditzel (Patterson & Ditzel, 1980) proposed the RISC (abbreviated as the Reduced Instruction Set Computer) and it is an efficient idea to design a smaller set of instructions, which make implementation of these most frequent performed tasks maximally efficient. The most common features of RISC architecture designs are a single instruction size, a small number of addressing modes, and no indirect addressing. These changes made it possible to develop successfully a new set of architectures with simpler instructions. The RISC-based machines focused the attention of designers on two critical performance techniques, the exploitation of instruction-level parallelism (initially through pipelining and later through multiple instruction issue) and the use of caches (initially in simple forms and later using more sophisticated organizations and optimizations).

The simplest version of computer architecture has to be the one instruction set computer (abbreviated as OISC sometimes also referred as the URISC or the ultimate reduced instruction set computer). The single instruction set computer is the penultimate reduced instruction set computer (Gilreath & Laplante, 2003). The idea of URISC is the opposite of a CISC, which incorporates many complex instructions as micro-programs within the processor. OISC is very simple, by examining its features and properties; we can gain greater insight than of RISC and CISC because such complex and varied instruction sets has a more fundamental and simplest features hidden. The URISC which was first proposed by (Parhami, 1987) is meant for educational purpose. It has been an inspiration and insight to the CISC (Complex Instruction Set Computer) and RISC (Reduced Instruction Set Computer). This simplified model of computer architecture is flexible with only a single instruction incorporated can be further expanded and implemented on hardware easily. The URISC uses only one instruction called the SBN instruction (Subtract and Branch If Negative). By using only the SBN instruction, the URISC is able to perform data addition and subtraction. Logical operations can be performed to execute data movement from one location to another.

The URISC consists of an Adder circuit as its sole ALU. Detailed operation of the URISC can be found in (Parhami, 1987). Figure 5 shows the schematic of the URISC architecture.

The 'Subtract and Branch if Negative' (SBN) processor was first proposed by Van der Poel (Gilreath & Laplante, 2003). With this primitive SBN instruction, the URISC is built from its basic processor. The basic operations of URISC are moving operands to and from the memory, with addresses corresponding to the registers. The arithmetic computation can be performed and the results are stored in the 2nd operand's memory location. Similarly, to execute URISC instructions, the Core subtracts the 1st operand from the 2nd operand, storing the results in the 2nd operand's memory location. If the subtraction results a negative value, it will 'jump' to the target address, else, it proceeds to execute the next instruction in the following sequence. Figure 5 shows the pseudo-code format of the SBN instruction written in programming.

In this paper, we use URISC as a platform and further expand it into a customized architecture called the MISC (minimal instruction set computer). The details of this expanded version of processing engine are further discussed in latter sections.

Wireless Identification and Sensing Platform (WISP)

Radio frequency identification (RFID) systems typically consist of small low-cost battery-free devices, called tags, which use the radio signal from a specialized RFID reader for energy harvesting and communication. The RFID tags are traditionally used as replacement for barcodes in such applications as supply chain monitoring, asset management, and building security (Buettner, et al., 2008; Peris-Lopez, et al., 2009). When the tags are queried, each tag responds to a unique identification number by reflecting energy back to the reader through backscatter modulation. The tags are often application-specific fixed-function devices that have a range of 10–50 cm for inductively coupled devices and 3–10 m for UHF tags. The core advantage of this technology is that they are not constrained by batteries and have potentially years of life span. Conventional applications are also benefiting from sensor-enhanced RFID tags. To date, there are several

Figure 4. The URISC architecture

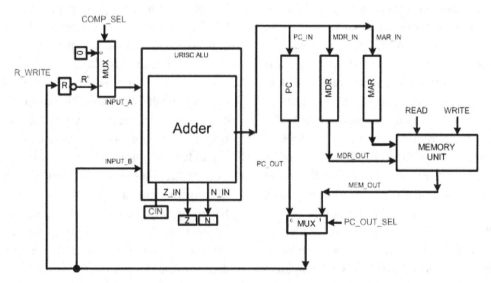

Figure 5. Pseudo-code format of the SBN instruction

```
SBN a,b,c ;
Mem[b] = Mem[b] - Mem[a];
if (Mem[b] < 0), goto c;
```

approaches for incorporating sensing capabilities into RFID. Active tags, which are a subclass of RFID tags, use batteries to power their communication circuitry, sensors, and microcontroller. Active tags can achieve high data and sensor activity rates from relatively long wireless range. However, the battery capacity, size and efficiency required by active tags are disadvantageous. In contrast, passive sensor tags receive all of their operating power from an RFID reader and are not limited by battery life. Some good examples of non-programmable passive tags with integrated temperature and light sensors, as well as an analog-to-digital converter (ADC) are shown in (Kocer & Flynn, 2006; Namjun, Seong-Jun, Sunyoung, Shiho, & Hoi-Jun, 2005). Their straight forward advantage is their suitability for applications in which neither batteries nor wired connections are feasible, for weight, volume, cost, or other reasons. Other source of power can be included such as solar or thermal can be used for passive tags but the only limitation for pure passive tags is the requirement of proximity to a RF reader.

Another problem for RFID tags is the configurability and computation limitation of the devices. Existing devices are generally □xed-function with respect to sensory inputs. They lack computational capabilities, limited by data path architecture. To overcome this problem, the WISP (also known as the Wireless Identification and Sensing Platform) is a sensing and computing device that is powered and read by off-the-shelf UHF RFID readers. WISPs have on-board microcontrollers that can be programmed to function as computing engine for security features, safe-guarding any information contained in the WISP.

WISP stands for Wireless Identification and Sensing Platform. The term "Identification" comes from "Radio Frequency Identification" (RFID). WISPs have similar capabilities of RFID tags, but also support sensing and computing. Like any passive RFID tags, WISP is powered and read by a standard off-the-shelf RFID reader, harvesting the power it uses from the reader's emitted radio signals. WISPs have been used to sense quantities such as light, temperature, acceleration, strain, liquid level, and to investigate embedded security. Most of the work on WISP so far has involved single WISPs performing sensing or computing functions. We think the next phase of WISP work will involve the interaction of many WISPs, and thus allow an exciting exploration of a new battery-free form of wireless sensor networking. Most people are familiar with RFID tags. Most common are passive RFID tags, where a battery-less IC device harvests power from a nearby RFID reader and uses it to respond to the reader with an identification number. Two broadly adopted standards for this technology are the Electronic Product Code (EPC) Class 1 Generation 1 and Class 1 Generation 2 standards, which operate in the Ultra High Frequency (UHF) bands. The standard is led by EPC Global.

According to (Buettner, Greenstein, Wetherall, & Smith), WISPs have the following features:

- Up to 10ft range with harvested RF power
- Ultra-low power MSP430 microcontroller
- 32K of program space, 8K of storage
- Light, temperature and 3D-accelerometers
- Backscatter communication to reader
- Reader to WISP communication (ASK)
- Real-time clock
- Storage capacitor (to sense without reader)
- Voltage sensor (measures stored charge)
- Extensible hardware (to add new sensors)
- HW UART & GPIO for external connections

- Works with select EPC Class 1 Gen 2 readers
- WISP software to sense and upload data
- Reader application to drive WISP
- Industry standard development tools
- Access to hardware design and source code

The objective to design tiny encryption module has always been a challenge. To meet the stringent cost and area design constraint requirement in the RFID tags, a small footprint hardware implementation for cryptography is highly desirable. Similar to any low-cost EPC Gen1 or Gen2 tags, the WISP provides a 16-bit ultralow-power microcontroller and thus, the ability to perform various form of computing operations, including sampling sensors, and reporting that sensor data back to the RFID reader.

With the nature of the WISP having resource constraint environment and environment adaptation requires on-field restructuring and reprogramming in software and hardware, design approaches often go towards low-area, low-cost and low complexity system designs. Sensing and data transfer mechanism usually occupies most of the area of the on-board FPGA microcontroller. This leads to a design for low complexity and low area system. In this chapter, the MISC is proposed to provide data security by fulfilling three criteria: low-area, low-complexity and re-configurable. The Anubis Cipher is proposed to be are implemented to the MISC processor to provided data security to the WISP. A pictorial illustration is shown in the Figure 6.

THE PROPOSED MINIMAL INSTRUCTION SET COMPUTER PROCESSOR FOR ANUBIS

In this section, we will discuss the implementation of the Anubis in detail. To come up with a minimalist design, the Anubis cipher has to be understood and efficiently implemented on the

Minimal Instruction Set Computer (MISC) processor. The MISC processor is a low-complexity computer architecture that provides a minimalist ideology of how a simple computer processor can be used to process and execute reduced instruction sets operations. One of the benefits of MISC processor is that it has only a minimum number of specific instructions that are sufficient to execute the specific encryption functions. Figure 7 shows the general MISC processor architecture implemented for the encryption system.

The MISC architecture can be broken down into several important design areas, from its ALU to its memory block. The ALU blocks can be seen and represented by alphabets A, B, and C in figure 7. Judging on what application the MISC has to be configured to function, the programmer himself can put in any ALU hardware blocks in any of the A, B, and C space. The MISC can also be configured to have 2 separate memory block for faster program execution, but in this chapter, we will on focus on the single memory architecture, which is also known as the von Neumann architecture. Similar to the SBN OISC, the MISC branching ability is inherited from the OISC. An N register is used to indicate a resultant negative is the outcome of a single instruction output. In other words, an SBN instruction can be used to branch to any memory item of the block RAM in the FPGA.

Figure 6. An illustration of WISP (Buettner, et al., 2008)

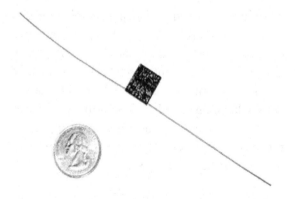

To differentiate what instruction the computer is processing, an op code is appended into the instruction sets. Note that the MISC instruction sets are 3-tuple, holding 3 bytes for each of the instructions. In figure 12, there are 2 registers, opcode1 and opcode2, both storing 1 bit of the appended op-codes (or function codes). With these values, the ALU output multiplexer will be triggered and will only store the intended data to the MDR register.

A Brief Introduction to MISC AES (Fundamental Building Block for MISC Anubis)

The MISC AES is a previous work that we have carried out and it has demonstrated satisfying results of hardware occupancy with competent results with other lightweight implementations. The results can be found in (Kong, Ang, Seng, & Adejo, 2010). In this section, we present the

Figure 7. General MISC processor proposed for Anubis encryption processing engine implementation in WISP

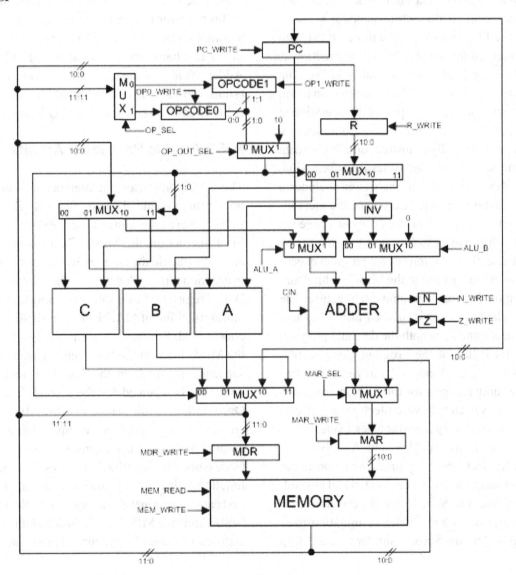

fundamental architecture that inspires the design and development of the MISC Anubis.

In order to perform AES computations onto the plain text, byte oriented transformations are studied and adapted from the AES encryption method. Instructions for transformations like Sub Bytes Transformation and Mix Columns Transformation have to be custom developed for the MISC AES Architecture. The four instructions are SBN (Subtract and Branch if Negative), XOR, xTime and Sub Bytes. The four MISC instructions are differentiated using the two MSBs of each of the instructions. The SBN instruction (Subtract and Branch if Negative) takes in two data values and subtracts them. If the output yields a negative value, the Program Counter will be added with the Target Address and a branch instruction is executed. The XOR operation takes in two data values and performs the XOR operation on them. The xTime instruction is a part of the operation in order to perform the complete Mix Column transformation. The Sub Bytes instruction takes in a data byte, and performs the Sub Bytes Transformation. In the MISC AES design, the design includes a 1024 x 10-bit Memory. The size of the memory varies on the demand of the designer if needed and once the memory is allocated, they have to be used efficiently as it is limited for program execution. The memory used in the MISC architecture is based on the Von Neumann Architecture. The total available memory is 1024 x 8-bit (512 bytes), which accommodates both the data and program codes. Each line of the Program Code section contains the Target Memory Addresses of the Data and the Jump Address for the SBN instructions. The memory used in this architecture can only be read or written at a particular clock cycle.

The similarity of AES and ANUBIS lies on the distinct grouping linear and non-linear blocks, making it easy to be divided and studied. For example, the Shift Rows is very similar to the Transposition and Cyclic Permutation used in Anubis. And the Sbox is similar to the P & Q box, and also the non-linear diffusion of Anubis is similar to the Mix Column executed in AES. And most importantly, the Key Addition in Anubis is a mimic of the Key Addition in AES, which are the exact same operations. Since the cryptographic operations of Anubis are very much similar to AES, therefore we are able to adapt this architecture for the Anubis implementation.

Minimal Instruction Set Computer Processor for Anubis

The MISC Anubis Processor together with 4 customized ALU for the Anubis cipher has the ability to perform any transformation in the Anubis algorithm. The MISC ALU that we present consists of 4 basic hardware blocks as the MISC ALU: Adder, XOR, xTimeAnu and Non-linear block (similar to the S-Box in AES, and in this case it is the tweaked s-box with P and Q boxes).

MISC Anubis Processor Architecture

The MISC only uses one memory unit to store both program and data for the ANUBIS. The MISC inherits the traits of the URISC. With the SBN instruction, the MISC is able to branch to any PC values in the memory unit and execute the instructions in any location of the memory unit. With 7 registers, 5 multiplexers, a single memory unit and 4 different ALU blocks, the MISC is being constructed. Similar to the structure of URISC, the MISC loads in the first memory address and subsequently loads in the first data item. This operation is repeated for the second data item. Once both data are loaded into the MISC, they are sent into the ALU for computation and the outputs will be chosen regarding to the Function Code embedded into the first address loaded. The function code is a 2-bit value, concatenated to the first data address in the memory unit. With the 2-bit MSB value, the MISC is able to determine which instruction is used for the current processor cycle and what data are stored back to the memory. The MISC Data Path is shown in Figure 8.

Most of Anubis' operations are byte-oriented. The MISC ANUBIS instructions are designed to be specific and comply only with the MISC ANUBIS Processor. Each instruction execution will result to byte-oriented processing onto the set of data addressed in the instruction's operand. Each instruction consists of three bytes: address of first operand, address of second operand, target address. Note that a 2-bit function code is used, occupying the 2-bit location of the first data address. The function codes are used to differentiate which instruction does the line of three bytes stands for. With this function code, the MISC is able to perform necessary operations in order to execute the correct command of data processing and storage.

A. MISC Anubis ALU

To break down the Anubis cryptographic primitives into instruction sets, firstly we would have to understand the respective functionalities of the Anubis processes. The SBN instruction is meant for the jumping instruction. The XOR instruction is meant for key addition and data location read-dressing. In the Anubis cipher, the linear diffusion and non linear layer is very similar to the Mix column and sub-bytes in the AES. The only difference is that the linear diffusion is an involution operation and the values of the matrix are different comparing with the mix column. The s-box in AES has the same size as the non-linear component in Anubis (8 bits in, 8 bits out). Since the Anubis is an involution cipher, the non-linear component for decryption is non-existent. The Anubis ALU that we present consists of the 4 main logic circuits: the Adder, XOR, xTimeAnu, and the non-linear block. From Figure 9, the AES Core Processor that consists of 4 main circuit blocks is shown. The Adder circuit takes in 2 inputs and adds them together. This Adder is essential in performing the adding operation of the SBN instruction. In the XOR Block, the circuit takes in 1 input and performs an XOR operation on to the data. The

Figure 8. The illustration of the architecture for MISC Anubis

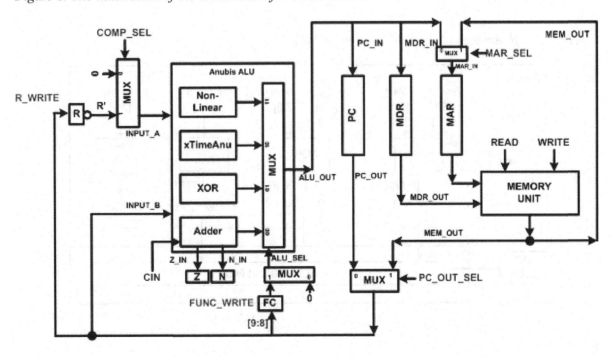

xTimeAnu Block is a part of the linear diffusion and key extraction transformation. As for the tweaked P and Q boxes, the block is represented as a single 'non-linear' block.

The circuit for the 'tweaked' P and Q non linear substitution box has been shown in section 1. Note that in (Good & Benaissa, 2006), the Xtime block used and designed was a reference to the $GF(2^8)$ reduction polynomial in AES. When we design this similar block for Anubis, the XOR points for the bit locations has to be re-routed. Figure 10 shows the redesigned xtime block that we call it the xTimeAnu.

B. Instruction Set

In order to perform Anubis computations on to the input plaintext such as, non-linear and XtimeAnu, a new series of instructions have to be developed. The MISC Instructions in the Table 4 are differentiated using the two MSB of each of the instructions. These instruction sets are the fundamental commands to the processor to operate accordingly to the Anubis algorithm. In the Table 4 below, we will show you the pseudo-code for the 4 customized instructions and their simplified formats.

From the pseudo-code in Figure 6, the SBN and XOR instruction takes in two input data items, whereas the xTime and Sub Bytes instruction takes in only one data item, whereas the Xtime-Anu and non-linear transforms the input data and maintains the data length of 8 bits.

C. The MISC Data Path Architecture

In the MISC Anubis architecture, 7 registers, 4 multiplexers (MUX), an Adder, an XOR Block, and xTimeAnu Block and a non-linear Block is used. At the top of the architecture design, the PC register stores the Program Counter (PC) value, which holds the subsequent program code in the memory to be read. The R register will store the first data read (Mem_A) from the memory. Memory Address Register (MAR) will stores the address of the memory, providing a reference to which memory location to be written or read. The Memory Data Register (MDR) is used to store the output computed by the Core ALU. The computation results can be either the results of the

Figure 9. The illustration of the Anubis processor ALU

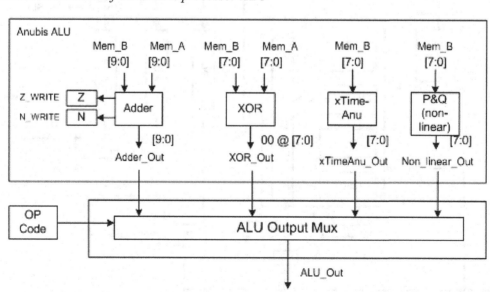

Adder, XOR, xTime or the Sub Bytes. The result will then be written back to the memory, replacing the value of the second data read (Mem_B). The Z and N registers indicate the arithmetic result performed by the Adder, is either a zero or a negative respectively. The Operation Code Register (OP) stores the Op Code of the current operation to be performed and it is used to select the output ALU. Figure 9 shows the data path of the architecture together with the custom ALUs for the MISC Anubis.

D. Memory Allocation

The advantage of MISC is that, all the components within this data path are reconfigurable, if the implementation method used is through FPGA. Since the WISP has a 32K program memory and an 8K data memory. It can be tested on a reconfigurable unit to verify that the Anubis design can be properly fitted into WISP. This architecture includes a 1024 x 10-bit Memory. The memory used in the MISC architecture is based on the Von Neumann Architecture. The total available

memory is 1024 x 10-bit (1280 bytes), which accommodates both the data and program codes. The data section lies at the address location of 0 to 127 (1024 bits), whereas the program section takes the location of 128 to 1024 (7168 bits). Each line of the Program Code section contains the Target Memory Addresses of the Data and the Jump Address for the SBN instructions. The memory used in this architecture can only be read or written at a particular clock cycle. Do note that, we have only tested Anubis with preloaded keys. The key evolution and key selection operations will require additional BRAMs.

E. Control Circuit

The Control Circuit is driven by a 4-bit Counter (C3C2C1C0). At each clock cycle, the control signals for particular control inputs are different. They are required to control the registers and store memory at any particular clock cycle.

During clock cycle 0, the value of the program counter (PC), is loaded into the MAR and at the same clock cycle, the Z register will be set accord-

Figure 10. The xTimeAnu circuit for the polynomial of $x^8 + x^4 + x^3 + x^2 + 1$ (0x11D)

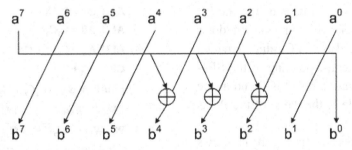

Table 4. Simplified instruction format with their respective operations

Operations	Function Code (2 MSB)	Instruction Format (3 bytes / instructions)
SBN	0 (00)	(00 @ Mem_A), (Mem_B), Target
XOR	1 (01)	(01 @ Mem_A), (Mem_B), Target
XTimeAnu	2 (10)	(10 @ 00000000), (Mem_B), Target
Non-linear (Tweaked)	3 (11)	(11 @ 00000000), (Mem_B), Target

ingly by the ALU Output to determine whether the PC has restart at 0x00. In the following clock cycles, the data is read from the memory location, addressed by the MAR which stored the value of PC. Subsequently, the read data is written back to MAR, storing the address of the data to be used for computation. These processes are repeated for a second data. The PC value will be increased by 1 after each data loading operation is done.

At clock cycle 1, the address for Mem_A is stored in the MAR, at the same time, the Op Code for the instruction is written to the OP Register. At clock cycle 2, the value of Mem_A is read and stored into the R register. At clock cycle 3, the incremented value of PC is stored in the MAR. At clock cycle 4, the address for Mem_B is stored in MAR. At the clock cycle 5, the value of Mem_B is read and sent to the ALU for computation. The Adder and the other hardware blocks will perform their individual operations from the two given inputs (Mem_A and Mem_B). At that particular clock cycle, depending on the value of the OP Register, the desired output will be chosen via an ALU MUX. At clock cycle 6, the output from the ALU is sent to the MDR Register for storage. With the arithmetic operations performed, clock cycle 7 will load the jump address from memory. Then the jump address will be added to the PC value at the same clock cycle. The jump address value will only be added to the PC value, provided that the OP value is corresponding to the SBN instruction and a negative result is found at the output. At clock cycle 8, the value of the PC is incremented.

The boolean expressions in Figure 11 shows the control signals to each bit of the 4-bit counter. If a branch occurs, the N register would have a value of 0. So, the PC register would just would take in take in the value of the jump address and increase by 1. Then the following instruction in the written program code will be performed.

A total of 9 clock cycles are required to perform one instruction written in the program code section. The control signals are produced by a com-

binational logic circuit during each clock cycles. The whole 9 clock cycles will repeat itself for until then end of the program reached. The 4-bit counter will restart once it reaches the value $C = 8$.

IMPLEMENTATION RESULTS AND DISCUSSIONS

In this section, we are presenting the result for MISC ANUBIS. The MISC ANUBIS Processor is designed and tested using the DK Design Suite software environment, which provides a Handel-C Hardware Descriptive language to ease the design process. A Celoxica RC10 board which houses the Spartan 3 XCS1500L-4 FPGA is used and on-board LEDs are used to observe the data memory items. To justify our proposed solution, we tried to make our program as small as possible. For the design simplicity, we didn't take account

Figure 11. The control signals to each bit of the 4-bit counter

$$\text{ALU\_B0} = \overline{C_2 C_0} + \overline{C_1} C_0$$

$$\text{ALU\_B1} = C_2 \overline{C_0}$$

$$\text{ALU\_A0} = C_1 C_0$$

$$\text{ALU\_A1} = \overline{C_2 C_0} + \overline{C_1} C_0$$

$$\text{CIN} = C_1 + C_0$$

$$\text{PMAR\_Write} = C_1 \overline{C_0} + \overline{C_2 C_1}$$

$$\text{DMAR\_Write} = C_1 \overline{C_0} + \overline{C_2} C_1 C_0$$

$$\text{PC\_Write} = \overline{C_2} \overline{C_1} C_0 + \overline{C_2} C_1 \overline{C_0} + C_2 \overline{C_1} \overline{C_0} N$$

$$\text{PMem\_Read} = C_2 \overline{C_0} + C_1 \overline{C_0} + \overline{C_2 C_1} C_0$$

$$\text{DMem\_Read} = C_1$$

$$\text{DMem\_Write} = C_2 \overline{C_0}$$

$$\text{R\_Write} = C_1 \overline{C_0}$$

$$\text{Z\_Write} = \overline{C_2} C_1 C_0$$

$$\text{N\_Write} = C_1 C_0$$

$$\text{MDR\_Write} = C_1 C_0$$

$$\text{Op\_Write} = \overline{C_2 C_1} C_0$$

$$\text{Op\_SEL} = C_1 C_0$$

of the memory required for the key expansion in ANUBIS and we assume that the key schedule is preloaded. As for the implementation results for MISC ANUBIS, it is shown in the Table 5 and the respective ALU component gate counts are shown in Table 6.

It is reported that WISP has scarce resources in terms of hardware and memory with 32K of program and 8K of data storage. With this, we can justify that the presented solution is taking up less resources that the WISP tag can offer. Although there is no direct comparison to our work as the basic unit to our design is the number of BRAMs, flip-flops and LUTs, but we can however confine our target to occupy as little resources as possible. In (Brumley, 2010), the author has presented the timing result in cycles per bytes. The implementation in (Brumley, 2010) is based on SIMD (single instruction multiple data) vectors. The machine used for benchmark-

ing is an Intel Core 2 Duo E8400 \Wolfdale" (45 nm) with 4GB of memory running Ubuntu 9.10, kernel 2.6.31-21, and GCC 4.4.1. The benchmark of comparison is none existent because what the author used is a fixed architecture with variable programming and the focus on instruction counts. Since we are dealing with a significantly small amount of program and data memory, our solution only requires a total of (1 x 12bits x 2k) 24kbits of memory.

Software based implementation of ciphers are usually not a benchmark for resource constrained devices as the design environment is totally different. We do believe that with efficient programming, the program memory occupancy can be significantly reduced. The Intel WISP uses a MSP430 microcontroller which is programmable. But micro-controller's architecture is fixed, not customized solely for security purposes as instruction sets are not optimized. Our solution has

Table 5. Implementation results for MISC ANUBIS

Components	Quantity	Total	Usage
No. of Slice Flip Flops	166	13,312	1%
No. of Occupied Slices	279	13,312	2%
Total No. of 4 Input LUTs	494	26,624	1%
No. of LUTs used a logic	438	495	~89%
No. of LUTs used a route-thru	55	495	~11%
No. of LUTs used a Shift Registers	1	495	~0%
No. of Bonded IOBs	44	221	20%
No. of BRAMs	3	32	9%
No. of GCLKs	4	8	50%
No. of DCMs	1	4	25%

Table 6. Gate counts on AES ALU components

ALU Block	AND	XOR	OR
Adder (10-bit)	20	20	10
XOR (8-bit)	-	8	-
xTimeAnu	-	3	-
Non-linear (P & Q boxes)	32	83 (4XNOR)	-

the ability to be configured by programmers for various optimizations. Optimization for codes size or design size is a good choice when it comes to have choices for efficient configurability. As for the implementation results from (Canni, et al., 2009), the KTANTAN cipher has been implemented in 462 GE while achieving encryption speed of 12.5 KBit/sec (at 100 KHz). And the KTANTAN48, which is the version that is recommend for implementation on RFID tags uses 588 GE. The KATAN64, which is the largest cipher among the candidates of the family, uses 1054 GE with a recorded throughput of 25.1 Kbit/sec (at 100 KHz). The implementation is done on CMOS on a logic process of 0.13um, with the size unit of gate equivalents. The platform we used to design MISC Anubis is on an FPGA with fundamental units of flip-flops, slices and LUTs.

In (Rouvroy, et al., 2004), the author has presented a set of comparisons to related works. By referring to Rouvroy et al, the comparison to our work is shown in the Table 7 below. Our version of AES has shown improvement as compared to the author's work. The comparison benchmark is to compare AES to AES and DES / 3DES to ANUBIS. The rational for comparison to this paper is due to the Rouvroy's aim to propose compact solutions for small embedded applications and the work covers AES and the weaker cipher, DES and 3DES. This is because our initial idea is to implement ANUBIS to be fitted as a level 1 or low-level security feature. DES and 3DES

has no direct comparison to ANUBIS in terms of cipher strength but our aim is to compare on the hardware level for ciphers deemed weaker than AES. From Table 7 below, we can observe that the Anubis has competent results are compared to DES and 3DES in terms for slices. The throughput is significantly lower but can be improved with efficient hardware design. Our justification is that, this version of Anubis can be implemented on a system that doesn't require quick real time processing or low latency systems with low clock rate, for minimal power consumption. There is another work from (Deshpande, Deshpande, & Kayatanavar, 2009) but there is no known data provided by the author to compare with.

Although there is no direct comparison to our work as the basic unit to our design, which is the number of BRAMs, flip-flops and LUTs, but we do hope that in future there will be related work to benchmark our findings.

On the other hand, the total instructions executed (excluding the key expansion for ANUBIS) are:

1. Transposition: (48 bytes / 3) * 10 = 160
2. Non-linear: (96 bytes / 3) * 10 = 320
3. Add key: (99 bytes / 3) * 10 = 330
4. Linear Diffusion: (633 bytes / 3) * 10 = 2000

This is amounted to the total of: 160 + 320 + 330 + 2000 = 2810 instructions. The total period for an ANUBIS encryption using MISC is 2810

Table 7. Comparison results with other Rouvroy et al's designs

	Rouvroy's AES (Rouvroy, et al., 2004)	Rouvroy's AES (Rouvroy, et al., 2004)	Rouvroy's DES (Rouvroy, et al., 2004)	Rouvroy's 3DES (Rouvroy, et al., 2004)	Our ANUBIS
Device	XC3S50-4	XC2S40-6	XC2S40-6	XC2S40-6	XCS1500L-4
Slices	163	146	189	227	166
Throughput (Mbps)	208	358	974	326	0.101
Block RAMs	3	3	0	0	3
Throughput/Area (Mbps/slices)	1.26	2.45	5.15	1.44	0.61kbps

x 9 cycles = 25290 cycles. The throughput for MISC ANUBIS is 101 kbps.

CONCLUSION

In this chapter, low complexity MISC design using Anubis cipher for WISP is presented. We have presented a processing engine to accommodate cipher operations with results showing minimal hardware utilization on an FPGA. With the small processing engine, a small and compact tweaked Anubis Processor can be realized. With a minimal amount of instructions used to perform encryption, the MISC Architecture implements on simple hardware components thus, reprogramming can be done by RFID readers and keys can be renewed or even other involution cipher can placed the existing ones in the WISP for added flexibility and configurability.

FUTURE RESEARCH DIRECTIONS

Some promising future research directions for strengthening the security mechanism for WISP can be: to find a unified processor for any block ciphers or other cryptographic primitives, improved reconfigurability for the microcontroller for application-specific design which will benefit the program size reduction with a more compact architecture, PKC (public key cryptography) for WISP and RFID devices, key management, key distribution and key renewal issues regarding RFIDs and an independent cryptographic platform in WISP specially for protecting the data integrity.

REFERENCES

Barreto, P. S. L. M. Rijmen. (2001a, 2008.11.19). *The Anubis block cipher*. Retrieved from http://www.larc.usp.br/~pbarreto/anubis.zip

Barreto, P. S. L. M. (2001b, 2008.11.19). *The Khazad legacy-level block cipher*. Retrieved from http://www.larc.usp.br/~pbarreto/khazad-tweak.zip

Biryukov, A. (2003). Analysis of involutional ciphers: Khazad and Anubis *Fast Software Encryption – 2003 Lecture Notes in Computer Science*, 4–5.

Bogdanov, A. (2008). Linear slide attacks on the KeeLoq block cipher . In Dingyi, P., Moti, Y., Dongdai, L., & Chuankun, W. (Eds.), *Information security and cryptology* (pp. 66–80). Springer-Verlag. doi:10.1007/978-3-540-79499-8_7

Bogdanov, A., Knudsen, L. R., Le, G., Paar, C., Poschmann, A., Robshaw, M. J. B., et al. (2007). PRESENT: An ultra-lightweight block cipher. *The Proceedings of CHES 2007*. Springer.

Brumley, B. B. (2010). *Secure and fast implementations of two involution ciphers*. IACR Eprint archive.

Buettner, M., Greenstein, B., Wetherall, D., & Smith, J. R. (2008). Demonstration: RFID sensor networks with the Intel WISP. *Proceedings of the 6th ACM Conference on Embedded Network Sensor Systems,* (pp. 393-394).

Canni, C., Dunkelman, O., & Knezevic, M. (2009). *KATAN and KTANTAN -- A family of small and efficient hardware-oriented block ciphers*. Paper presented at the 11th International Workshop on Cryptographic Hardware and Embedded Systems.

Chae, H.-J. A. Y., Daniel, J., Smith, J. R., & Fu, K. (2007, July). *Maximalist cryptography and computation on the WISP UHF RFID Tag*. Paper presented at the Conference on RFID Security.

Daemen, J., Knudsen, L. R., & Rijmen, V. (1997). *The block cipher square*. Paper presented at the 4th International Workshop on Fast Software Encryption.

Deshpande, A. M., Deshpande, M. S., & Kayatanavar, D. N. (2009, 4-6 June 2009). *FPGA implementation of AES encryption and decryption.* Paper presented at the International Conference on Control, Automation, Communication and Energy Conservation, INCACEC 2009.

Engels, D., Fan, X., Gong, G., Hu, H., & Smith, E. M. (2010). *Hummingbird: ultra-lightweight cryptography for resource-constrained devices.* Paper presented at the 14th International Conference on Financial Cryptograpy and Data Security.

Feldhofer, M., & Wolkerstorfer, J. (2007, 27-30 May 2007). *Strong crypto for RFID tags - A comparison of low-power hardware implementations.* Paper presented at the Circuits and Systems, 2007. ISCAS 2007. IEEE International Symposium on.

Frenger, P. (2000). The ultimate RISC: A zero-instruction computer. *SIGPLAN Notice, 35*(2), 17–24. doi:10.1145/345105.345111

Gilreath, W. F., & Laplante, P. A. (2003). *Computer architecture: A minimalist perspective.* Kluwer Academic Publishers.

Good, T., & Benaissa, M. (2006). Very small FPGA application-specific instruction processor for AES. *IEEE Transactions on Circuits and Systems. I, Regular Papers, 53*(7), 1477–1486. doi:10.1109/TCSI.2006.875179

Hell, M., Johansson, T., & Meier, W. (2007). Grain: A stream cipher for constrained environments. *International Journal of Wireless and Mobile Computing, 2*(1), 86–93. doi:10.1504/IJWMC.2007.013798

Israsena, P. (2006, 16-18 January). *Securing ubiquitous and low-cost RFID using tiny encryption algorithm.* Paper presented at the Wireless Pervasive Computing, 2006 1st International Symposium on.

Kocer, F., & Flynn, M. P. (2006). A new transponder architecture with on-chip ADC for long-range telemetry applications. *IEEE Journal of Solid-state Circuits, 41*(5), 1142–1148. doi:10.1109/JSSC.2006.872741

Kong, J. H., Ang, L. M., Seng, K. P., & Adejo, A. O. (2010, 5-8 December). *Minimal instruction set FPGA AES processor using Handel — C.* Paper presented at the 2010 International Conference on Computer Applications and Industrial Electronics (ICCAIE).

Law, Y. W., Doumen, J., & Hartel, P. (2006). Survey and benchmark of block ciphers for wireless sensor networks. *ACM Transactions in Sensor Networks, 2*(1), 65–93. doi:10.1145/1138127.1138130

Namjun, C., Seong-Jun, S., Sunyoung, K., Shiho, K., & Hoi-Jun, Y. (2005, 12-16 September). A 5.1-μW UHF RFID tag chip integrated with sensors for wireless environmental monitoring. *Proceedings of the 31st European Solid-State Circuits Conference,* ESSCIRC 2005.

Oren, Y., & Feldhofer, M. (2009). *A low-resource public-key identification scheme for RFID tags and sensor nodes.* Paper presented at Second ACM Conference on Wireless Network Security.

Parhami, F. M. A. B. (1987). *URISC: The ultimate reduced instruction set computer.* Department of Computer Science, University of Waterloo.

Patterson, D. A., & Ditzel, D. R. (1980). The case for the reduced instruction set computer. *SIGARCH Computer Architecture News, 8*(6), 25–33. doi:10.1145/641914.641917

Peris-Lopez, P., Hernandez-Castro, J. C., Tapiador, J. M., & Ribagorda, A. (2009). Advances in ultra-lightweight cryptography for low-cost RFID tags: Gossamer protocol . In Kyo-Il, C., Kiwook, S., & Moti, Y. (Eds.), *Information security applications* (pp. 56–68). Springer-Verlag. doi:10.1007/978-3-642-00306-6_5

Raddum, H. (2002). *The statistical evaluation of the NESSIE submission Anubis.*

Rolfes, C., Poschmann, A., Leander, G., & Paar, C. (2008). *Ultra-lightweight implementations for smart devices - Security for 1000 gate equivalents.* Paper presented at the 8th IFIP WG 8.8/11.2 International Conference on Smart Card Research and Advanced Applications.

Rouvroy, G., Standaert, F. X., Quisquater, J. J., & Legat, J. D. (2004, 5-7 April 2004). *Compact and efficient encryption/decryption module for FPGA implementation of the AES Rijndael very well suited for small embedded applications.* Paper presented at the International Conference on Information Technology: Coding and Computing, ITCC 2004.

Sample, A. P., Yeager, D. J., Powledge, P. S., Mamishev, A. V., & Smith, J. R. (2008). Design of an RFID-based battery-free programmable sensing platform. *IEEE Transactions on Instrumentation and Measurement, 57*(11), 2608–2615. doi:10.1109/TIM.2008.925019

Surhone, L. M., Tennoe, M. T., & Henssonow, S. F. (2010). *Anubis (Cipher).* VDM Verlag Dr. Mueller AG & Co. Kg.

Tae Youn, W., Ji Young, C., & Dong Hoon, L. (2008, 17-20 December). *Strong authentication protocol for secure RFID tag search without help of central database.* Paper presented at the IEEE/IFIP International Conference on Embedded and Ubiquitous Computing, EUC '08.

Technology, N. I. o. S. a. (2001). *Advanced encryption standard.*

Vaudenay, S. (2006). *RFID privacy based on public-key cryptography. ICISC 2006, LNCS* (pp. 1–6). Springer.

Woo Kwon, K., Hwaseong, L., Yong Ho, K., & Dong Hoon, L. (2008, 24-26 April 2008). *Implementation and analysis of new lightweight cryptographic algorithm suitable for wireless sensor networks.* Paper presented at the International Conference on Information Security and Assurance, ISA 2008.

Xiao, Y., Hu, F., & Kumar, S. (2007, Jan. 2007). *Towards a secure, RFID / sensor based telecardiology system.* Paper presented at the 4th IEEE Consumer Communications and Networking Conference, CCNC 2007.

ADDITIONAL READING

Batina, L., et al. (2006). *An elliptic curve processor suitable for RFID-tags.* Benelux Workshop Information and System Security (WISSec 06), 2006. Retrieved from http://eprint.iacr.org/2006/227.pdf

Canright, D., & Osvik, D. A. (2009). A more compact AES . In Rijmen, M. J. J. V. Jr., & Safavi-Naini, R. (Eds.), *Selected Areas in Cryptography* (*Vol. 5867*, pp. 157–169). Lecture Notes in Computer Science Springer. doi:10.1007/978-3-642-05445-7_10

Clavier, C., & Gaj, K. (Eds.). (2009). Cryptographic hardware and embedded systems. *Proceedings 11th International Workshop, Lecture Notes in Computer Science, Vol. 5747*, Lausanne, Switzerland, September 6-9, 2009.

Daemen, J., & Rijmen, V. (2002). *The design of Rijndael: AES--The advanced encryption standard.* Springer.

Daemen, J., Rijmen, V., & Barreto, P. S. L. M. (2002). *Rijndael: Beyond the AES.* Mikulasska Kryptobesdka 2002 - 3rd Czech and Slovak Cryptology Workshop, Dec. 2-3, 2002, Prague, Czech Republic.

Feldhofer, M., Wolkerstorfer, J., & Rijmen, V. (2005). AES implementation on a grain of sand . In *Proceedings on Information Security* (*Vol. 152*, pp. 13–20). IEEE Computer Society. doi:10.1049/ip-ifs:20055006

Gokhale, M., & Graham, P. S. (2005). *Reconfigurable computing: accelerating computation with field-programmable gate arrays*. Birkhäuser.

Hauck, S., & DeHon, A. (2008). *Reconfigurable computing: The theory and practice of FPGA-based computation*. Morgan Kaufmann.

Hung-Yu, C., & Chen-Wei, H. (2007). Security of ultra-lightweight RFID authentication protocols and its improvements. *SIGOPS Operation Systems Review*, *41*(4), 83–86. doi:10.1145/1278901.1278916

Juels, A. (2004). Minimalist cryptography for low-cost RFID tags. In *Proceedings of SCN'04, LNCS 3352*, (pp. 149–164).

Katz, J., & Lindell, Y. (2008). *Introduction to modern cryptography*. Boca Raton, FL: Chapman & Hall/CRC Taylor & Francis Group.

Mali, M., Novak, F., & Biasizzo, A. (2005). Hardware implementation of AES algorithm. *Journal of Electrical Engineering*, *56*(9-10), 265–269.

Oehler, R. R. (2003). *Reduced instruction set computer (RISC). Encyclopedia of computer science* (4th ed., pp. 1510–1511). Chichester, UK: John Wiley and Sons Ltd.

Parhami, B. (2005). *Computer architecture: From microprocessors to supercomputers* (pp. 151–153). Oxford University Press.

Philipose, M., Smith, J. R., Jiang, B., Sundara-Rajan, K., Mamishev, A., & Roy, S. (2005). Battery-free wireless identification and sensing. *IEEE Pervasive Computing / IEEE Computer Society and IEEE Communications Society*, *4*(1), 37–45. doi:10.1109/MPRV.2005.7

Poschmann, A., Leander, G., Schramm, K., & Paar, C. (2007). New light-weight crypto algorithms for RFID. In . *Proceedings of ISCAS*, *07*, 1843–1846.

Preneel, B. (2002). The NESSIE project: Towards new cryptographic algorithms. In *Information Security Applications, 3rd International Workshop, WISA 2002*, (pp. 16-33).

Sample, A. P., Yeager, D. J., Powledge, P. S., & Smith, J. R. (2007). *Design of a passively-powered, programmable sensing platform for UHF RFID systems*. IEEE International Conference on RFID 2007, March 26-28.

Satoh, A., Morioka, S., Takano, K., & Munetoh, S. (2001). A compact Rijndael hardware architecture with s-box optimization . In Boyd, C. (Ed.), *ASIACRYPT* (*Vol. 2248*, pp. 239–254). Lecture Notes in Computer Science. doi:10.1007/3-540-45682-1_15

Schneier, B. (1996). *Applied cryptography: Protocols, algorithms, and source code in C*. Wiley.

Smith, J. R., Sample, A. P., & Powledge, P. S. Roy, S., & Mamishev, A. (2006). A wirelessly-powered platform for sensing and computation. In *Proceedings of Ubicomp*, 2006, (pp. 495–506).

Uhsadel, L., Poschmann, A., & Paar, C. (2007). Enabling full-size public-key algorithms on 8-bit sensor nodes. *Proceedings of the 4th European Workshop Security and Privacy in Ad-hoc and Sensor Networks (ESAS 07), LNCS 4572*, (pp. 73-86). Springer-Verlag.

Van der Poel, W. (1952). A simple electronic digital computer. *Applied Scientific Research. B, Electrophysics, Acoustics, Optics, Mathematical Methods*, *2*, 367–400. doi:10.1007/BF02919783

Yeager, D., Prasad, R., Wetherall, D., Powledge, P., & Smith, J. (2008). *Wirelessly-charged UHF tags for sensor data collection*. IEEE International Conference on RFID 2008.

KEY TERMS AND DEFINITIONS

3-tuple: A sequence with 3 elements, in this case, 3 bytes

Branching Instruction: An instruction that changes the value of the Program Counter

Instruction Sets: A set of commands, understood by computer hardware

Involution: An involution is an operation whose inverse is the same as the forward operation.

Minimal Instruction Set Computer: A computer architecture with the compact and minimal number of instructions, sufficient to execute the desired operations.

Round Function: Iterated function execution

Xtime: An operation for multiplication by 2 over the $GF(2^8)$

Section 3
Applications

Chapter 7
Monitoring Sleep with WISP Tags

Enamul Hoque
University of Virginia, USA

Robert F. Dickerson
University of Virginia, USA

John A. Stankovic
University of Virginia, USA

ABSTRACT

This chapter presents a sleep monitoring system based on WISP tags. The authors show that their system accurately infers fine-grained body positions from accelerometer data collected from the WISP tags attached to the sides of a bed. Movements, duration, and bed entrances and exits are also detected by the system. The chapter presents the results of an empirical study from 10 subjects on three different mattresses in controlled experiments to show the accuracy of the inference algorithms. The authors also evaluate the accuracy of the movement detection and body position inference for six nights on one subject, and compare these results with two baseline systems. Preliminary data investigating the correlation between sleep stages from the Zeo and movement is also presented.

INTRODUCTION

RFID is an important technology that has already experienced great success in several different application areas. With the advent of adding sensing to RFIDs, as found in WISP tags, many new applications are possible. One promising area for WISP applicability is in smart homes. The tags may be used for applications designed to save

energy, automate homes, or to remotely monitor health. For monitoring many medical conditions, being able to assess the duration and quality of sleep plays an important role.

Because of its importance, many sleep-monitoring systems have been developed. These systems attempt to recognize sleeping disorders by providing healthcare providers with quantitative data about irregularity in sleeping periods and durations or the amount of agitation and restlessness experienced during the night. These

DOI: 10.4018/978-1-4666-1990-6.ch007

solutions vary in cost, comfort, and accuracy. In this chapter, we describe the main categories and characteristics of current solutions and then detail a new approach based on WISP tags.

The new system does not require any specific action from patients. In this system, we attach several WISP tags to the bed mattress and collect accelerometer data. Using the data we infer body positions, movements, and entries and exits from the bed. We compare the performance of our system with several baseline systems including using pressure pads, video, a popular iPhone based sleep monitoring application, and the Zeo.

BACKGROUND

To date, while there are many sleep monitoring systems there are very few low-cost, unobtrusive (comfortable) solutions. In this section we outline the major categories of solutions and describe their characteristics and limitations.

Physiological signals are regarded as the most accurate means to differentiate between awake and sleep phases such as light, REM, and deep sleep. The electroencephalogram (EEG) measures the frequency of brain waves to discern sleep and wake stages (Carskadon 1989). The electrooculogram (EOG) and electromyogram (EMG) are also standard technologies for sleep monitoring. The electrocardiogram (ECG) can be used to measure the heart rate, which is well known to decrease upon sleep onset. Some studies show that heart rate varies over different sleep stages (Redmond 2006, Shinar 2006) by use respiratory-derived features together with ECG-derived features for classifying different sleep stages automatically. These techniques have major limitations- they are costly since they require trained professionals in clinical environments to administer them and invasive since these techniques require equipment to be attached to patients, limiting their movement and causing discomfort. These physiological signals do not support monitoring body positions during sleep.

Temperature regulation in a body can also be used to monitor sleep quality. Skin temperature increases during sleep onset and decreases during wakeup (Krauchi 2004). But these temperature variations can only be measured under controlled laboratory conditions. (Yang 2006) uses an infra-red triangulation distance sensor to detect movements of different body parts without attaching any device to the body. But it does not provide any information about body position.

To overcome the limitations of the above techniques, there are many systems that enable sleep monitoring in home environments. Actigraphy (Sadeh 2002) is a commonly used technique for sleep monitoring that uses a watch-like accelerometer based device attached typically to the wrist. The device monitors activities and later labels periods of low activity as sleep. There are many commercial products like the Philips Actiwatch that are designed based on actigraphy. The Zeo is another commercial product for sleep monitoring in home environments. It is a headband that users need to wear each night so that it can detect sleep patterns through the electrical signals naturally produced by the brain. There is also an associated display that shows a person's sleep pattern for the previous night. These products are expensive and users need to wear the device.

Another method used for sleep monitoring is to instrument a mattress pad with sensors and passively infer body movements and sleep quality. The Bed Alarm Sensor Pad is such a commercial bed pressure-sensing pad that monitors change in body pressure on the pad to detect movements. In (Van der Loos 2001) the authors use pressure and temperature sensors laid out in a grid pattern in the mattress to determine quality of sleep. NAPS (Mack 2003, Mack 2006) is a low-cost physiological sensor-suite that can passively acquire important physiological and environmental characteristics. The NAPS suite allows subjects to simply lie on a mattress pad, embedded with vibration sensors, to obtain multidimensional data (e.g., body temperature, heart rate, respiration rate, positional mapping and movement). One might also use tiny sensor motes with accelerometers in place of vibration sensors. The main advantage of all these solutions is that users do not need to wear any device. But, in some cases batteries are needed and it may also be uncomfortable to sleep on a pad and thus, they can affect sleep quality. For patients with incontinence there is also a problem with keeping the pads clean. Also, pressure pads can only detect body movements and correlate them to sleep quality. In our WISP based solution, we can add additional sensors in the WISP tags to monitor other environmental parameters (e.g., temperature, light) that can be useful in analyzing the effect of external environment on sleep quality.

Audio and video signals can also be used to determine sleep quality. In (Peng 2006) a combination of heart rate, audio and video sensors is used to infer a sleep-awake condition. But such systems raise privacy concerns among the users.

SleepCycle is a popular iPhone based application that uses the accelerometer in the iPhone to monitor body movements and determine which sleep phase the user is in. The user just needs to put the iPhone in a suitable place on the bed. However, such iPhone based solutions can only monitor changes in accelerometer values in a certain place on the bed. In our experiments, we show that using WISP tags in three different places in the bed improves the accuracy of detecting body movements compared to using one/two WISP tags. Three WISP tags also enable us to monitor fine-grained body position. Moreover, the iPhone can accidentally fall off the bed and it needs to be connected to the charger for the whole night.

In summary, the advantages of our WISP-based sleep monitoring system are that users do not need to wear any device, they do not need to sleep on any mattress pads instrumented with sensors, no batteries are needed, the system is wireless, and it avoids privacy violations of video solutions.

WISP BASED SLEEP MONITORING SYSTEM

The Design

The sensing elements in our system are WISP tags shown in Figure 1. The device's antenna and power harvesting circuitry enable off the shelf EPC "Gen 2" RFID readers, shown in Figure 2 to power and read from it. To a RFID reader, a WISP appears as a normal RFID tag, but inside the WISP, the harvested energy is operating a 16-bit fully programmable ultra low-power microcontroller. The microcontroller can sample a variety of sensing devices including 3-dimensional accelerometers, lights, and temperature sensors. In our system, we only use 3D accelerometer readings. The WISP tags report these readings by encoding them as part of their identifiers that are read by a RFID reader.

We attach three WISP tags along the edge of the mattress. As an example of the accelerometer readings, the y-axis reading for Tag 1 is shown in Figure 3. Such simple information (from all three tags and all three dimensions) is used to differentiate between when the bed is empty, someone is lying on it, or someone is just sitting on the bed

Figure 1. WISP tag

Figure 2. Speedway reader

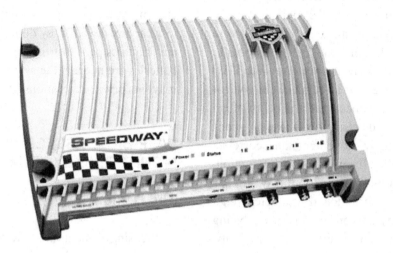

Figure 3. Accelerometer reading variation for empty/lying/sitting

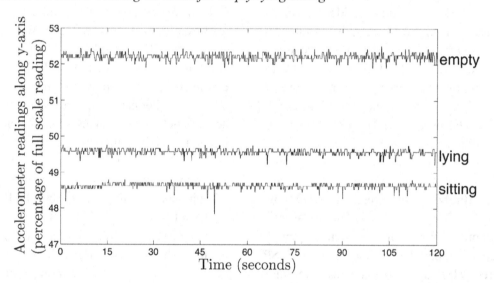

watching television or reading. When the bed is empty, the y-axis accelerometer of the tag is aligned perpendicular with respect to gravity, but when someone lays on the bed, because of the impact of the body on the mattress, the orientation changes. These orientations are different from the one when someone is just sitting on the bed.

Using the accelerometer readings, we distinguish four positions: lying on the back, stomach, left, or right sides (shown in Figure 4). In Figures 3 and 4 as an example, we show the accelerom-eter readings along the y-axis. Note that the readings along the z-axis(which is parallel to gravity) show similar variation. The readings along the x-axis do not show too much variation, but if we combine them with the readings along the y and z-axes, together they accurately differentiate among the four positions. The evaluation sections demonstrate the accuracy of this technique.

We are also interested in the amount of movement to detect restlessness and agitation. Each time someone moves on the bed, the accelerom-

Figure 4. Accelerometer reading variation for different lying positions

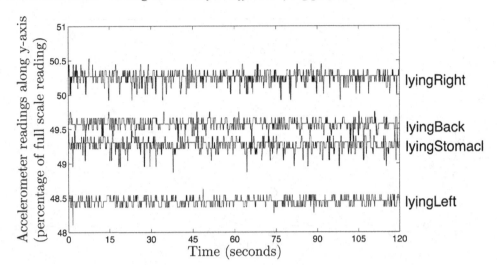

eter readings change rapidly. Using the change in these readings we detect each movement. The system monitors how many times a person tosses and turns during the course of the night and how many times the person leaves the bed. If someone is lying on the bed and does not move for a significant amount of time, then we assume the person is asleep. Frequency of movements is also different for different sleep stages and thus can be related to which sleep stage a person is currently in (Giganti 2008). Transitions between different sleep stages also correspond to change in frequency of body movements. Thus, based on a summary of movements made during each night, doctors can infer duration, quality of sleep and irregular sleeping patterns.

EVALUATION: CONTROLLED EXPERIMENTS

Our evaluation of using WISP for sleep monitoring consists of controlled experiments for body position and for movement as well as real overnight experiments. Note that some of these results and the associated discussion were previously reported in (Hoque 2010).

For our controlled experiments, 10 graduate students volunteered as subjects. The subject population was diverse in both height and weight (shown in Figure 5). All participants were volunteers, and were informed of the experimental procedures and the study's goals prior to participation. We conducted our experiments in three different beds to consider how different mattresses affect the measurements. Five subjects were evaluated on a twin-size bed in our University's medical testbed called Alarmnet (Wood 2008). Five other subjects participated in each of the other two beds that were in a graduate student's apartment. All three of them were twin mattresses. So, for each bed, five subjects participated in the experiments. For two of the three beds the participating subjects were the same.

For each experiment, we attached 3 WISP tags to the mattress of a bed. Figure 6 shows such a bed along with the positions of the tags. We placed the tags in such a way that when someone lies on the bed, there is one tag on each side of his body and one tag near the legs. We used two antennas for reading from the tags. The reader sends 10 read requests per second. The read rate from each of the tag was 4-7 reads per second during all our experiments. If we use one antenna, then read rate

Figure 5. Heights and weights of the subjects

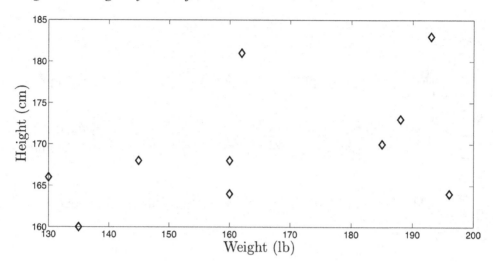

of one or more tag falls much lower. One disadvantage of using the WISP tags is that they need to be placed within 1-2 meters of the antenna of the reader. To meet the read-range requirement and to keep the equipment away from obstructing a resident's movement, we placed the antennas below the bed. The antennas were wired to the reader that was connected with the laptop.

Note that we also investigated the use of a fourth tag near the head. However, this tag did not improve the accuracy and so we eliminated it.

Figure 6. Experimental setup for the bed

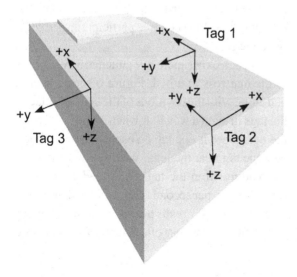

Each subject lay on the bed in the following four positions: on the back, on the stomach, on the left side and on the right side. These four positions are shown in Figure 7. Each subject also sat on the bed with his or her back on the wall and face towards the camera. This position resembles the way someone lies when watching television or reading a book while sitting in bed. For each position, we recorded data for two minutes. For each WISP tag, we obtained the acceleration along the x, y and z-axes. From the readings of all three tags, we get a 9-tuple. Note that, all three tags do not report their acceleration values synchronously. We combine the readings from the three tags within each second and construct each possible 9-tuple. We associate all the 9-tuples collected during these two minutes to that particular body position. We also recorded the readings from the tags when the bed was empty. We use the collected data to train our system.

After the training phase, the subject repeats the tasks again and we record data for 30 seconds for each position. Our system then classifies the new data based on previous training. For training and classification, we use the open source software "Orange Canvas" (Demsar 2004) which supports a number of classifiers. We decided to use the

Figure 7. Body positions while lying on the bed

| Left | Right | Stomach | Back |

Naive Bayesian Classifier. Note that for each subject, first we train our system based on the subject's training data and then classify his or her remaining data.

Controlled Experiments: Body Position

For each subject, we classify the collected data under three different settings. In the first setting, we test whether it is possible to differentiate between the bed being empty and someone lying on it (in any position). So we label all data collected during a subject lying on the bed in four different positions as lying. We do not include the data when the subject was sitting on the bed. In the second setting, we include the data for sitting and test whether it is possible to differentiate among the bed being empty, or someone lying on it (in any position), or someone sitting on it. In the last setting, we test whether it is possible to differentiate among all six cases: empty, lying on back, lying on stomach, lying on back, lying on left side, lying on right side and sitting. We name the above three cases as "set1", "set2" and "set3".

For each setting, first we train and classify based on the data collected from one tag only (tag no. 1, 2 or 3) of Figure 6 Then we use data from a combination of two of the three tags. Finally, we use data from all three tags. Our goal is to test how increasing the number of tags helps in reducing classification error. The results of our experiments on one of the three beds are summarized in Figure 8. Five of the 10 subjects participated

Figure 8. Average classification error for five subjects for one of the beds

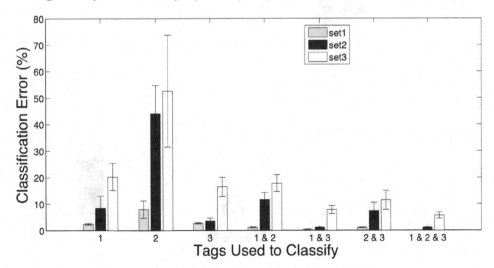

in the experiments on this bed. For each case, the y-axis shows the average of percentage classification errors for all five subjects. The error bars represent the standard deviations of the errors for each experiment.

As we see from Figure 8, if we increase the number of tags, the classification error decreases. When we use data from only one tag, the performance of tag 2 is the worst. This is expected, because it is placed near the leg, and so it fails to capture enough of the variation of body impact on the middle portion of the mattress for different positions. When we use data from any two of the three tags, we see that the combination of tags 1 and 3 performs best. This is because both of them are placed in the middle parts of the two opposite edges of the mattress. When we use data from all three tags, the error for "set1" becomes almost zero. For "set2" and "set3", average percentage errors are 1.06% and 5.64%, respectively. For the other two mattresses, we also observe similar trends, i.e., increasing the number of tags increases classification accuracy.

We also check how classification error varies over different mattresses. Figure 9 shows average classification error for all mattresses. Here

we calculate the average over the classification errors for all subjects that participated in the experiments on a particular bed. As we see from the figure, classification error for "set1" is almost zero for all mattresses. But for the other two sets, classification error is greater for mattress 3 than the other two mattresses. This mattress is the one that is in our university testbed. The testbed quality is different than the other two. It is hard and inflexible. So, the impact of the body weight does not change the orientation of the WISP tags immediately. As mentioned earlier, we classify the body positions for 30 seconds of data for each subject. Later we used the data from the last 20 seconds and the classification error went down significantly and was approximately same as the other two mattresses. So for such mattresses, we need to classify the body position after the body settles in to a new position. One implication of these results is that we could build a new mattress with the correct flexibility and embedded WISP tags that is especially targeted for those wanting or needing sleep monitoring.

Now, we analyze what body positions are misclassified most. Here, we consider misclassifications for "set3" only. For every mattress, the

Figure 9. Average classification error for all mattresses

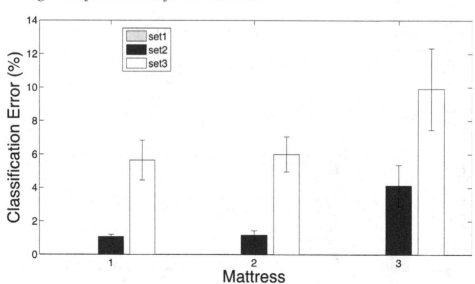

Figure 10. Average classification error for different body positions for all mattresses

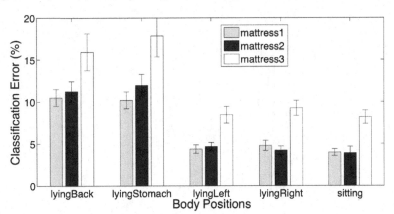

case when the bed is empty is classified correctly. For the other positions, the average of misclassifications for each mattress is shown in Figure 10. Here, we see that classification error is most prominent for the two body positions where a subject lies on back and on stomach. The reason is that sometimes one of these is classified as the other. For both these positions, the impact of body weight on the mattress remains almost same. For the other three positions, the classification error remains less than 10% for each mattress.

Note that, for these controlled experiments, the training period is only two minutes for each body position. For practical use, we need to train the system for longer periods. During our realistic overnight experiments, we train our system for several nights (about seven hours per night) and then run the system. The results are much better and are shown in the later section on realistic overnight experiments.

Controlled Experiments: Movement Detection

As we see from the earlier Figures 3 and 4, when a subject lies on the bed in a particular position or when the bed remains empty, the accelerometer values returned by the WISP tags remain within a noise level of a particular value. This is true for acceleration values along each of the three axes.

To find the maximum deviation in the readings, we calculate the derivative of all the readings when a subject remained in a particular position. The derivatives show that if the subjects remain in a particular position or if the bed is empty, the deviation remains in the interval $[+a, -b]$. The values of a and b vary for different tags, axes, and mattresses, but remain same for different subjects. We calculate these values from the data collected during the controlled experiments of the previous section.

If the subject moves to a new position or makes significant movements while remaining in the same body position, the derivative of the accelerations of all three tags along both y-axis and z-axis become higher than the corresponding $+a$ or lower than the corresponding $-b$. So during the movements, the derivatives of y and z acceleration values cross the threshold values ($+a$ and $-b$) several times. Figure 11 shows y-axis accelerometer readings during such a move. Here the values of both a and b are 1.

Our algorithm to extract movement events from derivatives of y and z-axes accelerations of the three tags is as follows: For each axis of each tag, we record timestamps when the reported reading is outside the interval $[+a, -b]$. We consider each of these moments a possible movement. Note that the three tags do not report values synchronously. We calculate the total number of

Figure 11. Accelerometer reading along y-axis during a movement

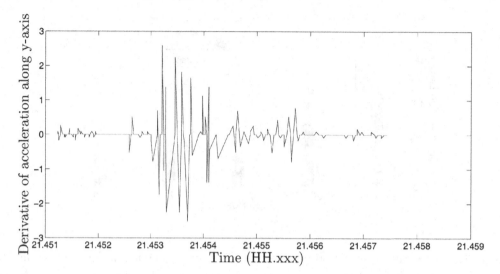

movements reported by the three tags within each two second time window. If the total number of movements within a time window is less than a predefined threshold, we consider those as discrete movements that do not affect sleep quality. We then cluster the other time windows, when a significant number of movements take place, using the DB-SCAN clustering algorithm (Ester 1996) to compute discrete movement events. The clustering also ensures that discrete movement events that happen within a short amount of time are combined as a period of restlessness. For each cluster, we set the movement level as the maximum of movement levels of all the time-windows belonging to that cluster.

Figure 12 shows the number of movements for each 2 second time window during 70 minutes of a controlled experiment. During the experiment, the subject got on the bed, laid there for 70 minutes during which he made several movements and

Figure 12. Number of movements per each two-second time-window during 70 minutes of controlled experiment

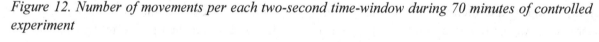

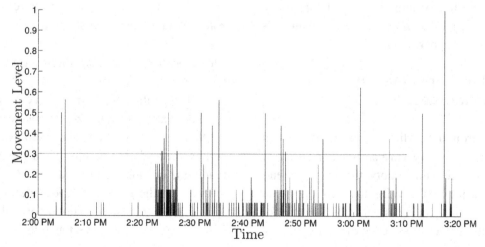

Figure 13. Discrete movement events during 70 minutes of controlled experiment

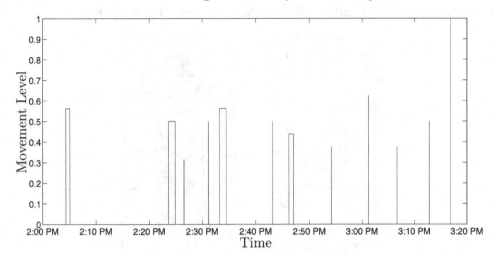

finally got off the bed. Some movements were from one body position to another and in some cases, the subject made significant movements while remaining in the same body position. We normalize the y-axis by dividing the number of movements for each time window by the maximum number of movements in any time window to get the movement level. We use 0.3 as the threshold to filter out the time windows where movement level is insignificant.

Figure 13 shows the discrete movement events as clustered by DB-SCAN. All the discrete movement events during the controlled experiment were successfully detected by our system. As we can see from Figure 13 some movement events span several minutes. During these movement events, the subject made a number of movements in quick succession. We comprehensively validate the performance of our movement detection algorithm by realistic overnight experiments that we present in the next section.

To evaluate the performance of our movement detection algorithm, we compare it with a baseline system that uses pressure pads to measure the movement levels. The pressure sensor we used was a USB-interface Multi-Platform Dance Dance Revolution (DDR) pad typically used in the popular DDR video game series. The configura-

tion of the pad is shown in Figure 14. Two pads were tiled to cover the area of a twin size bed. Data collected from the DDR pad is a bit-vector of size 16 representing which of the 16 buttons are activated. Our algorithm examines a time window, and takes the sum of the number of changes occurring in this bit-vector in that window. We chose a window size of two seconds, same as we did for WISP tags. After calculating the number of movements during each two-second-time window during the night, we clustered them in the same way as discussed in the previous section.

We also compare the performance of our system with an iPhone-based sleep monitoring application *SleepCycle* that uses accelerometer data to infer sleep quality. The application requires the iPhone to be placed on a suitable position of the bed (e.g., beside the pillow) all night and it collects data from the accelerometer of the iPhone for the whole night. Based on the data, it produces sleep quality related data that includes transitions between different sleep cycles. Durations of different sleep cycles over the course of the night are part of a person's sleeping pattern. So monitoring the transitions between sleep cycles helps in identifying irregular sleeping patterns. Our hypothesis is that transitions between the sleep cycles will correspond to higher number of

Figure 14. The DDR Pad has 8 binary contact buttons around the side, but the middle portion of the pad does not have a button. The authors tiled two pads can cover a twin size bed.

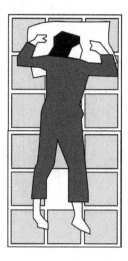

movements per time window. So from our overnight report of number of movements during each time window, we can infer the transitions between sleep cycles and the duration of each of them. We test our hypothesis in this section.

The study participant slept on the same bed for six nights. We collected and logged data from the DDR pads and the WISP devices simultaneously, and also placed an iPhone on the bed (beside the pillow) during each of these six nights. The *SleepCycle* application recorded sleep quality data and produced a report for each night. We also videotaped the sleeping period of the subject for each night after being given the subject's consent. We first validated the performance of the DDR pads in detecting movements during sleeping by comparing it with the video data for the first three hours of the recorded data for the first night. The validation result confirmed that the DDR pads could be used as ground truth to detect movements during sleeping. For evaluation, we use a cross validation approach. For each evaluation set, we choose five nights' data to train our system and evaluate the performance for the remaining night's data. So, there are six possible sets of training data. Thus, we have six sets of evaluation.

For each evaluation set, training of the movement detection algorithm includes calculating the thresholds of rate of change of acceleration values (i.e., values of **a** and **b**) along each axis for each tag and also the threshold to filter out the time windows where movement level is insignificant. During training, we consider movements detected by the DDR pads as ground truth. Training of the body inference algorithm includes training the Bayesian classifier with the accelerometer readings collected during the five nights with the corresponding body position. Collecting the actual body position for each time instant of each of these five nights is challenging. One option was to monitor the recorded video for each night and assign body positions accordingly. But this requires significant effort. To reduce effort, for each night, we watch the initial body position from the video and from then on we assume that unless there is a movement detected by the DDR pads, the position remains unchanged. When the DDR pads detect a movement, we fast forward to that time instant and see the new body position from the video and we continue in this way. Thus, we collect the ground truth for body position.

Figures 15 and 16 show the movement events during one night's evaluation (from the first evalu-

Figure 15. Movement determined by the system during one night's sleep of evaluation set 1

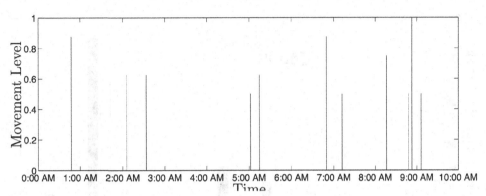

Figure 16. Movement determined from the DDR pad during one night's sleep of evaluation set 1

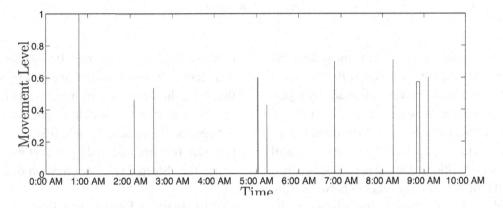

ation set) of the subject as detected by the WISP tags and DDR pads, respectively. If we compare these two figures, the first and last movements on both the figures represent the events when the subject got on and off the bed, respectively. Our system reported all movement events detected by the DDR pads. The timings of the movements are same in both figures. There was one movement that our system reported, but the DDR pads did not. It happened just after 7:00 AM in the morning. To investigate this incident, we fast forwarded to that specific time of the recorded video and observed that there was no significant movement during that time. So it was indeed a false positive.

Another notable difference occurred just before 9:00 AM in the morning, when our system reported two movement events and the DDR pads reported one movement event. However, the two

events reported by our system are very close to each other and can be considered a part of the same movement. The duration of the nine movements during this night that both systems detected are shown in Figure 17. From this figure, we see that there are no notable differences between the duration of movements calculated by both the systems. We present a summary of results and their implications for all six evaluation sets at the end of this section.

Figure 18 shows the report produced by the iPhone application *SleepCycle* to show the sleep quality for the same night as shown in Figures 15 and 16 The application shows various sleep stages like 'awake', 'deep sleep' and 'dreaming'. These sleep stages are irrelevant for our comparison. The application recorded data up to 8:00 AM in the morning. The vertical bars show when

Figure 17. Durations of the movements detected by our system and DDR pads

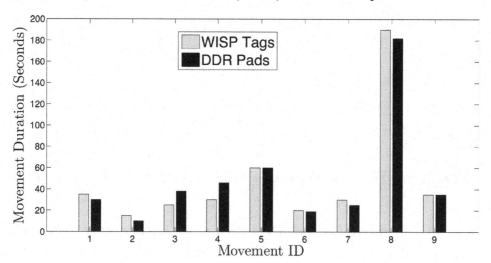

movement events are reported by our system. As we know, transitions between different sleep cycles correspond to movements made by a person. From the figure we can see that the timings of the movement events match to those of transitions between sleep cycles. There are no vertical bars for two transitions: one that happened between 7:00 and 8:00 AM and the other in between 4:00 AM and 5:00 AM. During the latter one, the subject was in deep sleep stage before and after the transition. So this is why there were no major movements. We explain the reason of lack of

movements during this transition at the end of this section. But this result proves our hypothesis that from the frequency of movements reported by our system, it is possible to infer transitions between sleep cycles. In addition, our system provides fine-grained body position monitoring which the "Sleep Cycle" application does not.

Evaluation of Body Position

Figure 19 shows the body positions as inferred by our system for the same night that was considered in Figures 15, 16, and 18. If we compare these four figures, we see that during each transition from one body position to another, there was a discrete movement event detected by our movement detection algorithm. Also, for the last three movements, the body position did not change. To ensure robustness against discrete erroneous classifications, we consider that the subject changed his body position if 20 successive instances are classified as the new body position. Also, if the movement detection algorithm detects that a movement is taking place, the body position is considered to be the same as it was before the movement until the movement is complete.

Figure 18. Sleeping quality report produced by the iPhone application

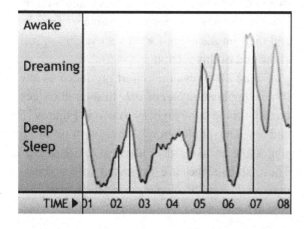

Figure 19. Body positions during one night's sleep of evaluation set 1

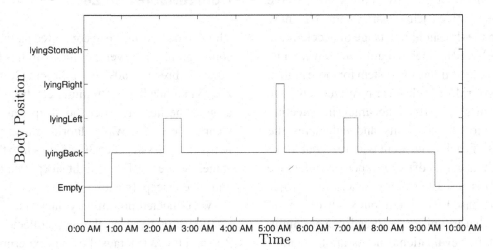

To evaluate the performance of the body position inference algorithm of our system, we generate 10 random instances of time for each night and check the subject's body position during each of those instances. For each night, the time instances are uniformly distributed over the course of the night. We define the accuracy of our inference algorithm to be the percentage of time instances when the body position inferred by our system match to the actual body position as seen from the recorded video data. We present the accuracy for each night as part of the summary of all results next.

Table 1 presents a summary of results for our six sets of evaluation. False negatives refer to the number of movement events that are detected by the DDR pads, but not by our system. Similarly, false positives refer to the number of movement events that are detected by our system, but not by the DDR pads. For each night, we define 'average error in movement duration' as the average of absolute differences between the movement durations calculated by our system and the DDR pads. Sleep cycle detection accuracy refers to the percentage of sleep cycle transitions (as shown by the iPhone application) that correspond to increased number of movements detected by our system.

From Table 1 we see that for each set, our system detected all the movement events detected by the DDR pads. Average error in calculating movement duration is less than six seconds

Table 1. Summary of results for six datasets

Evaluation Set	1	2	3	4	5	6
False Negatives	0	0	0	0	0	0
False Positives	1	0	0	1	0	0
Avg. Error in Movement Detection	6.9s	6.2s	2.2s	5.2s	4.1s	5.2s
Sleep Cycle Detection Accuracy	71.4%	75%	80%	75%	90%	80%
Body Position Inference Accuracy	100%	100%	100%	90%	100%	90%

for each night. But, for two nights, we observe one false positive each in our system. This may be due to the threshold in change of acceleration that we selected to filter insignificant movements. We believe by training the system for more nights, we can get rid of such false positives. Overall, our system shows 100% accuracy in detecting discrete movement events and calculates the durations of each movement with reasonable accuracy. The accuracy of the body position inference algorithm is at least 90% for all sets. Our evaluation was based on 10 randomly selected time instances that are uniformly spread over one night. More detailed evaluation is necessary to guarantee its performance. Therefore, we can say that, with proper training, our system performs as well as a system that uses pressure sensors and also is more comfortable for the users and completely unobtrusive. Moreover, our system provides fine-grained body position monitoring which no existing pressure sensor based sleep monitoring system provides.

We also compare our system with the popular iPhone based application "Sleep Cycle". Comparison results show that by only looking at the movement reports of our system, it is possible to identify most of the transitions between sleep cycles. Among the transitions that were not possible to identify, most of them were during deep sleep stages. The pressure sensors also did not identify them. So, these types of transitions do not correspond to significant body movements. We need to lower the value of the threshold for filtering out insignificant body movements which was set assuming the DDR pads' detected movements as ground truth. Therefore, we can say that, by training our system with the transitions detected by the iPhone application, it is possible to detect all the transitions between sleep cycles by our system.

Comparison with Zeo

The Zeo device has been compared against polysomnography in several clinical studies (Wright 2008, Fabregas 2009, Shambroom 2009). The Zeo, via a headband, monitors electrical signals around the head to infer the sleep stage of the wearer. Zeo's SoftWave algorithm uses a neural network to classify each 30 seconds into the sleep stages 'wake', 'REM', 'light sleep' (stages 1-2), and 'deep sleep' (stages 3-4).

We collected preliminary data to explore the relationships between the data collected by the Zeo and the WISP tags. For this experiment, we used a single sleeper on a twin-sized mattress. For seven nights, the subject wore both the Zeo headband and slept on a WISP instrumented bed. Eight WISP tags were used on the bed. The Zeo reported the sleep stage prediction results every 30 seconds, so correspondingly, we used a 30 second sliding time window on the WISP data and the movement levels on each accelerometer to extract the variation of the time signal. After the signal has been Z-normalized, values over a threshold of 2-sigma were classified as a 'movement event'. The result was binary data recognizing periods of movement in the sleep as shown in Figure 20.

Next we use a Naïve Bayesian classifier to classify the sleep stages. 15% of the data was used for testing, and the other 85% was used for training. We received poor classification accuracy especially with the REM, light sleep, and deep sleep. Next we trained a dynamic neural network using 70% as training data, and 20% for testing. The performance of the regression R values was 0.28 during testing with a mean squared error (MSE) of 0.64. In this very preliminary study with a single subject, there seems to be little correlation between the four sleep stages Zeo predicted and the features from the WISP tags.

Next, we limited the scope from predicting four sleep stages, to just two, sleeping vs. wake. Our dataset had 4285 frames for all the sleeping periods and 189 frames of wake periods. Our

Figure 20. One night's sleep period comparing the movement data with the Zeo

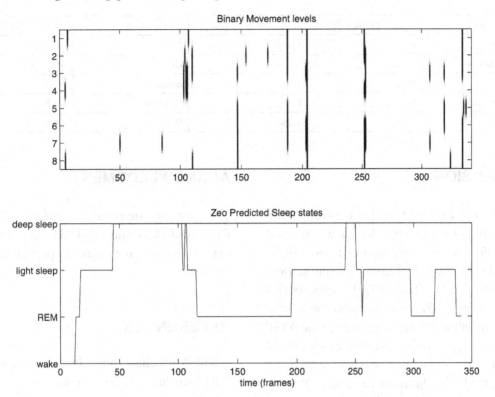

hypothesis was that wake periods receive more movement than sleep stages. We performed a one-tailed t-test on the two distributions assuming unequal variances. The null hypothesis was rejected with a p of 0.0017 that wake stages receive more movement than sleep stages with a mean of 1.59 vs. -0.07 standardized movements per frame. When the person first enters the bed and just prior to leaving the bed there is large period of movement. These preliminary results show that there could be a relationship with movement and consciousness. However, accurate segmentation of the time regions where someone drifted into sleep from being awake is a challenging and open problem. Future research once addressing this issue could then determine the ratio of the time sleeping vs. time spent in bed (sleep efficiency), which is an important metric when determining sleep quality. It will also be important to study both good and poor sleepers.

Previous research (Cole 1992) shows that wrist activity data collected when the user wears a wrist actigraph during sleep can be used to distinguish sleep from wakefulness with over 88% accuracy in a mixed sample including normal control, elderly individuals, sleep disorder patients, psychiatric patients and others. They do it using an automatic scoring algorithm that takes into account amount of movement in the current and adjacent time windows. A later study (Kogure 2011) formulates an automatic sleep/wake scoring algorithm that uses activity measurements obtained using a highly sensitive pressure sensor placed under the mattress. Results show that the pressure pad placed under a mattress or a futon can produce almost identical sleep/wake scores to actigraph. For our WISP based sleep monitoring system, similar medical studies need to be done to formulate an automatic sleep/wake scoring algorithm validated using a variety of patients.

Table 2. Qualitative comparison across devices

Device	Comfort	Privacy	Accuracy	Price
WISP Tag	High	Good	High	Moderate
Pressure Sensor	Medium	Good	High	Moderate
Camera	High	Very Low	Excellent	High
EEG/Actiwatch	Very Low	Good	Excellent	High
Zeo	Low	Good	High	High

CONCLUSION

In this chapter, we have described the use of WISP tags for monitoring sleep. We have compared our solution to other solutions and now briefly summarize the various sleep monitoring sensors and their tradeoffs in Table 2. We have shown that movement and body position can accurately be monitored with WISP tags. Furthermore, the WISP tags have a high comfort level since they are not on the sleeping surface and no device needs to be worn on the body. Our results also show that the recognition accuracy is similar to pressure sensors, but at a lower price and less intrusiveness. One main advantage of using WISP tags is that our system provides fine-grained body position monitoring which none of the existing systems offer. We plan to use our sleep monitoring system for monitoring restlessness of incontinence patients during sleep and also for depression studies where the goal is to monitor the sleep quality of patients suffering from clinical depression.

In the future we will consider the radiation caused by the reader since a long time period of exposure is expected, and further investigate the relationship between the amount and types of movement and sleep stages. We will also improve our algorithm to make our system usable in different scenarios e.g., in houses where there are pets, and where multiple people sleep in the same bed.

ACKNOWLEDGMENT

This work was supported, in part, by NSF grants ECCS-0901686 and IIS-0931972. The WISP tags were donated by Intel as part of the WISP challenge.

REFERENCES

1-800-Wheelchair. (n.d.). *Bed alarm sensor pad*. Retrieved June 16, 2011, from

Buettner, M., Prasad, R., Philipose, M., & Wetherall, D. (2009). Recognizing daily activities with RFID-based sensors. In *Proceedings of the 11th International Conference on Ubiquitous Computing* (Ubicomp '09) (pp. 51-60). New York, NY: ACM.

Carskadon, M., & Dement, W. (1989). Normal human sleep: An overview . In Kryger, M. H., Roth, T., & Dement, W. C. (Eds.), *Principles and practice of sleep medicine*. Philadelphia, PA: W B Saunders.

Chaudhri, R., Lester, J., & Borriello, G. (2008). An RFID based system for monitoring free weight exercises. In *Proceedings of the 6th ACM Conference on Embedded Network Sensor Systems* (SenSys '08), (pp. 431-432). New York, NY: ACM.

Cole, R. J., Kripke, D. F., Gruen, W., Mullaney, D. J., & Gillin, J. C. (1992). Automatic sleep/wake identification from wrist activity. *Sleep*, *15*(5), 461–469.

Demsar, J., Zupan, B., Leban, G., & Curk, T. (2004). Orange: From experimental machine learning to interactive data mining . In *ECML/PKDD* (*Vol. 3202*, pp. 537–539). Berlin, Germany: Springer. doi:10.1007/978-3-540-30116-5_58

der Loos, H. V., Kobayashi, H., Liu, G., Tai, Y., Ford, J., & Norman, J..... Osada, T. (2001). *Unobtrusive vital signs monitoring from a mulitsensor bed sheet*. In RESNA, 2001, Arlington, VA.

Ester, M., Kriegel, H., Sander, J., & Xu, X. (1996). A density-based algorithm for discovering clusters in large spatial databases with noise. In E. Simoudis, J. Han, & U. Fayyad (Eds.), *The Second International Conference on Knowledge Discovery and Data Mining (KDD-96)*.

Fabregas, S., Johnstone, J., & Shambroom, J. (2009). Performance of a wireless dry sensor system in automatically monitoring sleep and wakefullness. *Sleep, 32*, A388.

Giganti, F., Ficca, G., Gori, S., & Salzarulo, P. (2008). Body movements during night sleep and their relationship with sleep stages are further modified in very old subjects. *Brain Research Bulletin, 75*(1), 66–69. doi:10.1016/j.brainresbull.2007.07.022

Hoque, E., Dickerson, R., & Stankovic, J. (2010). *Monitoring body positions and movements during sleep using WISPs*. In Wireless Health' 10. San Diego, USA.

http://www.1800wheelchair.com/asp/viewproduct.asp?product_id=3060

Karlen, W. (2009). *Adaptive wake and sleep detection for wearable systems*. PhD thesis, Ecole Polytechnique Federale de Lausanne (EPFL).

Kogure, T., Shirakawa, S., Shimokawa, M., & Hosokawa, Y. (2011). Automatic sleep/wake scoring from body motion in bed: Validation of a newly developed sensor placed under a mattress. *Journal of Physiological Anthropology, 30*(3), 103–109. doi:10.2114/jpa2.30.103

Krauchi, K., Cajochen, C., & Wirz-Justice, A. (2004). Waking up properly: Is there a role of thermoregulation in sleep inertia? *Journal of Sleep Research, 13*(2), 121–127. doi:10.1111/j.1365-2869.2004.00398.x

Lexware Labs. (n.d.). *Sleep cycle*. Retrieved June 16, 2011 from http://www.lexwarelabs.com/sleepcycle/

Mack, D., Alwan, M., Turner, B., Suratt, P., & Felder, R. (2006). *A passive and portable system for monitoring heart rate and detecting sleep apnea and arousals: Preliminary validation* (pp. 51–54). Arlington, VA: Distributed Diagnosis and Home Healthcare.

Mack, D., Kell, S., Alwan, M., Turner, B., & Felder, R. (2003). *Non-invasive analysis of physiological signals (naps): A vibration sensor that passively detects heart and respiration rates as part of a sensor suite for medical monitoring*. In 2003 Summer Bioengineering Conference.

Oksenberg, A., & Silverberg, D. (1998). The effect of body posture on sleep-related breathing disorders: Facts and therapeutic implications. *Sleep Medicine Reviews, 2*(3), 139–162. doi:10.1016/S1087-0792(98)90018-1

Peng, Y.-T., Lin, C.-Y., & Sun, M.-T. (2006). Multimodality sensors for sleep quality monitoring and logging. *22nd International Conference on Data Engineering Workshops (ICDEW'06)* (p. 108).

Philips actiwatch. (n.d.). *Website*. Retrieved June 16, 2011, from http://www.actiwatch.respironics.com/

Redmond, S., & Heneghan, C. (2006). Cardio respiratory-based sleep staging in subjects with obstructive sleep apnea. *IEEE Transactions on Bio-Medical Engineering, 53*(3), 485–496. doi:10.1109/TBME.2005.869773

Sadeh, A., & Acebo, C. (2002). The role of actigraphy in sleep medicine. *Sleep Medicine Reviews, 6*(2), 113–124. doi:10.1053/smrv.2001.0182

Sample, A. P., Yeager, D. J., Powledge, P. S., Mamishev, A. V., & Smith, J. R. (2008). Design of an RFID-based battery-free programmable sensing platform. *IEEE Transactions on Instrumentation and Measurement, 57*(11), 2608–2615. doi:10.1109/TIM.2008.925019

Shambroom, J., Johnstone, J., & Fabregas, J. (2009). Evaluation of portable monitor for sleep staging. *Sleep*, A386.

Shinar, Z., Akselrod, S., Dagan, Y., & Baharav, A. (2006). Autonomic changes during wake-sleep transition: A heart rate variability based approach. *Autonomic Neuroscience, 130*(1-2), 17–27. doi:10.1016/j.autneu.2006.04.006

Sleep Foundation. (n.d.). *Sleep related problems.* Retrieved June 16, 2011 from http://www.sleep-foundation.org/articles/sleep-related-problems

Wisp: Wireless identification and sensing platform. (n.d.). Retrieved June 16, 2011, from http://www.seattle.intel-research.net/wisp/

Wood, A., Stankovic, J., Virone, G., Selavo, L., He, Z., & Cao, Q.,.... Stoleru, R. (2008). Context-aware wireless sensor networks for assisted-living and residential monitoring. *IEEE Network, 22*(4), 26–33. doi:10.1109/MNET.2008.4579768

Wright, K., Johnstone, J., Fabregas, S., & Shambroom, J. (2008). Evaluation of a portable, dry sensor-based automatic sleep monitoring system. *Sleep, 31*, A337.

Yang, Y., Shin, J., Jang, S., Lee, H., & Yoon, Y. (2006). Research of body movement during sleep with an infrared triangulation distance sensor, wavelets and neuro-fuzzy reasoning. *Proceedings of International Federation for Medical and Biological Engineering, 14*(7), 760–763.

Yeager, D., Holleman, J., Prasad, R., Smith, J., & Otis, B. (in press). Neuralwisp: A wirelessly powered neural interface with 1-m range. *IEEE Transactions on Biomedical Circuits and Systems.*

Zeo. (n.d.). *Personal sleep coach.* Retrieved June 16, 2011 from http://www.myzeo.com/

Chapter 8

WISP-Based Devices as Part of a Home Telecare Node

David Parry
Auckland University of Technology, New Zealand

Anne Philpott
Auckland University of Technology, New Zealand

Alan Montefiore
Auckland University of Technology, New Zealand

ABSTRACT

A systematic literature review of published sources that discuss radio frequency identification technology, ubiquitous health care, and dosage measurement was performed. The results were then critiqued. Methods of storing data and using Radio Frequency Identification (RFID) were studied. These results were used as an aid for developing a prototype system for monitoring medication dosages in a home health care environment. The combination of an RFID technology – the Intel Wireless Sensor Platform (WISPs) and the construction of a specific pill dispensing container in this prototype demonstrated that it is possible to use RFID technology to effectively and ubiquitously monitor and track drug taking compliance. With further refinements on the dispensing unit and optimization of the software this product could be manufactured and released to home care patients to help increase compliance and reduce health related issues. This could form the heart of a modular telecare data collection system. RFID-based devices that can store data in standardized formats may allow incremental development of home telecare systems in an economical fashion.

INTRODUCTION

Increasing health demands require investment in telecare. The "smart home" can support this but infrastructure issues and the rapid rate of change of technology act to make integrated compre-hensive systems unaffordable. In order to be proven clinically, particular applications need to be developed and deployed. Integrated telecare systems are complex and currently unaffordable and not scalable. Data storage and transmission also remain a major issue. However there are many conditions that need monitoring and assessment

DOI: 10.4018/978-1-4666-1990-6.ch008

in the patient's own homes, over long periods. In particular a high level of non-compliance relating to taking of medication can lead to compromised health benefits and wasted money. Improving patient compliance has the potential for improving issues related to Cardiovascular Disorder (CVD) and many other diseases. Current solutions either lack needed functionality or are much too costly to be used in a home care environment.

As the population of the western world ages, increased health costs and increasingly complicated drug regimes are becoming greater problems. In order to reduce health costs, or at least reduce the size of health cost increases, a large number of initiatives are taking place around the world in order to improve care in the home. To give a scale for the potential size of the problem, United Nations figures indicate that the number of people over 60 will triple over the next 50 years, with 2 Billion people over 60 by 2050(Population Division United Nations, 2002). Such large numbers of people with potential for chronic disease along with frailty may not want to be cared for in institutions, and care will shift increasingly to the home, supported by telecare technologies (UK Department of Health, 2005).

However, home telecare poses special challenges. The cost and intrusiveness of the infrastructure and technology needed to support it should be minimized. Each patient requires different care, so the monitoring and measurement toolset must be flexible, extensible and efficient. The equipment needs to be reliable and not impact on the lifestyle of the patient or other members of the household. Data management and storage remains one of the major challenges facing telecare, and the use of distributed data sources may form one answer to these problems.

The chapter will explain why WISP-based devices may offer a safe, secure and low-cost method of measuring compliance and may be incorporated into reminder systems as part of a home telecare node. The objective of this chapter is to outline the issues associated with the development and use of a home telecare node, the need to have common data standards and identify potential architectures based around WISP/wireless sensor devices and other RFID devices.

The example of a WISP-based drug compliance monitor (Montefiore, Parry, & Philpott, 2010) will be used as a guide to some of the potential problems and advantages of this approach.

Telecare

Telecare and assisted living technology have been recognized as a vital component of care for the increasing numbers of elderly and chronically sick people in western countries (UK Audit Commission, 2004) Not only is the percentage of the population classified as elderly increasing, but an increasing number of people are living with multiple chronic conditions. At the same time the potential pool of younger carers is decreasing. However the elderly of 2050 are likely to have greater awareness of the benefits of technology, and familiarity with handheld devices than a previous generation and such trends are likely to continue. Assisted living involves the use of technology to enhance the quality of life of such people, by allowing them to continue activities that might have been curtailed by ill-health or disability. Often these activities are very simple, and are taken for granted by the able-bodied. Telecare generally involves the remote monitoring or management of a disease process, in an effort to keep the patient as well as possible. This sometimes even applies to screening or preventative measures. Both approaches are attractive because of the potential to increase quality of life, and quality of care, but also to allow the user to retain autonomy and avoid institutionalization. There is also the potential to reduce costs of care, allow people to continue to live in remote areas and reassure relatives and informal caregivers. Although telecare is still largely at the pilot stage, benefits are beginning to be seen (Barlow, Singh, Bayer, & Curry, 2007).One of the issues in researching this area is that telecare is generally much more focused on the individual than conventional care, and needs to reflect the patient's lifestyle much more than systems based in an institutional setting. Many telecare devices have extremely specific functions, for example fall alert systems (Doughty, Lewis, & McIntosh, 2000), and their integration is becoming increasingly important. One approach to integration is the "intelligent home" or "smart home"(Inoue, Uemura, Minagawa, Esaki, & Honda, 1985; Stauffer, 1991). While the concept is not particularly new, smart homes have not been particularly intended for support of those with illness or disability. Large-scale integration of smart home features has not become popular, but more recently work has been done on the use of instrumented houses as technology test beds (Helal et al., 2005). Another study (Stefanov, Bien, & Bang, 2004), gives an overview of some requirements for a smart home. Appropriate interface design, context-awareness, standards for interoperability and, most importantly, usefulness are necessary for success.

In terms of interface design and usability, the principles of universal design for usability as developed by (Story, 1998) were used. These are; equitable use (i.e. designed for use by everyone), Flexibility in Use, Simple and intuitive interface, perceptible information, tolerance for error, low physical effort and appropriate size and space for approach and use.

The Need for Home Health Monitoring Devices

There have been many key steps in the development of information technology that allow us to ubiquitously monitor a patient's health as noted by (Choi et al., 2004). This is particularly beneficial for patients who have to live on their own and/or patients who can't make regular trips to the hospital. This paper noted that there is an increasing demand for automatic and unobtrusive monitoring devices for patient activities and drug usage.

New Zealand is following the general trend of the western world, with an ageing population. Current demographic trends from Statistics NZ show that New Zealand's population in the 65+ age bracket is going to increase by 30,000 in the next five years. (Statistics NZ - http://www.stats. govt.nz/) and this is true around the world. The issue of population ageing prompts the need for improving home based monitoring products as the elderly population often requires long-term care. These statistics, along with health care being one of the highest expenditures of the gross domestic product (GDP) justify the need for more ubiquitous and more affordable solutions to health care issues (Milenkovic, Otto, & Jovanov, 2006).

The home telecare hub will allow the integration of various independent devices, to provide information about both the current healthcare status of the patient, but also their use of treatment. By using WISP technology, independent, low-cost devices can be introduced and added to a low-infrastructure network to allow sharing of data as and when they are needed. By using this approach, the home health hub can be tailored to individual patient's requirements, and also act as a means of controlling appropriate sharing of data. This chapter will be relevant to people interested in home-based health informatics coming from both a clinical and technical background. The example of a WISP-based compliance monitor will be used to explore the use of these devices in this context.

RFID

Radio Frequency Identification (RFID) has become an integral part of our lives, in many application areas (Landt, 2005), perhaps the most familiar being the "proximity access card". RFID is a short-range radio technology that is best suited for communicating digital information about specific objects. In a typical scenario there are multiple RFID Tags/Transponders which are small and inexpensive and only one RFID Reader/

Interrogator. Information can be sent from the tags by generating, modulating and transmitting a radio signal the reader can detect. A number of modulation techniques can be used to enable read-only or read-write communication.

The use of passive RFID tags provides many benefits. They are inexpensive, disposable, and durable. These benefits allow the tags to be placed in many objects for gathering information for structural, medical and monitoring purposes. The major limitation of traditional Ultra High Frequency (UHF) is that they require a close proximity to a RFID reader to send data. This limits their functionality in scenarios such as supply chains or other applications where a reader is not always present. A common alternative to passive tags are active tags which use a battery to enable processing of sensor data when away from the reader. This however introduces new issues such as battery lifespan, recharging capabilities and hazardous chemicals such lithium are often used.

Wireless Identification and Sensing Platform (WISP)

WISP- (Sample, Yeager, Powledge, Mamishev, & Smith, 2008) is a new RFID tag technology that introduces a new power model that is different to standard passive or active tags. WISP's are similar to passive tags in the sense that they only use power harvested from the reader's RF signals, and like active tags they can continue collecting data away from readers. This ability is achieved by storing the harvested energy in a capacitor for later use. This is particularly useful for applications that have limited contact with readers. WISP-PDLs have been designed to run on the standardized EPC Gen 2 infrastructure. (www.epcglobalinc.org). This technology allows the use of commercially available readers that are compatible with EPC Gen 2 tags to collect tag data from the WISP (Figure 1). By allowing data capture and processing to continue while not within range of the reader, these tags are potentially an

extremely flexible means of collecting summary data from the environment. The WISP has three in-built sensors - temperature, capacitance and a 3D accelerometer. Data from these sensors, and a general purpose voltage sensor can be read when via the normal EPC gen2 air interface after loading a simple command program into the WISP onboard memory.

WISP's can use conventional UHF readers in this case the IMPINJ Speedway reader, a standard UHF Reader, connected to a PC via Ethernet. This reader has a normal operational range of around 1 metre with a 30cm square antenna. As with many other readers, multiple antennae can be attached to a single reader.

INTEGRATION OF WISP AND RFID DEVICES INTO A HOME TELECARE NODE

RFID technology can be used as the means of setting up a communications channel, although it is rarely thought of in this way. Compared to other communication methods, RFID techniques have distinct limitations and advantages compared to other modes, as communication is intermittent and often location dependent. Two main modes are available:

- Low cost opportunistic information passing
- Local storage linked to location and object presence

The first mode is especially useful is situations where a sensor needs to collect data for a long period, but this information does not need to be delivered to a central system continuously. One example of this is a temperature sensing device that is worn on the body to help predict fertility. This approach uses a HF RFID circuit to allow data to be transferred between the temperature sensing device and the processing unit only when in close proximity. This approach allows the temperature

Figure 1. Action of the WISP

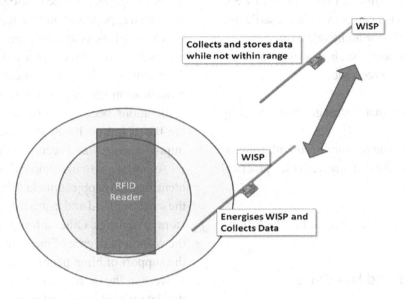

sensing unit to save power and prolong battery life by ensuring that it does not need to provide its own power for transmitting data. This mode of use utilizes the fact that in common use many devices can be located conveniently within range of a reader. Transmission of information may be the largest use of power in some sensor networks, so supplying this externally means a reduction in the cost, complexity and size of the sensor node.

The second approach is based on the concept of a very localized context-based information service. The fact that close physical proximity is required to communicate with the tags means that the location of physical objects, can be inferred from the fact that a tag can be identified. The identification can be mutual so that the presence of a particular reader- for example a mobile reader being held by an individual can be recorded onto a tag. This can provide a history of interactions that can be held within a number of different locations (tag, reader, central database). Like a set of POST-IT notes, such systems can provide localized information that is presented to the user only when it is relevant, or like a breadcrumb trail indicated where the object or user has been. Because more than one tag can be detected at a time, such approaches can also give a history of shared interactions, and possibly identify significant ones, or move from this to the identification of activity.

Deployment Strategies

There are two major approaches to setting up an RFID system. Either there are fixed tags and mobile readers, or mobile tags and fixed readers. Considering RFID as a communication channel it can convey information in two main ways. Firstly the data on the tag can be communicated to and from (in the case of read/write tags) the reader. Because of the very small storage space, such communication is very limited and only the extremely low cost of the tags makes it attractive compared to methods such as Bluetooth or Zygbee. However the ability to deploy very large number of tags with minimal maintenance also allows the communication of location. Compared with other approaches such as GPS or received signal strength lateration from Wi-Fi, RFID-based location systems are able to give a high degree of localization in particular areas with relatively low infrastructure costs. In the home environment it is suggested that a single RFID reader along with a combination of high-end (WISP) and low end (Passive RFID) tags can provide a versatile and affordable solution for home telecare.

In order to produce such a system three essential elements are required:

- A workable data storage and handling system
- Devices that can produce useful information
- Analysis tools that are reliable, practical and secure

These issues will be discussed in the following sections.

Data Storage and Handling

This chapter argues that for specialized applications such as home telecare, data storage on RFID tags for self-identification and distributed data storage may be preferable to a centralized middleware approach. Advantages of using tag-based descriptions include much reduced infrastructure costs, privacy, and flexibility. In particular this approach allows for the range of data stored to be tailored for particular applications.

Assisted living and telecare represent a very important area of development because of demographic and social changes in many nations. Methods of analyses of data from telecare devices will need development, along with training of clinical staff in the interpretation and use of this information. By implementing decentralized and standards-based data models, the data collected can be better controlled and used. Because home –based systems are always likely to be at the periphery of any data flow, such flexibility is important. The individuals involved in using assisted living and telecare devices are as diverse as any other sector of the population, if not more so because of the differing levels of (dis)ability. However the ingenuity of people in modifying and repurposing such systems can only be effectively harnessed if the infrastructure required

is not too great, and the consequences of errors and failed experiments not too harsh. By using the tags themselves to store relevant data, the system becomes more robust and less vulnerable to information security threats, and more amenable to modification and expansion as new devices and applications become available. A decentralized tag-based approach may increase scalability, and improve privacy and security (Figure 2)

Areas of particular clinical interest for self-identification of objects include information about the safety of food and instructions for the use and storage of drugs. Other information may include the color and degree of matching of clothing for the support of blind people when choosing what to wear in the morning. In many of these cases, the data would be provided by an external organization, which would not have full and continuous links to the user's database. The RFID tag data would therefore, need to be both readable and comprehensible to the user's system, without extensive handshaking. This situation is analogous to the situation on the WWW where for example, HTML pages are interpreted by a browser, despite there being no permanent connection between browser and server, and simple HTTP requests to obtain the data. The responsibility for interpreting the data received rests with the user, although the data provider – in this case the entity that writes to the tags is expected to do so according to a published, accessible standard. WISP devices are particularly suitable for this approach because of their large data storage, and processing capability.

Methods of Self-Identification

There has been a great deal of recent interest in self-identification via RFID, to bridge the gap between the real and virtual worlds (Römer, Schoch, Mattern, & Dübendorfer, 2004). RFID has been identified as a possible means of ensuring continuity of healthcare, by means of the storage of XML data on the tags, in a compressed format

Figure 2. Decentralised data storage

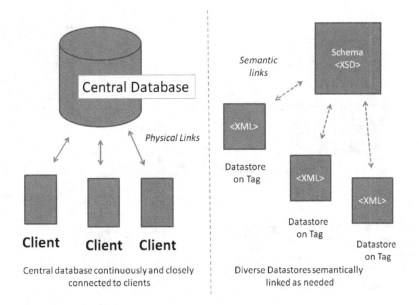

(Qiu, 2005). Self identification with data on the tag is attractive in the environment of assisted living for two main reasons, related to the environment of the home and the tasks being performed.

Firstly, the environment of the home is often not suited to sophisticated middleware applications. Network infrastructure in the home is often limited or non-existent, and even with the progressive fall in cost of Wi-Fi, and increasing use of home networks by individuals, the target group of users for this technology is unlikely to be able to easily set up and maintain the interconnected group of servers and handheld devices which are required. This may be one of the reasons that general home automation systems have not become popular.

Secondly, as RFID is increasingly used in the supply chain, it is likely that many objects will be tagged by third parties. By allowing self-identification standards to be created this data can be interpreted by end-users, and manufacturers could agree to provide free data space on the tags to allow self identification. Although this would increase the cost of the tags to the third parties, it may be an attractive marketing option for them.

The function of technology in object identification and characterization can be compared to the aim of the group of standards and approaches characterized as semantic web (Berners-Lee, Hendler, & Lassila, 2001) description and search. In the case of the web, the aim is to allow structured and shared representation of the contents of documents. For RFID self-identification, metadata about the tagged object is required. However in both cases, monolithic information systems such as single databases may not be appropriate. This is because the sheer variety of data that is significant for different users and objects implies too much complexity in a single centralized system. In addition, there is no central control of object production and labeling in either case.

In the case of the web, the use of XML is suggested because of its flexibility of representation, within a simple syntax. In addition the use of XML Schema and DTD's allow publication of both syntactic and semantic standards, which can be designed for different objects and domains. This approach allows extreme flexibility in the representation of data, but with links to "explana-

Figure 3. Distributed data storage

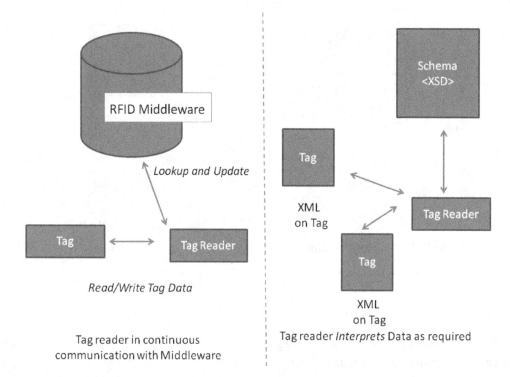

Read/Write Tag Data

Tag reader in continuous
communication with Middleware

XML
on Tag

XML
on Tag

Tag reader *Interprets* Data as required

tions" of the meaning of such data. The situation is shown diagrammatically in Figure 3.

The drawback is that if the schema changes, the documents will no longer be compliant. However validation of XML documents is a commonly performed process and the documents can provide a fraction of the data described in the schema, and the non-compliant parts can be ignored.

XML can be used in a number of ways in relation to RFID tags. An XML document can be associated with a tag ID, there can be partial storage of the data on the tag, with a link to a larger web-based document or all the data can be stored on the tag. Similarly, schema or DTD information can be stored on the tag, or in a web-accessible form. In the practical case of RFID applications in the home, the schema or DTD could be stored on the interrogation device, and only updated as required.

XML has the advantage that data is easily transformed into and out of this format, and indeed can be displayed by many web-browsers. Formatting information can be added with CSS or XSLT so that the user can easily extract the relevant data via interfaces such as browsers on mobile devices and voice-based tools.

Standardising Data Representation

The converse of the flexibility of XML is the fact that it requires common understanding by the users of a schema as to the semantics of the domain. This reflects the fact that XML is often used primarily to make documents machine-understandable. This is a problem if the understanding is not fully shared, or requires extensive development. One approach to solving this problem, is to use existing schemas, but these may be too restrictive and specific or effectively proprietary.

However in the medical domain Health level 7(HL7) (Dolin et al., 2006) is a well known set of standards that began as a medical messaging scheme, in a non-XML format. HL7 refers to the fact that it is intended to facilitate the application layer of the open systems interconnection (OSI) model (Zimmermann, 1980). The recent version 3 release including the clinical document architecture uses XML in the construction of documents. This standard introduces a much wider view, which sees HL7 documents as based around the semantics of the domain. This Chapter suggests that a message based approach, as used in other web services is an appropriate method for self identification because of the flexibility and robustness of this approach, where poorly formed or semantically incorrect messages, or in the case of our application, object identification documents can be ignored. This architecture also simplifies the use of multiple object tag interpretation schemes, for example if there is data that is relevant to only certain users. Overall a loosely-coupled scheme allows for a situation where the data associated with objects is large or even effectively unbounded, because the universe of tagged objects is so big. This approach supports not only the landmark and object tags that are known to the system but also tags produced and written to by third parties.

HL7 has already been used for the capture and analysis of clinical data collected at home (Garsden, Basilakis, Celler, Huynh, & Lovell, 2004)and this paper emphasizes the need for interoperability between health data systems, and the need for data to be presented in a meaningful way. The existence of many tools for dealing with HL7 and the body of knowledge in this area allow for rapid practical development. Because HL7 is an international standard, conformance with the standard can be checked, both in the documents themselves and in any writing or reading device. Within an HL7 document, various codes have traditionally been used in the medical domain such as LOINC. Recently SNOMED CT, a medical terminology, has been taken over by the international health terminology standards development organization (http://www.ihtsdo.org/) member countries are: Australia, Canada, Denmark, Lithuania, The Netherlands, New Zealand, Sweden, United Kingdom and United States. SNOMED was originally developed by the American College of Pathologists, and contains more that 600,000 concepts arranged in a hierarchy as an acyclic directed graph. SNOMED codes are between 6 and 12 bytes long, SNOMED allows extensions and has the concepts of attributes, which are themselves concepts, which can provide semantically meaningful relations between concepts. The terminology also contains synonyms and alternative descriptors for concepts which can also be coded, in short (6-12 byte) form. SNOMED CT is extensible, and codes for concepts that are not already in the system can be added by end users, although to be included in the release they would need to go through a review process.

Table 1. Comparison between distributed and central data storage

Central Data Storage	Distributed Data Storage
Requires continuous connection to network	May not require continuous connection
Vulnerable to unauthorized data download	Only data that is specifically exposed is available for download
May not be able to re-use data	Data may be reinterpreted
Requires all data collection systems to be compliant with the database	By using a standard data translation approach, new systems with different data formats can easily be integrated at the semantic level
Closely coupled, efficient but not easily extensible	Loosely coupled, less efficient but easily extensible with translation modules.

Producing Useful Information

To provide useful information to support telecare at home devices need to be able to collect data in the home environment without excessive input from the user. This means that tools that assist the user and can integrate into their normal life are likely to be more useful than those which merely collect data or impose costs in terms of changes to routine. This section describes some of the issues surrounding production of a prototype device.

Privacy and Security

Most nations have strict laws in terms of the privacy of medical information – for example in New Zealand these issues are dealt with by means of the Health Information Privacy code (New Zealand Privacy Commissioner, 2008). However, people may well wish to share some health information with their relatives, caregivers and friends. RFID-derived data is potentially extremely intrusive – revealing information about the details of one's private life. However by storing data on the tags rather than centrally, unauthorized access becomes harder and encrypting the data at source becomes possible. Ultimately such systems will need to be associated with personal health record systems – which are becoming more popular and widely available (Kaelber, Jha, Johnston, Middleton, & Bates, 2008), and are being developed with appropriate security and privacy measures (Homewood, Hall, Singh, Croft, & Parry, 2010).

APPLICATION AREAS

One of the important principles of this approach is that all data is potentially useful, the same data collected from location and movement sensing can be used in a number of ways. Three application areas seem particularly promising at the moment: activity recording and analysis, home automation, and medication adherence and compliance.

Activity Recording

One of the major concerns facing elderly people living alone is suffering a fall or other incapacitating event and being unable to summon help. Falls remain a common cause of morbidity and are difficult to prevent (Gillespie, 2006).Changes in levels of activity can indicate exacerbations of chronic disease – e.g. Chronic Obstructive Airways Disease (COAD). Some people with dementia may require warnings when they venture outside the home or begin wandering. Brain injured patients and others with some types of memory loss may have difficulty completing tasks that need several actions to be undertaken in a sequence, and require reminders and training to finish them.

Gross activity measurement has been performed by many groups. A recent paper (Suzuki et al., 2006) has used information fusion from a number of sensors to monitor activity, and wrist-based sensors have also been used for this purpose. Fall alarm systems have become popular and more sophisticated (Doughty et al., 2000). However, such systems do have their drawbacks. In particular the activity monitoring systems tend to measure gross activity, rather than whether the activity is directed and purposeful. Systems to notify carers of people straying are effective, but are single-use and tend to only perform this task (Altus, Mathews, Xaverius, Engelman, & Nolan, 2000).

Current measurement techniques of activity in the home are often inaccurate and cannot be used over long periods of time, because of drift, or awkwardness associated with accelerometers or pedometers, or the cost and difficulty of setting up dual isotope methods or direct observation. Other work has been done in this area (Smith et al., 2005), but this involved more complex tag design and seems more suited to laboratory studies, rather than in the home. To be useful in telecare, such a system has to be reliable, non-intrusive and cheap. Other sensing methods such as using Infra-red detectors (Buckland, Frost, & Reeves,

2006), or accelerometers are complementary, but are generally unable to give fine-grained data. As a by-product of the object location system described previously (McPherson, Parry, & Symonds, 2007) an activity monitoring system has been considered and this promises to improve on current techniques by incorporating the activity measurement into a useful tool – the object finder (Basrur & Parry, 2006)– to improve compliance. In addition the RFID system can give much greater detail about the actual activities being undertaken, as well as an estimate of the overall level of activity and the ability to generate a warning if activity ceases or appears to be disordered. In public health applications, such devices would provide valuable insights into the general population activity, and changes in response to interventions, such as medication or advice. The main experimental issues remain reliability of tag detection, user acceptance and interpretation of results.

Converting tag lists into energy expenditure estimates involves measuring the distance between detected tags, and multiplying by the energy use per meter travelled at the estimated speed, using standard estimation methods for the subject's weight and age. Note that this method can differentiate between distances travelled on a level surface and up and down stairs. It is also possible that energy used for other activities, e.g. getting out of chairs, may be able to be calculated.

Particular activity identification is more ambitious and involves interpreting the tag list data to infer an actual activity, for example making a cup of tea. In this example the subject may be detected being close to the sink, kettle and cupboard in sequence. Objects such as kettles, tea caddies and cups may need to be tagged to enable this part of the process. "Motifs" - sequences of movement or location - corresponding to particular activities may be able to be identified.

Home Automation

Various schemes have been put forward to support "smart appliances" such as refrigerators, washing machines or medicine cabinets. In each case these appliances would be equipped with RFID readers, and data already a written to tags on consumer goods would be used. The intelligent refrigerator may be used to identify food that is beyond its use date, order items that are out of stock or even suggest menu choices based on what is currently available in the kitchen. Similar systems could be devised to prevent washing clothes on the wrong setting or in the wrong combination. Medicine cabinets that provide reminders to take medication have also be proposed, and there are RFID-based approaches to counting the number of pills dispensed, or even providing warning s if it appears that the wrong combinations of medicine are being taken. As RFID is being increasingly used in the supply chain, such information could be potentially made available ad minimal extra cost. Compared to other approaches to home automation, each application can be added with relatively little cost and added infrastructure, especially if multi-aerial readers are being used. By storing relevant data on the tag, access to a central database, and hence a LAN technology such as WIFI is not required, although this can be added, if for example including a URL that can be accessed for more information.

Medication Adherence or Compliance

A high level of non-compliance in regards to taking prescription medicine is a cause for compromised health benefits and wasted money. A large review in 2001 (Vermeire, Hearnshaw, Van Royen, & Denekens, 2001) indicated that 30%-50% of patients fail to comply completely with treatment. Compliance issues have become a major focus for research and many interventions have taken place to help reduce the impact of non-compliance. Improving patient compliance has a great highest potential for improving issues related to Cardiovascular Disorder (CVD) and many other important chronic diseases. Some authors are unhappy with the term "compliance" (Anon, 1997), preferring "adherence" or "concordance", but compliance is still widely used in the literature.

There are a number of ways that people can fail to comply with prescribed treatment, from whether they actually fill the prescription to when and whether they actually take the medication according to the prescribed regimen. This study is concerned with "secondary non-compliance", that is when patients have physical possession of the medication but do not follow the instructions exactly. Studying secondary non-compliance is difficult as this generally occurs within the patient's own home and it is not reflected in statistics that are routinely collected e.g. in terms of prescriptions filled.

A CASE STUDY: WISP-BASED DEVICES FOR COMPLIANCE MONITORING

WISP and RFID offer a number of intriguing possibilities for assisted living and telecare. A case study is given of a WISP-based pill counting device for compliance measurement.

History of Drug Compliance Monitoring Devices

There have been many developments in information technology intended to allow ubiquitous monitoring of a patient's health in their own homes. A large scale review (Pare, Jaana, & Sicotte, 2007) identified that the main areas of monitoring that had been used included pulmonary conditions, diabetes, hypertension, and cardiovascular diseases. Demonstration of benefit is still elusive(Martin, Kelly, Kernohan, McCreight, & Nugent, 2008),

but acceptance of this technology is high (Alwan et al., 2006).

These approaches may be particularly beneficial for patients who have to live on their own and/or patients who can't make regular trips to the hospital. (Choi et al., 2004) Note that there is an increasing demand for automatic and unobtrusive monitoring devices for patient activities and drug usage.

Previous Approaches to Compliance Monitoring and Reminding

To accurately be able to monitor how many and how often a patient is taking pills unobtrusively is the key functionality of a compliance monitor. One particular approach for monitoring drug dosage is via the placement of needed drugs into set compartments.. Many companies and pharmacists offer this type of packaging for pills. However it does not ensure the patient is taking the drugs. Another downfall is that patients often have to load the pills themselves each week which can lead to errors and incorrect dosages. This style of system offers no tracking/monitoring of any type so in the event of non-compliance there is no way of notifying a care-giver.

More recent systems (such as the system described in (Ng Nghee Guan, 2009)) work along the lines of an unobtrusive dispenser and dosage monitor.

Another dosage system is the MEMS 6 Monitor developed by Aardex Group. The MEMS Monitors main objective is to record the dosing history of a users medication in clinical trials. The MEMS Monitor uses specially developed micro circuitry and a crafted lid to track the opening of the container. Currently the monitors are designed to support one type of drug at a time for one user. The user or care giver can place the Monitor on a special reader which transfers the obtained data to a host computer. (Ardex Group). Figure 4 shows images of the MEMS 6 Monitor with LCD dosage display and the MEMS

Figure 4. The MEMS monitor (http://www.aardexgroup.com/aardex_index.php?group=aardex&id=85)

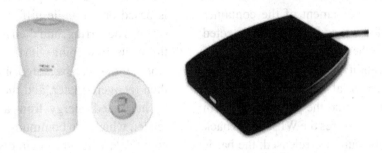

Although the MEMS Monitor seems to be adequate for recording dosages there are, however, some flaws in the design. The system cannot detect multiple pills coming out of the container. This functionality is vital when monitoring patients as it informs a care giver that the patient might be over dosing and also gives the care giver an indication when the patient is nearly be out of medication.

Another flaw in the MEMS Monitor is that the user is required to place the monitor directly on the reader when wanting to upload dosage information. Using RFID technology monitoring the patient's dosages in real-time, linked to an internet connection would be able to give caregivers instant alerts if there is a problem with the medication. RFID technology will also remove the step of having to specifically place the container on the reader as the RFID container will be able to be detected from a much longer range.

With both these implementations manufacturing costs are a likely issue for patients requiring home based care. RFID technology can provide a cheap and reliable alternative without the need for specialised development equipment.

Prototype Requirements

The prototype pill counter was built based on the following requirements Each pill being dispensed would be detected, and movement of the container that did not result in a pill being dispensed would not result in a false count. Because of limitations in the development environment for the WISP on-board computer, all data collection occurred within range of the RFID reader, with minimal on-bard processing. Finally pill dispensing should be based on a simple operation or set of operations.

Prototype Development

An approach based on a commercial confectionary dispenser was used after some unsuccessful designs. We used a commercial mint dispenser, and we make no claim on the existing mechanical design which is protected by copyright.

This device includes a rotating "catcher" that is moved via a gear when a button is pressed, converting linear motion to rotation. As the catcher rotates through the pill reservoir, it picks up a single pill and transfers it to the dispensing location. As the button is released the pill is taken into the dispensing area and falls from the device. The key design elements are that the device will not work if not held close to upright, and that the device can only dispense one pill per button press.

This design could not utilize a single WISP for movement. The incorporation of two WISP tags allows the application to detect differences between their values, so when the button is pressed one of the WISPs rotates with respect to the other. The difference in values is detected by the system and is the difference is great enough it can be inferred that a pill was dispensed. The angle "noise" of the accelerometer is +/- 0.6 of a degree when stationary so the amount of rotation has to be greater than 2 degrees to reliably determine that the button was pushed. Any gross movement of the container would be cancelled out as it would be detected by both devices at the same time.

To achieve the ability to rotate the rotating wisp WISP we built an internal mechanism that forces the WISP to rotate around the point specified in Figure 4. The spring ensures the WISP locks back into place when the button is released; the bar is used for holding one side of the WISP up while the other drops. A second WISP tag is placed anywhere on the container and calibrated to return the same values during gross movement of the whole devices as the rotating tag. This design can handle being dropped and shaken around as the detection is based on differences rather than individual movements.

Some considerations taken into account were isolating the sensors from elements that might affect reliability of results and distinguishing between actual dispensing of pills and general handling of the pill container.

Data Interpretation and Analysis

Software to interpret this data was vital but was fairly simple, involving modules to calibrate the two WISPs, a module to read the unique Tag ID's and record the angle as derived from the onboard sensors, and a calculation module to calculate the difference between the two angles and count

Figure 5. How the internal tilting mechanism works (only one WISP shown for clarity)

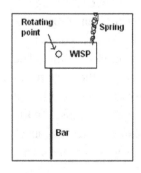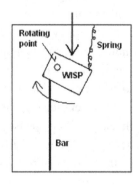

pills dispensed. The "pills dispensed" algorithm is based on an angle difference of more than 6 degrees occurring, and then returning to less than this value. In keeping with the philosophy of "local storage" data can be stored on WISP- devices and then recovered later, and the device is powered by harvested energy from a commercial RFID reader, which can communicate the recorded pill count data when the small dispenser is in range.

RESULTS

For testing the system as a whole (including container, static and dynamic tags, software and users) a number of tasks were used to ensure the system provided reliable results in all situations. The data collected from each test included a test number, software detected result (binary), actual result (binary), tilt value of static tag, tilt value of dynamic tag and the tilt difference. When evaluating the results the software detected result and the actual result pair will be studied to see how closely the samples agree.

The results from the tests showed that the software and container were performing as expected. With a 100% success in detection under general usage it can be seen that the device performs reliably. Because the dispenser only works within a certain range of angles, (close to the vertical) the system was able to detect when the dispenser was being used correctly, even if the button to cause rotation was being pressed in an orientation when the device would not dispense.

After developing and using the WISP there are only two shortcomings that could be identified. The first was the in-ability to utilise the in-built EEPROM. This was mainly due to a lack of sample code from the WISP developers as it was proving difficult to control voltages to the memory. (Yeager, et al., 2008). The second issue was related to noise from the 3D accelerometer. This produced a 'flicker' of 0.6 degrees when the WISP was at rest. This made specifying the boundary conditions for detection difficult as there was a possible 0.3 degree buffer each side of the actual limit. The manufacturing company however, has released a 'drop-in' successor to the ADXL330 accelerometer with a newer model that reduces noise and power consumption. (Analog Devices).

CONCLUSION

RFID and WISP devices offer exciting possibilities in the area of home telecare. Data storage and processing would benefit from being based around a multilevel model, with data and process at the lowest level on the WISP device, collection via the RFID reader, and interpretation of activity, or clinical needs via a local or remote PC. The processing load is actually very low and mobile devices such as phones or tablets will be easily able to cope with this.

The use of the WISP was ideal for the development of this prototype. The WISP had all the capabilities of standard RFID tags, but also supported in-built sensors and processing. WISP's are standard EPC gen1 or gen2 tags but use the harvested power from standard readers to power a low power MSP430 microcontroller and a range of sensors. The only sensor used in this project was the 3D accelerometer. With up to 3 metre range and the ability to add a capacitor for 'away from reader' capabilities, the WISP avoided many of the issues that would have been faced using standard RFID tags.

The combination of RFID technology (WISPs) and the construction of a specific pill dispensing container in this prototype has proved to demonstrate that it is possible to use RFID technology to effectively and ubiquitously monitor and track drug taking compliance. With further refinements on the dispensing unit and optimizations in the software this product could be manufactured and released to home care patients to help increase compliance and reduce health related issues.

RFID is a technology that is being adopted by many large corporations for internal use, but consumer applications seem increasingly inevitable. Costs are being lowered, for both tags and readers and specialized applications such as payment systems via NFC are becoming more popular. Using a limitation of technology, the short range of RFID data transmission – systems can be designed to gain a benefit in terms of localization. RFID is extremely convenient and simple to use compared to other techniques, and may be particularly useful for people with disabilities or the elderly. It is also fun to use, and is very attractive to hobbyists and artists, who have developed many diverse applications. Ultimately the last few inches between the digital and the physical may be the most important niche for this technology.

DISCUSSION

This paper argues that for specialized applications such as assisted living object location and activity monitoring, data storage on RFID tags for self-identification is preferable to a centralized middleware approach. Advantages of using tag-based descriptions include much reduced infrastructure costs, privacy, and flexibility. In particular this approach allows for the range of data stored to be tailored for particular applications. The use of a messaging standard reflects the physical reality where the tags are detected and read in an unsynchronized way. The use of HL7 and SNOMED including extensions along with data compression may allow existing protocols and validity checking software to be used, and support the use of RFID in the clinical domain.

There are disadvantages to this approach. There will be a requirement to store index data about objects and locations in a database for many applications. The amount of data storage on tags and the accuracy and reliability of reading and writing data remain issues to be resolved. By following a layered model as described in the OSI standard, even if the application layer format changes, most of the reader hardware and software in the lower layers are unaffected.

To be effective, object self-identification via RFID will require standardization in terms of both the structure and semantics of the data. However, just as the World Wide Web has seen novel applications arise through the development of simple standards, incremental development, based around competing models may be more successful than more monolithic solutions. Examples on the web such as the development of collaborative bookmarking, and tagging of applications such as Google maps, may demonstrate ways in which the semantic web, and the "Internet of things" can come together. However, the success of the web in its earliest form of HTML and HTTP suggests that extremely simple, effectively asynchronous and loosely coupled data transfer can give enormous benefits. The success of SMS text messaging also suggests that there are roles for apparently less rich and less high-quality communication methods, if there are compelling advantages for the user. Assisted living and telecare represent a very important area of development because of demographic and social changes in many nations. Methods of analyses of data from telecare devices will need development, along with training of clinical staff in the interpretation and use of this information. By implementing decentralized and standards-based data models, the data collected can be better controlled and used. Because home –based systems are always likely to be at the periphery of any data flow, such flexibility is important. The individuals involved in using assisted living and telecare devices are as diverse as any other sector of the population, if not more so because of the differing levels of (dis)ability. However the ingenuity of people in modifying and repurposing such systems can only be effectively harnessed if the infrastructure required is not too great, and the consequences of errors and failed experiments not too harsh. By using the tags themselves to store and collect relevant data, the system becomes more robust and less vulnerable to information security threats, and more amenable to modification and appropriation by the users themselves.

ACKNOWLEDGMENT

INTEL generously supplied hardware via the WISP Challenge project, and AM was supported by an AUT Summer studentship.

REFERENCES

Altus, D. E., Mathews, R. M., Xaverius, P. K., Engelman, K. K., & Nolan, B. A. D. (2000). Evaluating an electronic monitoring system for people who wander. *American Journal of Alzheimer's Disease and Other Dementias*, *15*(2), 121–125. doi:10.1177/153331750001500201

Alwan, M., Dalal, S., Mack, D., Kell, S., Turner, B., & Leachtenauer, J. (2006). Impact of monitoring technology in assisted living: Outcome pilot. *IEEE Transactions on Information Technology in Biomedicine*, *10*(1), 192–198. doi:10.1109/TITB.2005.855552

Anon., (1997). Editor's choice. Compliance - A broken concept. *British Medical Journal*, *314*(7082), 1.

Barlow, J., Singh, D., Bayer, S., & Curry, R. (2007). A systematic review of the benefits of home telecare for frail elderly people and those with long-term conditions. *Journal of Telemedicine and Telecare*, *13*, 172–179. doi:10.1258/135763307780908058

Basrur, P., & Parry, D. (2006). "Where are my glasses?": An object location system within the home. *Bulletin of Applied Computing and Information Technology*, *4*(2).

Berners-Lee, T., Hendler, J., & Lassila, O. (2001, May). The Semantic Web. *Scientific American*, 29–37.

Buckland, M., Frost, B., & Reeves, A. (2006). Liverpool Telecare Pilot: Telecare as an information tool. *Informatics in Primary Care, 14*(3), 191–196.

Choi, J. M., Choi, B. H., Seo, J. W., Sohn, R. H., Ryu, M. S., Yi, W., et al. (2004, 1-5 September). *A system for ubiquitous health monitoring in the bedroom via a Bluetooth network and wireless LAN.* Paper presented at the 26th Annual International Conference of the IEEE Engineering in Medicine and Biology Society, IEMBS '04.

Dolin, R. H., Alschuler, L., Boyer, S., Beebe, C., Behlen, F. M., & Biron, P. V. (2006). HL7 clinical document architecture, release 2. *Journal of the American Medical Informatics Association, 13*(1), 30–39. doi:10.1197/jamia.M1888

Doughty, K., Lewis, R., & McIntosh, A. (2000). The design of a practical and reliable fall detector for community and institutional telecare. *Journal of Telemedicine and Telecare, 6*(Supp. 1), S150–S154. doi:10.1258/1357633001934483

Garsden, H., Basilakis, J., Celler, B. G., Huynh, K., & Lovell, N. H. (2004). *A home health monitoring system including intelligent reporting and alerts.* Paper presented at the 26th Annual International Conference of the IEEE Engineering in Medicine and Biology Society, IEMBS '04.

Gillespie, L. G., Robertson, M. C., Lamb, S. E., Cumming, R. G., & Rowe, B. H. (2006). *Interventions for preventing falls in elderly people.* Cochrane Database of Systematic Reviews.

Helal, S., Mann, W. H., King, J., Kaddoura, Y., & Jansen, E. (2005). The Gator Tech smart house: A programmable pervasive space. *Computer, 38*(3), 50–60. doi:10.1109/MC.2005.107

Homewood, A., Hall, A., Singh, A., Croft, F., & Parry, D. (2010). *Are personal health records secure? A security Audit of "Google health."* (Poster). Paper presented at the Health Informatics New Zealand, Wellington.

Inoue, M., Uemura, K., Minagawa, Y., Esaki, M., & Honda, Y. (1985). A home automation system. *IEEE Transactions on Consumer Electronics, CE-31*(3), 516–527. doi:10.1109/TCE.1985.289966

Kaelber, D. C., Jha, A. K., Johnston, D., Middleton, B., & Bates, D. W. (2008). A research agenda for personal health records (PHRs). *Journal of the American Medical Informatics Association, 15*(6), 729–736. doi:10.1197/jamia.M2547

Landt, J. (2005). The history of RFID. *IEEE Potentials, 24*(4), 8–11. doi:10.1109/MP.2005.1549751

Martin, S., Kelly, G., Kernohan, W. G., McCreight, B., & Nugent, C. (2008). Smart home technologies for health and social care support. *Cochrane Database of Systematic Reviews, 2008*(4). doi:10.1002/14651858.CD006412.pub2

McPherson, K., Parry, D., & Symonds, J. (2007, December 4). *Radio frequency identification (RFID) for assisted living: Testing the aura object location (AOL) model.* Paper presented at the 18th Australasian Conference on Information Systems, Toowoomba, Queensland, Australia.

Milenkovic, A., Otto, C., & Jovanov, E. (2006). Wireless sensor networks for personal health monitoring: Issues and an implementation. *Computer Communications, 29*(13-14), 2521–2533. doi:10.1016/j.comcom.2006.02.011

Montefiore, A., Parry, D., & Philpott, A. (2010). *A radio frequency identification (RFID)-based wireless sensor device for drug compliance measurement.* Paper presented at the Health Informatics New Zealand, Wellington.

New Zealand Privacy Commissioner. (2008, 2008). *The New Zealand health information privacy code*. Retrieved 1st May 2011, from http://privacy.org.nz/health-information-privacy-code/

Ng, N. G. (2009). *Programmable medication alert & dispenser*. Singapore: SIM.

Pare, G., Jaana, M., & Sicotte, C. (2007). Systematic review of home telemonitoring for chronic diseases: The evidence base. *Journal of the American Medical Informatics Association, 14*(3), 269–277. doi:10.1197/jamia.M2270

Population Division United Nations. (2002). *World population ageing: 1950-2050*. New York, NY: Author. Retrieved from http://www.un.org/esa/population/publications/worldageing19502050/

Qiu, R. G. (2005). *An Internet computing model for ensuring continuity of healthcare*. Paper presented at the Systems, Man and Cybernetics, 2005 IEEE International Conference on.

Römer, K., Schoch, T., Mattern, F., & Dübendorfer, T. (2004). Smart identification frameworks for ubiquitous computing applications. *Wireless Networks, 10*(6), 689–700. doi:10.1023/B:WINE.0000044028.20424.85

Sample, A. P., Yeager, D. J., Powledge, P. S., Mamishev, A. V., & Smith, J. R. (2008). Design of an RFID-based battery-free programmable sensing platform. *IEEE Transactions on Instrumentation and Measurement, 57*(11), 2608–2615. doi:10.1109/TIM.2008.925019

Smith, J. R., Fishkin, K. P., Jiang, B., Mamishev, A., Philipose, M., & Rea, A. D. (2005). RFID-based techniques for human-activity detection. *Communications of the ACM, 48*(9), 39–44. doi:10.1145/1081992.1082018

Stauffer, H. B. (1991). Smart enabling system for home automation. *IEEE Transactions on Consumer Electronics, 37*(2), xxix–xxxv. doi:10.1109/30.79314

Stefanov, D. H., Bien, Z., & Bang, W.-C. (2004). The smart house for older persons and persons with physical disabilities: Structure, technology arrangements, and perspectives. *IEEE Transactions on Neural Systems and Rehabilitation Engineering, 12*(2), 228–250. doi:10.1109/TNSRE.2004.828423

Story, M. (1998). Maximizing usability: The principles of universal design. *Assistive Technology, 10*(1), 4–12. doi:10.1080/10400435.1998.10131955

Suzuki, R., Ogawa, M., Otake, S., Izutsu, T., Tobimatsu, Y., & Iwaya, T. (2006). Rhythm of daily living and detection of atypical days for elderly people living alone as determined with a monitoring system. *Journal of Telemedicine and Telecare, 12*(4), 208–214. doi:10.1258/135763306777488780

UK Audit Commission. (2004, 12th February). *Assistive technology independence and well-being 4*. Retrieved 1 November 2006, from http://www.audit-commission.gov.uk/reports/national-report.asp?CategoryID=&ProdID=BB070AC2-A23A-4478-BD69-4C19BE942722

UK Department of Health. (2005, 2nd March 2007). *Our health, our care, our say*. Retrieved 1st July 2007, from http://www.dh.gov.uk/en/Publicationsandstatistics/Publications/PublicationsPolicyAndGuidance/DH_4127453

Vermeire, E., Hearnshaw, H., Van Royen, P., & Denekens, J. (2001). Patient adherence to treatment: three decades of research. A comprehensive review. *Journal of Clinical Pharmacy and Therapeutics, 26*, 331–342. doi:10.1046/j.1365-2710.2001.00363.x

Zimmermann, H. (1980). OSI reference model--The ISO model of architecture for open systems interconnection. *IEEE Transactions on Communications, 28*(4), 425–432. doi:10.1109/TCOM.1980.1094702

ADDITIONAL READING

McDonald, H. P., Garg, A. X., & Haynes, R. B. (2002). Interventions to enhance patient adherence to medication prescriptions: Scientific review. *Journal of the American Medical Association, 288,* 2868. doi:10.1001/jama.288.22.2868

Olivieri, N. F., Matsui, D., Hermann, C., & Koren, G. (1991). Compliance assessed by the medication event monitoring system. *Archives of Disease in Childhood, 66,* 1399. doi:10.1136/adc.66.12.1399

Osterberg, L., & Blaschke, T. (2005). Adherence to medication. *The New England Journal of Medicine, 353,* 487. doi:10.1056/NEJMra050100

Ostrom, J. R., Hammarlund, E. R., Christensen, D. B., Plein, J. B., & Kethley, A. J. (1985). Medication usage in an elderly population. *Medical Care, 23,* 157. doi:10.1097/00005650-198502000-00006

Parry, D., Houliston, B., & Symonds, J. (2008, October 2008). *RFID-based self-description to support low-cost telecare and assistance at home.* Paper presented at the 13th International Symposium for Health Information Management & Research (ISHIMR), Auckland.

Parry, D., Symonds, J., & Briggs, J. (2007). RFID solutions to support home telecare information flows. *Health Care and Informatics Review Online,* October 2007.

Parry, D. T., & Narayanan, A. (2010). RFID enabled smartcards as a context-aware personal health node. *Healthcare Informatics Review Online, 14*(2).

Shalansky, S. J., Levy, A. R., & Ignaszewski, A. P. (2004). Self-reported Morisky score for identifying nonadherence with cardiovascular medications. *The Annals of Pharmacotherapy, 38,* 1363. doi:10.1345/aph.1E071

Symonds, J., Parry, D. T., & Ayoude, J. (Eds.). (2009). *Auto-identification and ubiquitous computing applications: RFID and smart technologies for information convergence.* Hershey, PA: IGI Global. doi:10.4018/978-1-60566-298-5

Szeto, A. Y. J., & Giles, J. A. J. (1997). Improving oral medication compliance with an electronic aid. *IEEE Engineering in Medicine and Biology, 16,* 48. doi:10.1109/51.585517

Vurgun, S., Philipose, M., & Pavel, M. (2007). A statistical reasoning system for medication prompting. In J. Krumm, G. D. Abowd, A. Seneviratne, & T. Strang (Eds.), *UbiComp 2007: Ubiquitous Computing, 9th International Conference,* Vol. 4717. Springer Verlag.

Wendel, C. S., Mohler, M. J., Kroesen, K., Ampel, N. M., Gifford, A. L., & Coons, S. J. (2001). Barriers to use of electronic adherence monitoring in an HIV clinic. *The Annals of Pharmacotherapy, 35,* 1010. doi:10.1345/aph.10349

KEY TERMS AND DEFINITIONS

Compliance: The degree to which patients comply with the prescribed regimen of medications, including whether they take the medication, at the right time and in the right quantities.

HL7: Health level 7 - a message-based system for transmitting clinical information.

Radio Frequency Identification (RFID): A technique for auto-identification of objects using active or passive tags that supply information often including a unique identification number. RFID can work without a line-of-sight, and can often involve reading multiple tags at the same time.

Semantic Web: A method of making data sources – often XML-based – interpretable by machines so that meaningful information can be discovered and shared on the web.

SNOMED: Systematized Nomenclature of Medicine - A controlled clinical vocabulary, including thousands of concepts. Designed to allow interoperability of clinical systems and clinical information analysis.

Telecare: Healthcare that is delivered at a distance – using digital communications to monitor and assist patients, often in their own homes.

XML Schema: A means of standardising the content and format of XML documents that may be accessed via the web.

XML: Extensible markup language, a means of encoding the meaning of data items as well as the data content in a simple format, using user-defined tags.

Chapter 9
Real–Time Traceability with Sensing in RFID Applications:
Design Issues

Ana V. Alejos
University of Vigo, Spain

Iñigo Cuiñas
University of Vigo, Spain

José Antonio Gay Fernández
University of Vigo, Spain

Manuel García Sánchez
University of Vigo, Spain

ABSTRACT

Traceability and embedded sensing are analyzed in this chapter by three main approaches: firstly, a Wireless Sensor Network; secondly, a Sensor Area Network; and, finally, a Wireless Identification and Sensing Platform. This chapter presents an introduction to the "RFID F2F" action, and its application to the wine sector briefly describing a wine pilot developed in Spain. The traceability system resulting of the WSN and RFID integration is sketched and concisely described. The current deployment of this pilot is commented. In a second block, this chapter introduces the accomplishment of an RFID tracking through a mesh of individual active radiofrequency (RF) barriers composed by active emitter and receiver nodes/tags that cover only small individual areas. The result is a Sensor Area Network (SAN). Finally, the authors of this chapter discuss the Wireless Identification and Sensing Platform technology. WISP chips have the capabilities of RFID tags □compliant with EPC Class-1 Generation-2 standard– but they also support embedded sensing and computing. WISP technology is shown as the next step forward in the design of pervasive devices. The chapter discusses the main features of the emerging computational RFID technologies.

DOI: 10.4018/978-1-4666-1990-6.ch009

INTRODUCTION

Due to the extension of food economy, different proposals appear to improve the quality of the products and the information received by the consumers. Even more, consumers look for specific or differentiated products, with high quality standards and guaranteed origin. Such requirements claim for new traceability systems that could be useful for the producers and distributors (as current traceability standards) but also for the final consumers. Indeed, clients want to know more and more about the goods they are buying.

Passive UHF RFID resulted in an optimal alternative to barcodes for industrial supply-chain tracking applications. By simply applying a cheap and inexpensive electromagnetic label named RFID tag on the product, it is possible to know plenty of information about the goods to be purchased. However, once covered the basic tracking needs, it is desirable to incorporate additional data to the RFID tagged items. To collect data externally to the supply-chain, such as origin production conditions or transport temperature and humidity conditions, sensoring devices are needed to be present in the items or surroundings jointly to the RFID tags. The integration of both elements in an embedded sensing RFID network would be an optimal approach for unobtrusive monitoring and traceability, providing more confidence in the product.

Embedded sensing is analyzed in this chapter by three main approaches: firstly, a Wireless Sensor Network; secondly, a Sensor Area Network; and, finally, a Wireless Identification and Sensing Platform. Then, the chapter is structured in three main blocks, one per approach, as follows. A first block is related to the WSN solution. We offer a brief introduction to the "RFID F2F" action, and its application to the wine sector depicting one wine pilot in Spain along with the individuated requirements of a well-performing traceability system. The block diagram of the whole traceability system is sketched and concisely described. Then,

the environment where the WSN were deployed is presented and after that, the main elements of the WSN are showed. The propagation model estimated from experimental measurement in the vineyard is used for the deployment of in an actual wireless network with sensors to collect data of foil and soil temperature, ambient humidity and solar radiation.

In a second block we discuss the SAN configuration, followed by a block dedicated to the WISP technology. We also discuss other platforms derived from the WISP technology and depict pros and cons of the so-called computational RFID devices. Finally, some conclusions and key terms are presented to close this chapter.

Previous Related Work

The European Action "RFID F2F" is focused on the test of a system to solve the mentioned situation, by using radio frequency identification (RFID) and Wireless Sensor Networks (WSN). Indeed, despite the joint use of such technologies has already been investigated (Catarinucci 2008; Catarinucci 2009; Catarinucci 2010), a global system merging both traceability and sensor data is still missing in literature.

A wireless sensor network was deployed in a vineyard, and the maximum distance between installed nodes is necessary to be previously estimated. Therefore, some propagation studies have been conducted in order to analyze the behavior of such specific radio channel at the frequency band assigned to these wireless networks: 2.4GHz. Propagation studies in rural environments, plantations and cultivated areas should consider the presence of vegetation in the propagation channel and its possible effect on the wireless communication.

Although there are several research works related to propagation under these conditions (LaGrone 1961; Richter 2005) and even an International Telecommunication Union recommendation (ITUR 2007), most of them are focused on a classical master-slave (or base station to mobile terminal) wireless communication configuration, in which the base station has a prominent height over the coverage area.

The herein proposed sensor application is intended to be deployed as a peer to peer collaborative network where the transmitter and the receiver are both at similar heights. And there is a lack in the research literature for such a configuration (Hashemi, 2008).

Some previous work related to the deployment of wireless sensor network (WSN) in forested scenarios has been reviewed. For instance, Nükhet and Haldun (Nükhet 2009) has shown the importance of using these WSNs in the analysis of forest fire propagation, but a radio propagation study is indicated to be needed in order to optimize the deployment of these WSNs. Hefeeda and Bagheri (Hefeeda 2007) deployed a WSN in order to analyze the forest fire propagation, but no study was done regarding the radio propagation conditions in these forested environment.

We also present in this chapter one specific tracking system, the sensor area network (SAN) (Alejos 2010), as a cost affordable and reliable alternative to classical WiFi-based tracking systems. The technique here introduced presents a double potential: acting as a location system for RFID tracking applications, but also as a WSN.

A SAN (Alejos 2010) is composed of a mesh of individual sensors or reader nodes placed in specific and strategic spatial positions; those sensors or reader nodes can register the user moving tags as they come in and out of their operation zone or coverage area. The coverage area is divided in single nodes or cells that can follow the movement of the user tags. The capabilities (transmitter/receiver/transceiver) of each element (cell node/moving tag) have to be defined a priori.

For intended use in RFID tracking, this operation principle offers one important additional option to warrant the privacy of the tags. If the detection capacity for the SAN nodes drops on the active moving tags, a tagged asset penetrating a cell can decide if it reveals its presence in the network. Actually, any generic RFID reading network may implement this node detection capability as a complementary tool to identify specific tags or reader nodes, and as a way to estimate the movement direction of tag subsets.

For a SAN system, the propagation phenomena can limit the provided benefits and, thus, constrain the reliability of signal strength estimation. Attenuation and signal fading, time dispersion and frequency fading caused by the multipath propagation or by people moving in the surroundings of reader terminals/nodes/tags, are undesired impairments, sometimes characterized by time-varying features, which make the compensation of the associated effects extremely difficult. This is a key problem because the success of tracking systems based on Received Signal Strength indicator (RSSI) estimation depends on its ability to keep a received signal power similar to LOS conditions despite the impairments produced by the propagation channel. The receiver front-end sensitivity largely constraints the capacity of a user moving tag to detect the coverage areas, namely cells.

In this research work, we analyze and discuss the influence of four hardware architectures to accomplish the node detection capability in the ISM unlicensed frequency band of 2.4GHz. They have been analyzed in combination with two options for the transmitted signal: continuous wave (CW) or modulated wave.

We have deployed one single SAN cell with a detection capacity relegated to the moving tag. Experimental measurements have been conducted in actual scenarios showing a good accuracy in the position estimation, as well as an adequate robustness against interferences produced by neighbour *WiFi* networks and multipath impairments. Different transmitted signals have been tested, some of them proposed for the RFID standard (Alejos 2010), in combination with the examined receiver front-end architectures.

The final approach introduced in this chapter for wireless sensing is the Wireless Identification and Sensing Platform (Ahson 2008). A WISP is a sensing and computing device that is powered and read by off the shelf UHF RFID readers. WISPs have on board microcontrollers that can sample a variety of sensing devices, creating a wirelessly-networked, battery-less sensor device. WISP devices can auto-power from analog light sensors to temperature sensors to strain gauges by harvesting the energy coming from the RFID UHF reader. The computing features of WISPs have also been used by RFID security researchers.

The WISP technology offer powered passive tags dotted with sensing and computation capabilities. The WISP has been widely tested showing promising results in supply-chain traceability, healthcare, and other applications. The WISP chips have also the capabilities of RFID tags ☐compliant with EPC Class-1 Generation-2 standard. WISP technology is the next step forward in the design of pervasive devices. However, the intermittent feature of the WISP communication produced by the harvested energy is one challenge to be solved by developing specific communication protocols.

BACKGROUND

The project "RFID from Farm to Fork" is a CIP-Pilot action involved within the 7th Frame Work of the European Union. The main objective is the use of only one system to perform the complete traceability, recording data at each stage of the food chain, from the producer point to the sale point. These data could be useful to determine the perfect condition of the final product, but also to control the process during the elaboration. Thus, both final consumers and producers would take advantage of such systems.

The final consumers could access different data about the whole process experienced by the product they are buying, just by moving the object (labeled with a RFID tag) in the vicinity of a RFID reader, which can be installed in the supermarket or even as an application at each personal smart phone. The individual identification of the product allows the software to obtain a complete traceability report from a central database, and to bring the consumer this information.

Each producer along the supply chain could use the identification by radio frequency to control its production and storage, and to know some previous information of its ingredient matters. The project involves both the design of the complete system and its tests at different stages of the chain: fishing companies, wine producers, food transporters, and final users, in order to define the actual interest of the system, its performance, and its advantages and disadvantages.

The project has shown the ability of RFID technologies to make a return on investment for SMEs in the food industry, as well to provide large information to the consumers (RFID-F2F 2011). The opportunities for such a return on investment arise from the increment of productivity due to authentication, quality control, wastage reduction, and energy optimization. Until now, these advantages have been demonstrated in large organizations, which have control over most or all of the value chain and are in a position to make an end-to-end investment. Vice versa, the "farm to fork" traceability system has not yet been adopted by independent SMEs, which only participate in one stage of the value chain.

By linking RFID and sensor network technologies with a European wide database, i.e. EuroFIR (European Food Information Resource) (EuroFIR 2011), which can store the exact history of any food product, SMEs will be given the opportunity to optimize their own business process to maximize return. In addition, a pan-union resource will be created allowing producers to demonstrate the quality and freshness of their product, which will have the effect of increasing consumer confidence as well as producer margins (Cuiñas 2011).

TRACEABILITY BASED ON WSN AND RFID

The grapes involved in the production of Ribeiro wines have to come from a strictly delimited area, controlled by the Denomination of Origin "Ribeiro" council.

So, a requirement is to assure the location of the vineyards by means of precise methods. Concretely, the vineyards of the pilot under study are planted on the lands along the Miño river. The control of different weather parameters is also a must. The region is very rainy in winter, but it is also very warm in summer. The high humidity levels, in co-ordination with the sunny time, could lead to problems with plagues (i.e., mildew) or different illness which are the main problem during the growing of the grapes. The installation of a wireless sensor network (WSN) with adequate sensors would help the farmers to develop tailor made strategies to keep the health of their plants.

During the wine manufacturing, at the winery, a fast and comfortable way to control all the movements of liquids among the different barrels, across filters, cleaners, and other machines is one of the claims of the winery managers. The proposal of a RFID based traceability system, operated by a handheld RFID reader appears to be a good solution for this task. Water resistant RFID tags will be attached to each machine and each barrel, and a three steps standard event will be defined to describe the movement of liquids from barrel A (first step) to barrel B (second step) across the machine M (third step).

Some chemical data, the outcomes of the periodic analysis, would be added manually to the database at different stages along the manufacturing of the wine.

The management of the bottle stock by individually RFID tagging the bottles appear to be more complicated: there were some reading problems around the bottles and, which is no less important, the cost of each tag is still too high to be economical assumable by the winery.

Traceability Model

The block diagram of the traceability system based on the integration of RFID and WSN technologies is depicted in Figure 1, which contains all the wine processes from the vineyard to the delivery towards final sellers, indicating which technology is used at each stage, as well as the most significant data to be collected at each point.

Figure 1. RFID system model

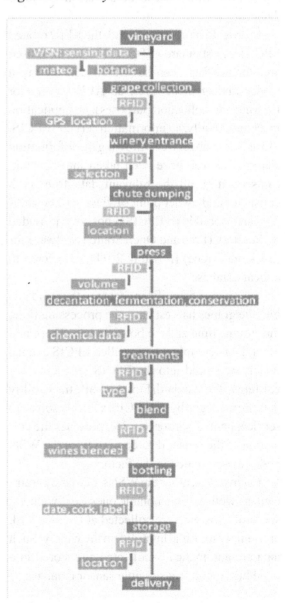

As indicated in the picture, the deployment of such a traceability system requires the joint use of several technologies: identification technologies (RFID) within the winery, WSNs for the botanic and climatic data measurement, and a software infrastructure for the management of all kind of data. The different stages of the process are indicated in dark blue, along the vertical line. On the left traceability technologies are shown, whereas yellow blocks are referred to RFID-based actions and light blue ones to WSNs. Finally, light red indicates the data to be stored in the local database.

In order to be compliant with the RFID related EPCGlobal standard (EPC 2011), the proposed architecture has been structured considering a reader management layer, an ALE server (for filtering and collection purposes), an application layer and, finally, an information service, EPCIS. Moreover, web services allowing the information sharing via web have been taken into account. Consequently, all the relevant data from both consumer and owner point of view, will be gathered and stored in an EPCIS repository, provided by Fosstrak (Free and Open Source Software for Track and Trace) (Fosstrak 2011), connected to a local database.

Examples of significant data are the wine origin, the grapes harvest date, the processing time, the storage time and the bottling date. These and other data determine the so called EPCIS events, which are stored into the EPCIS repository according to the standard. Finally, while traceability data are managed by the EPCIS, a further software service, named Sensor Service, provides the collection of the sensor data coming from the WSN nodes disseminated in the factory.

For instance, as for the WSNs, climatic parameters as well as some botanic ones (leaf wetness, soil moisture, etc.) are collected at the vineyard, and temperature and humidity in the winery. Such data are automatically collected and stored in a local but remotely accessible sensor database.

At the winery stage, instead, among different identification technologies, passive RFID plays a very relevant role, as it guarantees the automatic event tracking and must be preferred to other identification solutions, not suitable in unclean environments such as vineyards and winery. In addition to passive RFID, barcodes and QR codes could also be used. Indeed, both of them can be printed on the bottle label and allow an easy management in the retail site on the one hand, and a direct query to the traceability data, even by means of smart phones, on the other.

The development of a user-friendly capturing application capable to collect and appropriately store in the EPCIS repository all data related to the events is another important aspect.

Once a wine bottle is produced and its QR or barcode is read, data stored in both EPCIS repository and sensor database are made accessible through a proper webpage which summarizes all the relevant information.

Consequently, the consumer is allowed to know the history of each specific wine bottle, in terms of harvesting date, harvesting typology, time in barrels, temperature, humidity and intensity of rainfall during grapes growth and many others.

RADIO-PROPAGATION ANALYSIS

Before installing the wireless sensor network, it was necessary to study the maximum distance between consecutive nodes. There are some propagation studies in rural environments at 2.4GHz. For instance, Cuiñas in (Cuiñas 2010) presented a study on the propagation in mature forest at 2.4GHz. Furthermore, Gay-Fernandez in (Gay-Fernandez 2010) described the main parameters to consider when deploying a wireless sensor network in a scenario with vegetation. However, for the wine pilot case, since that the wireless sensor nodes were going to be deployed at a mean height of 3m over the ground, and the vineyard grows up to 2m, the propagation environment seems to be quite different from the scenario presented in the aforementioned literature.

Since the propagation analysis could not be performed in the selected vineyard due to the advanced status of the vineyard harvest, two experimental field measurements were performed in grasslands and scrublands, in order to obtain a general propagation expression valid for the vineyard environment by extrapolating data from these two different environments.

Experimental Measurements

A separate transmitter and receiver configuration has been used during both measurement experiments. Therefore, large distances between transmitter and receiver could be reached in order to check how the signal strength attenuation with distance is. The transmitter equipment consisted of a signal generator Rohde-Schwarz SMR, which fed an omnidirectional wide band antenna, Electrometrics EM-6865. A portable spectrum analyzer Rohde-Schwarz FSH-6 was used at the receiver equipment with an omnidirectional antenna, similar to the one used in the transmitter end.

The data were collected along two different radials for each environment. Each radial consisted of 25 points and 150m for the grassland environment, and 16 points and 32m for the scrubland case. The number of power samples per gathered trace in the spectrum analyzer at grass and scrub lands was 301 and 3010 respectively. Three different heights were analyzed for the transmitting and receiving antennas: 0.9, 1.2 and 1.6 meters. Both antennas were placed at the same height in

our analysis, in order to simulate the best conditions for a peer to peer propagation setup.

Radio-Propagation Model

The objective of the data processing is the analysis of the results by means of a regression to estimate the power decay trend as a function of distance. The attenuation of the received power seems to fit a linear equation given by $P_{Rx}=P_0-10n\log_{10}d$, where d is the distance between transmitter and receiver in meters, P_0 is the received power, in dBm, at 1m from the transmitter, P_{Rx} is received power, in dBm at a distance of d meters away from the transmitter, and n is a factor that shows the power decay rate with distance in dB/m.

If the previously explained linear fitting regression is applied to the collected samples, we obtain the parameters shown in Table 1 for grassland and scrubland. This table shows the attenuation factors n_1 and n_3, obtained for the first and second slope regression; the mean error in the estimated received power versus the real measured values if the power decay model is used; and the cut-off point for the two regressions.

Figures 2 show the fitting results for both environments. All the values of received power shown in these figures have been normalized versus a transmission power of 0dBm, in order to easily use with another transmitting power value. According to the wireless nodes datasheet, the transmission power is +3dBm and their sensitivity is -101dBm. So, considering the threshold given by the sensitivity and the parameters shown in Table 1 for the theoretical linear fitting model,

Table 1. Grassland vs scrubland regression data

Height (m)	n_1		n_2		error (dB)		point (m)	
	grass	scrub	grass	scrub	grass	scrub	grass	scrub
0.90	1.75	2.36	4.13	4.63	1.44	2.61	22	13
1.20	2.07	2.20	3.55	5.18	1.20	1.60	37	13
1.60	2.04	1.88	3.61	5.58	1.70	1.23	85	13

Figure 2. Propagation model for grassland (left) and scrubland (right) cases

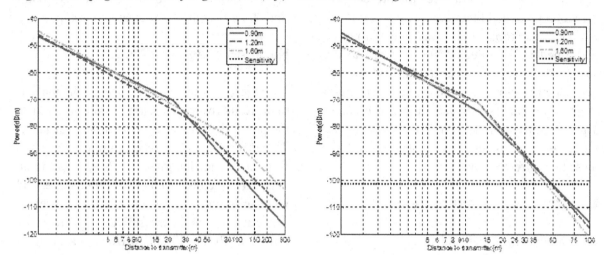

Table 2. Maximum distances between nodes

Height (m)	Grass	Scrub
0.90	123m	48m
1.20	162m	48m
1.60	254m	44m

an estimation of the maximum distance between nodes was done for both environments.

Figures 2 include a dotted line at -101dBm by mean of which the maximum range coverage can be estimated at the point that the regression lines intercept the sensitivity threshold. Table 2 summarizes the maximum achievable distances between nodes for each environment and antenna height. In consequence, when deploying the wireless sensor network, the nodes should be separated a maximum distance of 250m if there is Line of Sight (LoS) between them, and a maximum of 48m if there are scrubs or trees between them. The antenna heights considered for grasslands and scrublands experiments could likely represent the vineyard situation. There, the antennas would be higher over the ground, but the distance to the canopies would be similar than the one chosen for the performed measurements.

The Crossbow Eko pro series kit of Figure 3 was the equipment selected to implement the WSN nodes. This kit is a wireless agricultural and environmental sensing system used for precision agriculture, microclimate studies and environmental research. The Eko system can be complemented with different sensors, such as soil moisture, ambient humidity and temperature, leaf moisture, soil water content and solar radiation. All of them have been used in the WSN deployment under study.

Figure 4 illustrates the main components of the deployed WSN which includes also a GPRS modem to connect the WSN to the F2F central database acting as a bridge towards the RFID tags. The WSN includes six Eko nodes, one Eko base station, and several sensors plugged into each Eko node. Following we describe in detail each item of the WSN architecture and how they were interconnected:

Figure 3. Eko node (left) and Eko base station (right)

Figure 4. System architecture

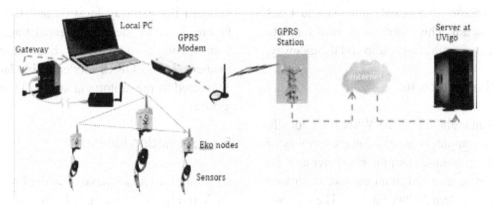

1. WSN nodes:

The Eko nodes (Figure 3 in yellow) are a fully integrated, outdoor, solar-powered wireless sensing device that allows users to deploy a multi-point monitoring solution that provides real-time data from their environment. These nodes can reach an outdoor coverage range up to 3.2Km or 2 miles, depending on the surrounding environment and the chosen node hardware configuration. These Eko nodes integrate a Memsic's IRIS processor radio board and an antenna, powered by rechargeable batteries and a solar cell. Each Eko node can accommodate up to 4 different sensors.

2. Sensors:

Six of these Eko nodes were deployed in this test, each one with four different sensors plugged in. The number assigned to each kind of sensor in the WSN has been fixed according to the requirements of the vineyard owner.

3. Base station: gateway and base radio

The Eko base station, shown in Figure 3 right picture, consists of two hardware components: the Eko base radio and the Eko gateway. The Eko gateway consists of an embedded network device providing an Ethernet connection, so that a local computer can be connected to check, view or copy all the WSN collected data.

Figure 5. WSN nodes distribution

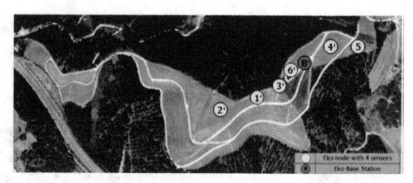

The Eko base radio is a fully integrated system that enables the connection between nodes, sensors and gateway. The base radio integrates another IRIS processor/radio board, antenna and USB interface board. This interface is used for data transfer between the base radio and the gateway.

4. WSN architecture:

Sensor data gathered on the WSN were locally stored in a computer connected to the gateway of the Eko base station. Both the computer and the gateway were installed in an outdoor enclosure requiring an external power supply. The location of this enclosure is represented as a red circle in Figure 5.

The data stored in the local PC should be transmitted to a remote server at the University of Vigo. By this way, all the sensor data could be available in real time outside the vineyard. For this sensor data transmission, a GPRS modem was needed. Another choice could have been to connect the Eko base station and the winery building by some radio technology; however this was not possible due to the absence of line of sight between the gateway location and the winery. Figure 3 shows the implemented transmission system, composed by the Eko base station and a Siemens TC-65 GPRS modem. This latter was connected to the laptop via an RS232-serial interface.

5. WSN nodes location:

Up to six Eko nodes have been deployed inside the Vitivinícola's vineyard. Each one with four different sensors plugged in. The distribution of the nodes along the vineyard has been done so that each one was located in a different variety of grape, according to the vineyards owner suggestion. The correspondence between node location and variety is indicated in Table 3. This table also summarizes the distances from each node to the base station.

Table 3. Node location and environment

Node	Grape variety	Distance to BS (m)
1	Godello	165
2	Albariño	345
3	Treixadura	80
4	Treixadura	200
5	Loureira	295
6	Godello	105

As per suggestion of the vineyard's owner, all the Eko nodes are able to measure ambient temperature and humidity, and the same parameters for the soil. Additionally, the leaf moisture appears to be quite important, so this sensor has been also connected to each node. Since that solar radiation and soil water content sensors seem to provide less important data, they have been equally distributed over the WSN.

6. WSN configuration:

The node distribution is shown in Figure 5, where Eko nodes are represented as yellow circles, the base station location is shown with a red circle, and the Vitivinícola del Ribeiro central building is represented with a red square.

Table 3 shows the final network configuration according to Figure 5 and the received power value gathered during December 2010 for the first link of the data path. The second column in Table 3 indicates the consecutive node in the path of the data transmission from the source node towards the base station. These nodes are usually called "father" node. The third column indicates the distance between each node and its father. The last column shows the received signal strength indicator (RSSI) in dBm between each node and its father. These values show that almost all the radio links between one node and its father present a large value of received power likely due to LoS propagation conditions. The only radio link with some problems is the one between nodes 3 and 6. This link seems to have very low signal strength likely due to a small terrain elevation between these nodes.

Figure 6 presents different data gathered by the sensors plugged-in to the Eko nodes. For instance, upper left chart of Figure 6 shows the evolution of the ambient and soil temperature, in °C, during December 2010. According to this plot, the average ambient temperature was 6.28°C with a standard deviation of 4.67°C, while the soil average temperature was 7.26°C with a standard deviation of only 2.63°C. The upper right plot of Figure 6 represents the ambient humidity of node 2 during the same month. These data reveals that the average ambient humidity is around 90% with

Figure 6. Sensoring information from Node 2: ambient and soil temperature (°C); ambient humidity (%); soil water content (%); solar radiation (W/m²)

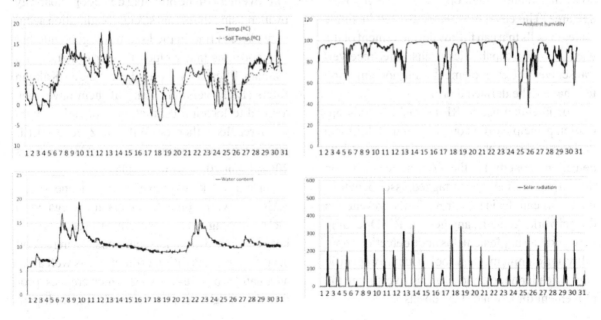

a standard deviation of 10%. Bottom left chart of Figure 6 shows the soil water content present at the node 2 location. Peaks at day 7 and 10 indicate they were rainy days, followed by a 12 days period almost without rain. Solar radiation, in Watts per square meter, present at node 2 is plotted in bottom right chart of Figure 5.

TRACEABILITY BASED ON SENSOR AREA NETWORKS

Tracking systems based on *ad-hoc* designs can offer larger accuracy levels than WiFi-based solutions, but they also present important disadvantages, which, on many occasions, are related to the system modularity and the difficult maintenance and calibration. In this chapter we present one of these specific tracking systems, the SAN, as a cost affordable and reliable alternative to classical WiFi-based tracking systems. The technique here introduced presents a double potential: acting as a RFID tracking application, but also as a WSN.

A SAN is composed of a mesh of individual sensors or reader nodes placed in specific and strategic spatial positions; those sensors or reader nodes can register the user moving tags as they come in and out of their operation zone or coverage area. The coverage area is divided in single nodes or cells that can follow the movement of the user tags. The capabilities (transmitter/receiver/transceiver) of each element (cell node/moving tag) have to be defined a priori.

For intended use in RFID tracking, this operation principle offers one important additional option to warrant the privacy of the tags. If the detection capacity for the SAN nodes drops on the active moving tags, a tagged asset penetrating a cell can decide if it reveals its presence in the network. Actually, any generic RFID reading network may implement this node detection capability as a complementary tool to identify specific tags or reader nodes, and as a way to estimate the movement direction of tag subsets.

For a SAN system, the propagation phenomena can limit the provided benefits and, thus, constrain the reliability of signal strength estimation. Attenuation and signal fading, time dispersion and frequency fading caused by the multipath propagation or by people moving in the surroundings of reader terminals/nodes/tags, are undesired impairments, sometimes characterized by time-varying features, which make the compensation of the associated effects extremely difficult. This is a key problem because the success of tracking systems based on RSSI estimation depends on its ability to keep a received signal power similar to LOS conditions despite the impairments produced by the propagation channel. The receiver front-end sensitivity largely constraints the capacity of a user moving tag to detect the coverage areas, namely cells.

In this research work, we analyze and discuss the influence of four hardware architectures to accomplish the node detection capability in the ISM unlicensed frequency band of 2.4GHz. They have been analyzed in combination with two options for the transmitted signal: continuous wave (CW) or modulated wave. In this scenario, the Kalman filtering technique can be applied to mitigate the propagation impairments resulting in improvement of the node/tag detection capability by increasing the signal strength estimation accuracy. We can find in the literature a good number of algorithms that include the performance of a Kalman filter, as well as the extended Kalman filters (EKF) version. Many of them have been applied to location estimation applications with good results either on WiFi or Zigbee (IEEE 802.15) based systems (Jiang 2003; Timmons 2004; Egan 2005; Simon 2006).

In our work, we have deployed one single SAN cell with a detection capacity relegated to the moving tag. Experimental measurements have been conducted in actual scenarios showing a good accuracy in the position estimation, as well as an adequate robustness against interferences produced by neighbour *WiFi* networks and multipath impairments. Different transmitted signals have been tested, some of them proposed for the RFID standard (Alejos 2010), in combination with the examined receiver front-end architectures.

SAN Architecture

The SAN system herein considered constitutes a mesh of individual RF barriers, each of which works according to the same principle of a laser barrier. The main advantage of the radio version of a barrier is the robustness against the environmental conditions, such as dust, particles in the air or humidity. Basically, an RF barrier consists of a transmitter that sends information, modulated or not, and a receiver that can detect and measure some parameters of that signal in order to detect the barrier presence. Actually, both ends of the barrier can present additional capabilities, as transceivers that decide to act in a barrier mode under specific circumstances.

If the parameter selected to estimate the presence of a barrier is the signal amplitude, the range or spatial limits of the RF barrier or cell are given by a limit value that must be determined in accordance with the sensibility of the receiver architecture. If the parameter level measured by the tag is greater than the delimiting value, the tag can consider itself located within the area delimited by the RF barrier. This happens under ideal propagation conditions: if the multipath and the interferences do not cause important impairments in the travelling signal.

Although the physical perimeter of the RF barrier is delimited in terms of signal parameters, the shape of the coverage area depends on the antenna pattern footprint, but it can be basically considered circular or ellipsoidal. This consideration will be taken into account when dimensioning the cell range.

The operating principle of the SAN system will be as follows: a tag (moving user) will send information required from a reader (static cell node) while it stays within a restricted area.

Besides, the tag can also be equipped with an additional feature. As an RFID system that can preserve the privacy of the tag, the tag knows that it is placed in a specific region, and, then, decides if a transmission must be performed or not, independently of the query performed by the reader; that constitutes an intelligent choice. This last performance would require an active tag, a smart location algorithm and some additional hardware on the tagged asset.

The hardware architecture to be implemented in the receiver end of a moving tag in order to detect the RF barrier presence will depend firstly on whether the transmitted signal is modulated or not. Basically, in this scene, the main options for a receiver end consist in a heterodyne front-end, a power detector and an envelope detector. We have also added one more alternative given by a heterodyne receiver provided with an envelope detector at the non-zero Intermediate Frequency (IF) output.

As following described, the combinations resulting from the transmitted signal type and the receiver end architecture options have been used to determine the coverage area provided by an RF barrier and the accuracy reached by each combination in the detection of the barrier presence. This detection has been fixed in terms of measured radio range. If we choose the power level as signal parameter to monitor, it can be translated into a receiving distance value or radio range. According to the transmitted power and the budget link, the theoretical radio range can be compared with the measured range. From this comparison the decision on the cell presence is taken.

The following expression (1) has been applied to obtain the theoretical coverage area or radio range, assumed as a circular shape zone of radio r in meters:

$$r \leq \sqrt{10^{(P_{Tx}+G_{Tx}+G_{Rx}-32.4-20 \cdot log_{10}f-S)/10} - h^2} \qquad (1)$$

Figure 7. SAN geometry (Alejos 2010)

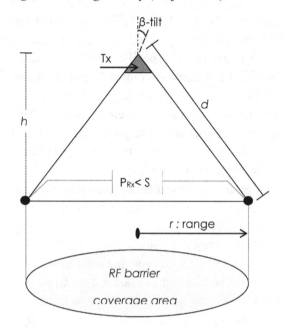

The above range parameterization was derived taking into account the condition to determine the limit of the barrier in terms of power (2), the free-space propagation model (3) and the cell geometry setup described in Figure 7:

$$P_{Rx} \leq S \qquad (2)$$

$$P_{Rx} = P_{Tx} + G_{Tx} + G_{Rx} - 32.4 - 20 \cdot log_{10} f - 20 \cdot log_{10} d \qquad (3)$$

where P_{Tx} is the power of the transmitted carrier in dBm; G_{Tx} and G_{Rx} are the transmitting and receiving antenna gain in dBi; f is the frequency expressed in GHz; h is the transmitter antenna height in meters; and S is the receiver sensitivity in dBm. The graphical illustration of the geometry assumed for the RF barrier is shown in Figure 7. The possibility of using antennas with no isotropic radiation patterns, and even with mechanical or electronic tilt degree β, can be considered.

A fit using a log-distance model ($P_{Rx}=P_0-10n\log_{10}x$) was discarded after finding that the

power-decay factor n was around 2, as found in free-space conditions. This model should be considered for other values of n, and so the range r (1) would be expressed as in (4):

$$r \leq \sqrt{10^{2(P_0-S)/10n} - h^2} \qquad (4)$$

where P_0 should be estimated for a reference point within the coverage area, and would include the transmitter and receiver antenna gains. The coverage area is assumed to be illuminated by the main antenna lobe, within the region of the 3dB beam width; under that condition, antenna gains can be considered as constant. For any other case, a correction factor must be included. Another equally important aspect to consider is the depolarization of the antennas that can be easily incorporated in (1) and (4) as a loss factor.

SAN Architecture

Two kinds of transmitted signals have been tested. In the not modulated case, a CW single carrier is transmitted at 2.45GHz; in the second case, a GMSK modulation is applied. In the latter, the modulation will help to mitigate part of the interferences in the largely radio-electrically polluted 2.4GHz ISM band. The modulated transmission option is aligned to the standard developed for RFID systems working in this frequency band (Alejos 2010).

During all the experimental measurements, the transmitter end consisted of a generator and an isotropic antenna placed at the room ceiling at a height h of 3m. The power of the transmitted signal was changed according to the receiver scheme used. The transmitter set-up has been depicted in Figure 8.

Measurements have been carried out in an open area inside an actual room, free of furniture, and under quiescent conditions. Some of the few elements of the scenario, such as the ceramic floor, are supposed to present large reflection coeffi-

Figure 8. Diagram block for CW transmission end

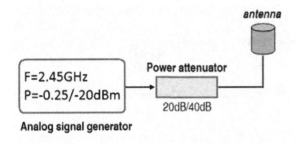

1. Single carrier CW transmission

A CW transmission is applied to send a single carrier at 2.45GHz. The first receiver scheme used is based on a spectrum analyzer. In this case, the signal received by a stub omnidirectional antenna is visualized in a spectrum analyzer configured to determine the peak power level of the received carrier using the RMS detector. In this configuration the spectrum analyzer is equivalent to a heterodyne receiver. The sensitivity of this configuration is -100dBm.

The second tested receiver consisted of measuring the power of the signal on a supposed 500KHz bandwidth channel. The channel power measurement option given by the spectrum analyzer was employed. In this configuration the spectrum acts as an RMS power meter. The sensitivity was -100dBm.

The third receiver architecture used consisted of an ad-hoc receiver as shown in Figure 9. We used a heterodyne receiver with output at an IF of 70MHz. This IF was applied to an envelope detector and the peak amplitude of the voltage signal present at its output was measured by an oscilloscope. The sensitivity of this receiver architecture, -35dBm, is given by the detector. In this third case, the power of the transmitted carrier was increased from -20.25dBm to -0.25dBm to compensate the lower value of sensitivity achieved by this receiver scheme.

cients. Regardless of this fact, distortions did not seem to affect the registered signal.

We delimited the maximum measured range area by finding the point in which the received power stopped being greater than the noise, which prevented distinguishing both signals. This point is called the barrier limit. The received power measurements were performed along two perpendicular diameters of a 9m-radius circumference, which was centered in the transmitting antenna placement; those measurements were taken in every meter of the diameter at ground level.

Following we describe the different receiver schemes implemented for each kind of transmitted signal described below. In Table 4 we summarize the combinations detailed in terms of the main parameters of the experiment performed, such as transmitted power P_{Tx} and receiver sensitivity S. Later, we discuss the results obtained from the actual measurements with the described combinations.

Table 4. Receiver architectures and delimiting ratio for SAN (Alejos 2010)

Transmission & Transmitted Power P_{Tx}		Receiver architecture	Sensitivity S(dBm)	Measurement procedure	Range r (m)	Delimiting Ratio = P_{Rx_out} / P_{Rx_in}
CW	CW @ -40.25dBm	Heterodine front-end	-100	Peak Power level	9	-90 / -100
	CW@ -40.25dBm	RMS power meter	-100	RSSI	8	-83 / -89
	CW@ -20.25dBm	Heterodine front-end & Envelope Detector	-35	Peak voltage level	9	-19 / -31
Mod-ulated	GMSK @ -0.25dBm	RMS power meter & Envelope Detector	-100	RSSI	7	-52 / -62

Figure 9. Diagram block for super-heterodyne receiver end with envelope detector (Alejos 2010)

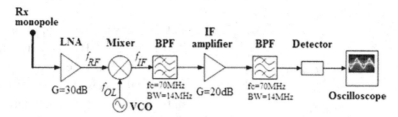

Figure 10. Diagram block for the GMSK transmitter

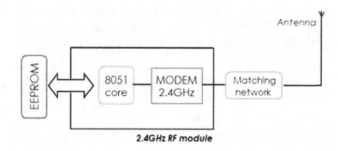

2. Modulated transmission

A GMSK signal was transmitted with a centre frequency of 2.45GHz. The data rate selected was 250Kbps, with a Gaussian filter of bandwidth-time product 0.3 $B \cdot T$, where B is the used bandwidth for transmission and T the selected bit pulse. The product $B \cdot T$ is related to the filter -3dB bandwidth and data rate by BT=-3dB-cutoff-frequency/data-rate. The bandwidth of the transmitted signal is 2MHz.

An advantage of this case is the double functionality offered as a RF barrier and a wireless information system. A block diagram of this GMSK transmitter can be seen in Figure 10. The transmitter version for a SMA connectorized antenna can be observed in Figure 11.

The receiver scheme considered for the modulated transmission case consisted of the combination of a superheterodyne downconverter and an envelope detector. The signal was captured by an omnidirectional stub antenna, amplified and later on downloaded by a mixer to an IF of 70MHz. This IF signal was amplified

again and bandpass filtered before being driven to an envelope detector. The voltage amplitude signal offered by the Schottky diode detector was observed in an oscilloscope and its RMS value was calculated. This scheme based in an envelope detector corresponds with many actual systems. The block diagram is the same as shown in Figure 9.

Figure 11. Implementation of GMSK transmitter with SMD chip antenna

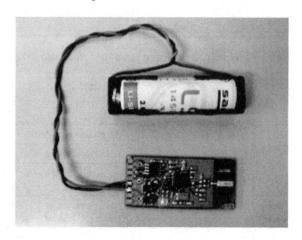

The sensitivity value of the Schottky detector is -35dBm, as indicated above; thus, a transmitted power of -0.25dBm was used again.

In the four front-end cases, the analysis performed consisted of determining the range obtained through practical power measurements. The power levels of the signals observed at the output of each receiver scheme have been compared with the ideal value given by the free space model as plotted in Figure 12. In Table 4 the range achieved for the different cases are summarized. For the first and second receiver cases, the theoretical and the measured radio ranges are very close. In the other two cases, we observe a measured range inferior to the theoretical one of 9m.

However, this comparison, according the measured range, does not seem to be enough to determine the goodness of a receiver hardware scheme. This fact drives us to define a new parameter in order to estimate the accuracy in the estimation of the RF barrier limit. Additionally, the power of the transmitted signal differs from

one end-architecture to another, then we have defined a new parameter in order to achieve a more meaningful comparison of the different architectures' behavior. This parameter takes into account the power level measured inside the area delimited by the barrier and the power found outside the limits. We have named this parameter Delimiting Ratio (DR) and it is given by the ratio of the averaged highest power measured within the barrier and the noise power level measured in the barrier limit. This parameter constitutes a measurement of the receiver accuracy to determine the presence of the RF barrier.

In Table 4, we also summarize the values of the DR parameter obtained for the different receiver architectures considered. A larger value of DR indicates that the system has larger reliability to detect the barrier and a more optimal robustness against interferences and multipath. According to this criterion, the combination of a CW single carrier and a receiver architecture formed by a heterodyne receiver and an envelope detector

Figure 12. Comparison of measured ranges

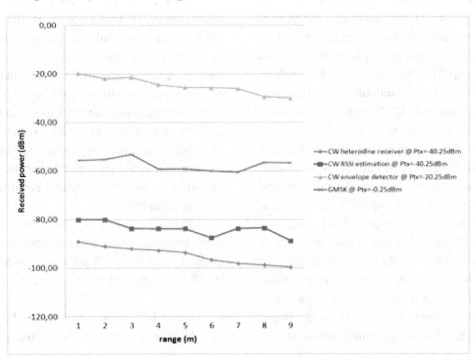

Figure 13. WISP tag

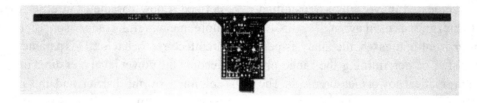

seems to be a suitable candidate for a practical implementation of a SAN-tag; however, this scheme presents a disadvantage due to the limited range detection of the envelope detector, which requires a larger transmitted power.

WIRELESS IDENTIFICATION AND SENSING PLATFORM

In the above sections we have described two needs shared by many present applications involving traceability: sensing and privacy. Both needs would require a smart tag dotted able to perform a minimum computation that additionally is intended to obtain without the presence of a battery item. The power harvesting capability of the passive RFID must be reinforced to support the extra features of the intended new RFID smart tags. This is the principle of the so-called computational RFID technologies which follow the EPC Gen 1 or Gen2 standards to result compatible with commercial RFID readers, so limiting the challenge to the tag end.

One of these technologies is the Wireless Identification and Sensing Platform (WISP) which merge the features of RFID tags with sensing and computing capabilities (Chae 2007; Sample 2008; Yeager 2008; Yeager-Sample 2008). Like any passive RFID tag, WISP is powered and read by a standard off-the-shelf RFID reader, harvesting the power it uses from the reader's emitted radio signals. WISPs can be used to sense signals such as light, temperature, acceleration, strain, liquid level, and to investigate embedded security.

Most of the work on WISP so far has involved single WISPs performing sensing or computing functions. The next phase of WISP work should involve the interaction of many WISPs as a new battery-free form of wireless sensor networking. A WISP tag is shown in Figure 13.

Most of passive RFID tags have a battery-less chip that harvests power from a nearby RFID reader and uses it to respond to the reader with an identification number and other additional information. Two broadly adopted standards for this technology are the Electronic Product Code (EPC) Class 1 Generation 1 and Class 1 Generation 2 standards, which operate in the Ultra High Frequency (UHF) bands. The standard is led by EPCGlobal.

WISPs are powered by harvested energy from off-the-shelf UHF RFID readers. To a RFID reader, a WISP is just a normal EPC gen1 or gen2 tag; but inside the WISP, the harvested energy is operating a 16-bit general purpose microcontroller. The microcontroller can perform a variety of computing tasks, including sampling sensors, and reporting that sensor data back to the RFID reader. WISPs have been built with light sensors, temperature sensors, and strain gauges. Furthermore, WISPs can write to flash memory and perform cryptographic computations.

WISP is a project of Intel Research Seattle with significant input from students and faculty of the University of Washington (WAS, USA). WISPs offer the following features:

- Up to 10ft range with harvested RF power
- Ultra-low power MSP430 microcontroller

- 32K of program space, 8K of storage
- Light, temperature and 3D-accelerometers

- Backscatter communication to reader
- Reader to WISP communication (ASK)
- Real-time clock
- Storage capacitor (to sense without reader)
- Voltage sensor (measures stored charge)
- Extensible hardware (to add new sensors)
- HW UART & GPIO for external connections
- Works with select EPC Class 1 Gen 2 readers
- WISP software to sense and upload data
- Reader application to drive WISP
- Industry standard development tools
- Access to hardware design and source code

Since WISP communicates using the EPC-global Class1 Generation-2 (C1G2) protocol, it can be used with commercial RFID readers within a 10 feet range. This allows the WISP to leverage existing infrastructure and to inter-operate with standard tags. The WISP writes sensor data to flash memory, and the data is collected using the C1G2 READ command which allows the reading of arbitrary tag memory locations. Even without its sensing capabilities, the Intel WISP can be used as an open and programmable RFID tag; for instance, the RC5 encryption algorithm was implemented on the Intel WISP.

One of the significant challenges of incorporating microcontrollers, sensors, and peripherals into passive RFID technology is the ability to manage the large power consumption of these devices. The resulting power consumption is significantly larger than typical passive RFID tags. Under these conditions the harvester cannot continuously supply power to the WISP during a single reader query.

One method used to overcome this challenge is to use a large storage capacitor (on the order of microfarads) to accumulate charge over multiple EPC queries. This allows for short bursts of power

to activate and measure sensors and communicate at long distances where received power is minimal.

If the single query power requirements are not met, the WISP sleeps for several reader transmission cycles. This allows more time for charge accumulation. The approach of duty cycling is often used in low power applications; however this presents a challenge for RFID networks when the WISP is not necessarily able to respond to each reader query.

However, WISP is not the only computational RFID platform emerging at present provided of batteryless computation and sensing capabilities. We can mention also the Moo project (Zhang 2011) derived from the WISP technology. The UMass Moo is a passive Computational RFID that harvests RFID reader energy from the UHF band, communicates with an RFID reader, and processes data from its onboard temperature sensor and accelerometer. Its function can be extended with its general-purpose I/Os, serial buses, and 12-bit ADC/DAC ports. The Moo provides a RFID-scale, fully programmable, batteryless sensing platform. The programs execute on an MSP430 microcontroller. The Moo 1.0 derives from the open source Intel DL WISP 4.1.

Given the increasing capacity of microcontrollers to reduce their power consumption, it is expected that more platforms give arise with more complex and innovative features. At present, the tag requires a board that can be seen as a limitation not only in terms of physical dimensions but also from the point of view of costs. Even that the new computational devices result compatible with the commercial RFID readers by using the EPC standard we should not forget that the power harvesting requires additional cycles to fulfill a reader query. This fact can slow down the whole tracing process to reach even critical values for a query cycle, introducing a trade-off between application requirements and update rate.

In the same line, a key parameter for maximizing the read distance of the WISP is sleep current consumption rather than active current consumption. This would lead to increase the query cycle duration in order to improve the link range. This trade-off is clearly subordinated to the application specific requirements.

We cannot forget that the WISP tags can be more exposed to the usual privacy risks considered for the passive RFID tags (are tag cloning, tag snatching, tag data manipulation, and disruption) or that new threads should be considered given that a circuitry is accessible and the information flow maybe sniffed directly from the source.

CONCLUSION

Three different approaches are discussed as an option to implement traceability and sensing together with RFID. Firstly, an actual integration of RFID and WSN is discussed. Previously to its implementation, a complete measurement campaign was developed to model the propagation channel of the radio links among the WSN nodes. This propagation model has been used for planning the actual installation in a vineyard in the Northwestern Spain. The Eko technology from Memsic has been selected for this deployment. Up to six Eko nodes were set up into the vineyard, to cover an area of approximately 6 Km^2.Four different sensors have been plugged into each Eko node, to collect different ambient and soil parameters, like humidity, temperature, solar radiation, moisture, etc. These sensor data are transferred to a central RFID database, so that the data can be incorporated to the traceability systems, including the RFID tags.

In second place we introduced the SAN approach to provide localized tracking and sensing, both in a simple architecture. The most interesting outcome regarding this block has been the study of the receiver hardware influence that could be extrapolated for the proper design of RFID readers. Experimental measurements were done to estimate the influence of the transmitted signal type and the receiver end architecture in the detection of

the RF barrier presence. The parameterization of the coverage area of a SAN cell in terms of power is derived for both free-space and log-distance propagation models.

Outcomes here described show a promising performance for this wireless network design, which has not received enough attention in literature. SAN tracking features can find an interest application such as the one described in (Sharma 2009) to guide users towards specific items in commercial surfaces. We conclude that the SAN approach offers some advantages over classical tracking systems, which are based on Wireless Sensor Networks (WSN), especially in the multipath impairment mitigation, such as a controlled power emission, and the chance to warrant privacy regarding the exchange of RFID information.

Finally, the WISP technology was briefly described as an optimal sensor embedded wireless option; however, the short range turns it into not applicable for all the cases. It is evident that wireless sensing can bring the advantages of RFID technology to wireless sensor networks, and vice versa. However, as shown in the pilot described in this chapter, even that it cannot be expected to fully replace WSNs for all applications, they do open up new applications, especially for long-life, or inaccessible devices cases (Buettner, 2008).

As the traditional RFID usage model is very different from that of WSNs, RSNs face at least two substantial research challenges to integrate the two technologies (Buettner, 2008):

1 WISPs are powered only when they are in range of an RFID reader. This can result in the WISP losing power in the middle of a task. Additionally, when tags are powered the cost of transmission is essentially zero, which allows them to be re-tasked. Thus, techniques must be developed that enable task completion in the face of intermittent power.

2 Radio sensor networks will be highly asymmetric in terms of their communication abilities because they build on RFID. This complicates protocols designed to gather and process sensor data. The standard RFID strategy of identifying and then communicating with each device is wasteful as most devices may not have relevant data. Instead, query languages and protocols need to be developed, ideally in a manner that is compatible with the C1G2 standard.

REFERENCES

Ahson, S., & Ilyas, M. (2008). *RFID handbook: Applications, technology, security and privacy.* CRC Press. doi:10.1201/9781420055009

Alejos, A., García Sánchez, M., Cuiñas, I., & Garcia Valladares, J. C. (2010). Sensor area network for active RTLS for RFID tracking applications at 2.4GHz. *Progress in Electromagnetics Research, 110,* 43–58. doi:10.2528/PIER10100204

Barth, A., Collin, J., & Mitchel, J. C. (2008). Robust defenses for cross-site request forgery. In *15th ACM Conference on Computer and Communications Security,* (pp. 75-88).

Buettner, M., Greenstein, B., Sample, A., & Smith, J. R. (2008). *Demonstration: RFID sensor networks with the Intel WISP.* In 6th ACM Conference on Embedded Networked Sensor Systems.

Catarinucci, L., Cappelli, M., Colella, R., & Tarricone, L. (2008). A novel low-cost multisensor-tag for RFID applications in healthcare. *Microwave and Optical Technology Letters, 50,* 2877–2880. doi:10.1002/mop.23837

Catarinucci, L., Colella, R., & Tarricone, L. (2009). *A cost-effective UHF RFID tag for transmission of generic sensor data in wireless sensor networks.* IEEE Transaction on Microwave Theory and Techniques – Special Issue on RFID Technology.

Catarinucci, L., Colella, R., & Tarricone, L. (2010). Sensor data transmission through passive RFID tags to feed wireless sensor networks. In *Proceedings of IEEE MTT International Microwave Symposium.*

Chae, H. J., Yeager, D. J., Smith, J. R., & Fu, K. (2007). Maximalist cryptography and computation on the wisp UHF RFID tag. In *Proceedings of the Conference on RFID Security.*

Crossbow Technology Inc. (2009). *Eko PRO series user's manual, rev. C.*

Cuiñas, I., Catarinucci, L., & Trebar, M. (2010). RFID from farm to fork: Traceability along the complete food chain. In *Proceedings of Progress in Electromagnetic Research Symposium.*

Cuinas, I., Gay-Fernandez, J. A., Alejos, A., & Sanchez, M. (2010). A comparison of radioelectric propagation in mature forests at wireless network frequency bands. In *European Conference on Antennas and Propagation* (pp. 1-5).

Egan, D. (2005). The emergence of ZigBee in building automation and industrial controls. *Computing & Control Engineering Journal, 16*(2), 14–19. doi:10.1049/cce:20050203

EPCglobal. (2005). *EPC radio-frequency identity protocols class-1 generation-2 UHF RFID protocol for communications at 860MHz- 960MHz version 1.0.9.*

EPCGlobal Standard. (n.d.). *Web page.* Retrieved October 15, 2011, from http://www.epcglobalinc.org

EuroFIR (European Food Information Resource). (n.d.). *Website.* Retrieved October 15, 2011, http://www.eurofir.net

Fosstrak (Free and Open Source Software for Track and Trace). (n.d.). *Web page*. Retrieved October 15, 2011, http://www.fosstrak.org

Gay-Fernandez, J. A., Garcia Sanchez, M., Cuiñas, I., Alejos, A. V., Sánchez, J. G., & Miranda-Sierra, J. L. (2010). Propagation analysis and deployment of a wireless sensor network in a forest. *Progress in Electromagnetics Research*, *106*, 121–145. doi:10.2528/PIER10040806

Hashemi, H. (2008). Propagation channel modeling for ad hoc networks. In *European Microwave Week*, (pp. 39- 43).

International Telecommunication Union (2007). Attenuation in Vegetation. *ITU-R Recommendation 833-6*.

ISO/IEC 18000. (2010). *Information technology AIDC techniques – RFID for item management – Air interface*.

Jiang, T., Sidiropoulos, N. D., & Giannakis, G. B. (2003). Kalman filtering for power estimation in mobile communications. *IEEE Transactions on Wireless Communications*, *2*(1), 151–161. doi:10.1109/TWC.2002.806386

Karabuk, T., Hefeeda, M., & Bagheri, M. (2007). Wireless sensor networks for early detection of forest fires. In *IEEE International Conference on Mobile Adhoc and Sensor Systems*, (pp. 1-6).

LaGrone, A., & Chapman, C. (1961). Some propagation characteristics of high UHF signals in the immediate vicinity of trees. *IRE Transactions on Antennas and Propagation*, *9*(5), 487–491. doi:10.1109/TAP.1961.1145049

Nükhet, S., & Haldun, A. (2009). The importance of using wireless sensor networks for forest fire sensing and detection in Turkey. In *5th International Advanced Technologies Symposium*, (pp. 1323-1327).

RFID-F2F (RFID From Farm to Fork). (n.d.). *Website*. Retrieved October 15, 2011, http://www.rfidf2f.eu/

Richter, J., Caldeirinha, R. F. S., Al-Nuaimi, M. O., Seville, A., Rogers, N. C., & Savage, N. (2005). A generic narrowband model for radiowave propagation through vegetation. In *Vehicular Technology Conference*, Vol. 1, (pp. 39- 43).

Sample, A. P., Yeager, D. J., Powledge, P. S., & Smith, J. R. (2008). Design of an RFID-based battery-free programmable sensing platform. *IEEE Transactions on Instrumentation and Measurement*, *57*(11), 2608–2615. doi:10.1109/TIM.2008.925019

Simon, D. (2006). *Optimal state estimation: Kalman, H-infinity, and nonlinear approaches*. John Wiley & Sons. doi:10.1002/0470045345

Timmons, N. F., & Scanlon, W. G. (2004). Analysis of the performance of IEEE 802.15.4 for medical sensor body area networking. In *First Annual IEEE Communications Society Conference on Sensor and Ad Hoc Communications and Networks*, (pp. 16-24).

Yeager, D., Prasad, R., Wetherall, D., Powledge, P., & Smith, J. (2008). Wirelessly-charged uhf tags for sensor data collection. In *Proceedings of IEEE RFID*.

Yeager, D. J., Sample, A. P., & Smith, J. R. (2008). Wisp: A passively powered UHF RFID tag with sensing and computation. In Syed, M. I., & Ahson, A. (Eds.), *RFID handbook: Applications, technology, security, and privacy*. CRC Press.

Zhang, H., Gummeson, J., Ransford, B., & Fu, K. (2011). Moo: A batteryless computational RFID and sensing platform. Tech report UM-CS-2011-020, Department of Computer Science, University of Massachusetts Amherst.

ADDITIONAL READING

Corbley, K. P. (November, 2008). WiFi inadequate as real-time asset and patient tracking solution. *Wireless Design Magazine*, 30-31.

Geok, T. K., Reza, A. W., & Tan, C.-P. (2008). Objects tracking utilizing square grid RFID reader antenna network. *Journal of Electromagnetic Waves and Applications, 22*(1), 27–38. doi:10.1163/156939308783122724

Holleman, J., Yeager, D., Prasad, R., Smith, J. R., & Otis, B. (2008). An energy-harvesting wireless neural interface with 1-m range . In *Proceedings of Biomedical Circuits and Systems*. NeuralWISP. doi:10.1109/BIOCAS.2008.4696868

Liu, H.-Q., & So, H.-C. (2009). Target tracking with line-of-sight identification in sensor networks under unknown measurement noises. *Progress in Electromagnetics Research, 97*, 373–389. doi:10.2528/PIER09090701

Liu, H.-Q., So, H.-C., Lui, K. W. K., & Chan, F. K. W. (2009). Sensor selection for target tracking in sensor networks. *Progress in Electromagnetics Research, 95*, 267–282. doi:10.2528/PIER09070802

NASA. (2007). *Sensor area network for integrated systems health management*. SBIR contract NNX08CC91P.

Ransford, B., Clark, S., Salajegheh, M., & Fu, K. (2008). Getting things done on computational RFIDs with energy-aware checkpointing and voltage-aware scheduling. In *USENIX Workshop on Power Aware Computing and Systems*.

Sample, A. P., Yeager, D. J., Powledge, P. S., Mamishev, A. V., & Smith, J. R. (2008). Design of an RFID-based battery-free programmable sensing platform. *IEEE Transactions on Instrumentation and Measurement, 57*(11), 2608–2615. doi:10.1109/TIM.2008.925019

Sample, A. P., Yeager, D. J., & Smith, J. R. (2009). A capacitive touch interface for passive RFID tags. In *Proceedings of the 2009 IEEE International Conference on RFID*.

Sharma, N., & Youn, J.-H. (2009). RFID-based direction finding signage system (DFSS) for healthcare facilities . In Turcu, C. (Ed.), *Sustainable radio frequency identification solutions*. In-Tech. doi:10.5772/8014

WISP. (n.d.). *Wiki*. Retrieved October 15, 2011, http://wisp.wikispaces.com/

Yeager, D. J., Holleman, J., Prasad, R., Smith, J., & Otis, B. (2008). Neural WISP: A wirelessly-powered neural interface with 1-m range. *IEEE Transactions on Biomedical Circuits and Systems, 3*(6), 379–387. doi:10.1109/TBCAS.2009.2031628

Yeager, D. J., Sample, A. P., & Smith, J. R. (2008). WISP: A passively powered UHF RFID tag with sensing and computation . In Ahson, S. A., & Ilyas, M. (Eds.), *RFID handbook: applications, Technology, security, and privacy*. CRC Press.

Zhao, J., Zhang, Y., & Ye, M. (2006). Research on the received signal strength indication location algorithm for RFID system. In *Proceedings of the International Symposium on Communications and Information Technologies*, (pp. 881-885).

KEY TERMS AND DEFINITIONS

Air Interface: The operating system of a wireless network. Technologies include mobile (AMPS, TDMA, CDMA, GSM) and wireless (WiFi) standards.

Antenna: Device for transmitting and receiving radiofrequency signals. The size and shape of antennas are generally determined by the frequency of the signal they manage.

Attenuation: Loss of signal strength measured in decibels (dB), due to interference, propagation impairments and just range or distance between transmission and receptor.

Base Station: Central radio transmitter/receiver that communicates with mobile telephones within a given range (typically a cell site).

Broadband: Transmission facility having a bandwidth (capacity) sufficient to carry multiple voice, video or data channels simultaneously. Broadband is generally equated with the delivery of increased speeds and advanced capabilities, including access to the Internet and related services and facilities "that provide 200 kbps upstream and downstream transmission speeds" (per the FCC's Fourth Annual Report to Congress on the "Availability of Advanced Telecommunications Capability in the United States," September 2004).

Gateway: Internetworking system capable of joining together two networks that use different base protocols. A network gateway can be implemented completely in software, completely in hardware, or as a combination of both. Depending on the types of protocols they support, network gateways can operate at any level of the OSI model.

GPRS: Packet based technology approach that enables high-speed wireless Internet and other GSM-based data communications, offering a very efficient use of available radio spectrum for data radio transmission.

Node: Any device connected to a computer network. Nodes can be computers, personal digital assistants (PDAs), cell phones, or various other network appliances. On an IP network, a node is any device with an IP address.

Power Harvesting: Also named energy harvesting is a process to collect and store energy derived from different ambient sources, such as radio propagated electromagnetic signals. This energy can be used to power autonomous and portable battery-free devices.

Sensor: Electro-mechanical device designed to be able to collect information from different parameters or variables, such as temperature, pressure, humidity, movement, light, sound, touch, and others, which are outside the system. A sensor measures a physical quantity and converts it into a signal which can be read by an observer or by an instrument.

Chapter 10
From the Farm to Fork:
Information Security Accomplishment in a RFID Based Tracking Chain for Food Sector

Ana V. Alejos
University of Vigo, Spain

Iñigo Cuiñas
University of Vigo, Spain

Isabel Expósito
University of Vigo, Spain

Manuel García Sánchez
University of Vigo, Spain

ABSTRACT

In this chapter the authors present their information model implemented for one pilot developed in the "RFID from Farm to Fork" (F2F) project which looks for the extension of RFID technologies throughout the complete food chain. They describe the privacy assessment proposed by the European Union that allows the evaluation of the privacy and security impact for a RFID application under study. The main privacy risks have been identified and described by the related EU Directives concerning RFID technology. The authors describe the questionnaire elaborated by the EU to assess the privacy robustness level of a RFID application, and they showcase a real wine pilot deployed in Spain. In this chapter, the authors also examine the privacy risks in the middleware communication with both RFID reader and back-end system. The EPCIS has been the Open Source middleware solution adopted in the F2F project. For the F2F pilot deployed in the wine sector, the authors describe the privacy impact assessment questionnaire designed for this case. Finally, they discuss the threads on the RFID tags, the advantages provided by the WISP technology in this regard and its repercussion on the risk questionnaire.

DOI: 10.4018/978-1-4666-1990-6.ch010

INTRODUCTION

The production and distribution of food is one of the largest and most important activities in each country all over the world. Because of this extension of food economy, different new proposals appear to improve the quality of the products and the information received by the consumers.

The RFID technology is largely suitable to ensure traceability along the complete production chain, enabling also to collect information from the final costumer. As in other information systems, RFID is not exempt of security threats and vulnerabilities, then it is required to define and implement strategies to assure privacy, data and information protection, as well as security in applications, for all users along the production chain and final selling points.

Even that the specific security and privacy risks are strongly dependent on the RFID application, the main points susceptible of risks are the tag, the reader, the air interface, the middleware communication with the RFID reader and the back-end system. As we explain in this chapter, both regulations and assessment are required to prevent security and privacy risks.

This chapter presents the works regarding the privacy and information security developed along the project "RFID from Farm to Fork", a CIP-Pilot action involved within the 7th Frame Work of the European Union. Our proposal looks for the extension of RFID technologies throughout the complete food chain from the farms where cows, fishes, sheep, grapes, etc. are grown to the final consumer at the supermarkets, including all intermediate stages: transports, factory, processes, and storage. The main objective is the use of only one system to perform the complete traceability, recording data at each stage. These data could be useful to determine the perfect condition of the final product, but also to control the process during the elaboration. Thus, both final consumers and producers would take advantage of such systems.

The final consumers could obtain different data above the whole process undergone by the product they are buying, just by moving the object (labeled with a RFID tag) in the surroundings of a RFID reader, which can be installed in the supermarket or even as an application developed for personal smart phones. The individual identification of the product allows the software to obtain a complete traceability report from a central database, and to bring the consumer this information. Each of the producers along the chain could use the identification by radio frequency to control his production and storage, and to know some previous information of his ingredient matters. The project involved both the design of the complete system and its tests at different stages of the chain: fishing companies, wine producers, food transporters, and final users, in order to define the actual interest of the system, its performance, and its advantages and disadvantages.

The project showcases the ability of RFID technologies to make a return on investment for small and medium sized enterprises (SMEs) in the food industry, as well to provide large information to the consumer. The opportunities for such a return on investment arise from the following:

- Opportunities to create markets for premium products (organic, etc.) if technology can address authentication, condition monitoring and quality control.
- New opportunities are created to increase quality, reduce wastage, reduce energy used for refrigeration, reduce chemical usage for preservatives, optimize carbon use, etc.
- Impact on competitiveness and productivity gains.
- Potential for new markets for food producers in the regions.
- Increased productivity through reduced wastage.
- Authentication of origin, process and transport of products.

Until now these advantages have been demonstrated in large organizations, which have control over most or all of the value chain and are in a position to make an end-to-end investment. However, they are not available to independent SMEs, which only participate in one stage of the value chain. By linking RFID and sensor network technologies with a Europe wide database, which can store the exact history of any food product, SMEs will be given the opportunity to optimize their own business process to maximize the return. In addition, a pan-union resource will be created which will allow producers to demonstrate unequivocally the quality and freshness of their product, which will have the effect both of increasing consumer confidence, and increasing producer margins.

The purpose of this chapter is to describe some aspects about the privacy and security in RFID applications, as well as an assessment to help to reach the desired level of security. The first step that must be taken for the security analysis of a system is the identification of potential targets. A target can be an entire system (if the intent is to completely disrupt a business), or it can be any section of the overall system (from a retail inventory database to an actual retail item). Therefore, it is necessary to ensure the security of an RFID system to identify how their assets are being protected and how they might be targets.

When evaluating and implementing security on RFID, the data protection is very important, but sometimes attacks to physical assets can be more dangerous. The data may never be affected, even though the organization could still suffer tremendous loss (Sanghera 2007). The following example in the retail sector (Thorton, 2010) can illustrate the importance of data protection. If an individual RFID tag was manipulated so that the price at the sale point was reduced, the store would suffer a loss of the retail price, but with no damage to the inventory database system. The database was not directly attacked and the data in the database was not modified or deleted, and yet, a fraud was committed because part of the RFID system had been manipulated.

According to the model described in Figure 1, we can identify the following components on a RFID system which could experience attacks to each one of these elements: RFID tag; link tag/reader; RFID reader; link reader/middleware; and middleware/backend systems.

The above model was developed in the frame of the "From the farm to fork" (F2F) project described in the next section. The following sections of the chapter present the privacy concerns relative to the use of radio frequency identification along an actual RFID deployment. We also put emphasis on the current legal environment on

Figure 1. RFID system model

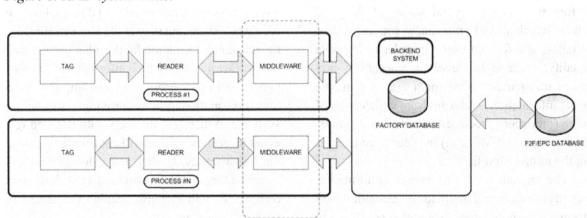

privacy protection and on the use of the EU Directive 95/46/EC which define a code of conduct and best practices, which can help to manage information security and privacy measures throughout the whole RFID-enable business process.

We explain the questionnaire for the privacy impact assessment (PIA) that has been prepared in accordance with the "Privacy and Data Protection Impact Assessment Framework for RFID Applications", adopted on 2011 by the European Data Protection Commission. It aims to uncover the privacy risks associated with an RFID application and to address them.

The middleware EPCIS used for the link between reader and back-end is described and analyzed in terms of privacy and security. The PIA questionnaire is answered for the pilot developed for the F2F project implemented in the wine sector. Finally, the WISP technology and its advantages provided for the RFID tags are also discussed in this chapter explaining the effect on the PIA questionnaire. Terms from EU Directive 95/46/EC related to data protection are also incorporated and defined.

BACKGROUND

In the F2F project, funded by the EU 7th Framework in 2009, an information traceability system was implemented. This project aimed at identify the information flow considering all details of information in the several stages of the food chain, which might be relevant by processing, for retailers and for end users, in order to improve quality, reach higher levels of traceability and more transparency. One main question maybe what information is relevant from which point of view (producers, processers, retailers, consumers, health experts, SME, etc.) in order to consider it in the information flow.

The ultimate goal is to have the information readily available and in a ready-to-be-understood format at the point of need for each user of the new

system. To achieve this, the relevant information needs to be fed into the information chain at some point in order that it can be transported along with the foodstuffs on the RFID tags to the users.

Possible "locations" to include the information into the information chain are food producers (i.e. farmers producing grapes, meat, milk etc.), food transporters (i.e. the ones in charge of transport of the primary foodstuffs to the place of food "refining"/"procession"), food processors, food transporters (to point of sale), food retailers/seller.

For each case, the point of upload of the information that makes the most sense needs to be identified, but in any case, the information needs to be available in a standardized format to enable its sustainable use. Furthermore protocols, privacy and security requirements has been identified.

After having identified and collected all necessary information, the information modeling and architecture has been developed and evaluated as well as then tested in the second project's phase by implementation of pilots in specific food sectors. This project had shown strong contribution and collaboration with the following issues: systems integration and databases construction, especially in mapping of technology to information model.

Overall Benefit of Privacy and Security

By means of an information traceability system the single users in the food chain would not only have access to the information of direct relevance for each of them, but they could also benefit from the relevant information for the other users. Once farmers are aware of which information is important for all other users, they can improve their own information process providing the whole system and therefore all users with detailed relevant information. This will contribute to ensure the transparency, quality and value of the new system. Once quality is provided, confidence and compliance will increase, leading the access to premium markets.

Considering that traceability systems aim to catch and recognize identifiers in the course of operation, for example recognizing their location and time until the post sale stage (end consumer's household), one major issue in this systems should be to ensure privacy and security fulfilling all related requirements. This project also intended to explore different existing techniques to assure privacy of all system users and system's security. Existing technologies allow considering new ways to implement privacy. So, from this point of view they define a technology to obtain that aim: privacy. It is a transformative technology: enhance the security of personal data, if collected/used.

In order to reach this goal in relation to end consumers, guidelines and rules for administrators and users must be defined (consumer notice, consumer choice, etc). Regarding the system's security the use of smart tags, key encryption units, cryptographic protocols as well as closed identification must be investigated. In addition, issues concerning product items authentication must be analyzed and levels of security and granularity of the security features should be defined.

In the next section we explain the "Privacy and Data Protection Impact Assessment Framework for RFID Applications", which has been adopted the 11th of February 2011 by the Working Party 29 of the European Data Protection Commission. The principles described on it aim to help to uncover the privacy risks associated with an RFID application and to address them; and they are sufficiently general to be applicable to all RFID applications.

PRIVACY AND DATA PROTECTION IMPACT ASSESSMENT FRAMEWORK FOR RFID APPLICATIONS

The European Commission ("the Commission") issued a recommendation dated 12 May 2009 on the implementation of privacy and data protection principles in applications supported by Radio Frequency Identification ("RFID recommendation"). In that recommendation, the Commission established a requirement for the endorsement by the Article 29 Data Protection Working Party of an industry-prepared framework for Personal Data and Privacy impact assessments (PIA) of RFID applications. These assessments are commonly referred to as privacy impact assessments, or PIAs. This RFID application PIA framework ("the framework") addresses that requirement.

The benefits of conducting PIAs for RFID Applications are numerous. These include helping the RFID Application Operator:

- To establish and maintain compliance with privacy and data protection laws and regulations;
- To manage risks to its organization and to users of the RFID application (both privacy and data protection compliance-related and from the standpoint of public perception and consumer confidence); and
- To provide public benefits of RFID applications while evaluating the success of privacy by design efforts at the early stages of the specification or development process.

The PIA process is based on a privacy and data protection risk management approach focusing mainly on the implementation of the EU RFID recommendation and consistent with the EU legal framework and best practices.

The PIA process is designed to help RFID application operators uncover the privacy risks associated with an RFID application, assess their likelihood, and document the steps taken to address those risks. These impacts (if any) could vary significantly, depending on the presence or lack of personal information processing by the RFID application. The PIA framework provides guidance to RFID application operators on the risk assessment methods, including adequate measures to mitigate any likely data protection or privacy impact in an efficient, effective and proportionate manner.

Finally, the PIA framework is sufficiently general to be applicable to all RFID applications, while allowing for particularities and specificities to be addressed at sectorial or application type level.

The PIA framework is part of the context of other information assurance, data management, and operational standards that provide good data governance tools for RFID and other applications. The current framework could be used as a basis for the development of industry-based, sector-based, and/or application-based PIA templates. As in the implementation of any theoretical document, the PIA framework may require clarification of its application of terms, as well as guidance on practices that should be based on practical experience, that may help in its implementation.

In the context of RFID technology, the following taxonomy applies:

- A *Privacy Impact Assessment (PIA)* is a process whereby a conscious and systematic effort is made to assess the privacy and data protection impacts of a specific RFID application with the view of taking appropriate actions to prevent or at least minimize those impacts.
- The *Framework* identifies the objectives of RFID application PIAs, the components of RFID applications to be considered during PIAs, and the common structure and content of RFID application PIA Reports.
- A *PIA Report* is the document resulting from the PIA process that is made available to competent authorities. Proprietary and security sensitive information may be removed from PIA reports before the Reports are provided externally (e.g., to the competent authorities) as long as the information is not specifically pertinent to privacy and data protection implications. The manner in which the PIA should be made available (e.g., upon request or not) will be determined by member states. In particular, the use of special categories of

data may be taken into account, as well as other factors such as the presence of a data protection officer.
- *PIA Templates* may be developed based on the framework to provide industry-based, application-based, or other specific formats for PIAs and resulting PIA reports.

These and other terms, such as *Users* and *Individual*, are for the purpose of this PIA framework also described in Key Terms at the end of this chapter.

The execution and reporting, where appropriate, of PIAs are in addition to other obligations that the RFID application operators may have under specific applicable laws, regulations, and other binding agreements. RFID application operators should have their own internal procedures to support the execution of PIAs

Internal Procedures

RFID Application Operators should have their own internal procedures to support the execution of PIAs, such as the following:

- *Scheduling of the PIA process* so that there is sufficient time to make any needed adjustments to the RFID application and to make the PIA report available to the competent authorities at least six weeks before deployment.
- *Internal review of the PIA process (including the initial analysis) and PIA reports* for consistency with other documentation related to the RFID application, such as system documentation, product documentation, and examples of product packaging and RFID tag implementation. The internal review should provide a feedback loop to address any impacts collected after the Application is implemented and to accommodate results from prior PIAs.

- *Compilation of supporting artifacts* (that may include results of security reviews, controls designs and copies of notices) as evidence that the RFID application operator has fulfilled all of the applicable obligations.
- *Determination of the persons and/or functions within the organization who have the authority for relevant actions* during the PIA process (e.g., completion of the PIA initial analysis and PIA report, signing the PIA report, maintaining applicable documents, and any separation of duties for these functions).
- *Provision of criteria for how to evaluate and document whether the application is ready or not ready for deployment* consistent with the Framework and any relevant PIA template.
- *Consideration/Identification of factors that would require a new or revised PIA report* is warranted. Criteria should include: significant changes in the RFID application, such as material changes that expand beyond the original purposes (e.g., secondary purposes); types of information processed; uses of the information that weaken the controls employed; unexpected personal data breach[1] with determinant impact and which wasn't part of the residual risks of the application identified by the first PIA; defining of a period of regular review; responding to substantive or significant internal or external stakeholder feedback or inquiry; or significant changes in technology with privacy and data protection implications for the RFID application at stake. Material changes that would narrow the scope of collection or use would not trigger per se the need for a revised PIA. Throughout the lifetime of the RFID application, a new or revised PIA Report would be warranted if the RFID applica-

tion changes in level as described in the *Phase I: Initial Analysis* subsection.

- *Stakeholder Consultation.* Opinions and feedback from relevant stakeholders related to the RFID application under review should be appropriately considered as part of the PIA review of potential concerns and issues. Consultations should be appropriate to the scale, scope, nature, and level of the RFID application. Within companies, individuals are designated with responsibility for overseeing and assuring organizational or departmental privacy. These individuals are essential participants in the PIA process to the extent that they are involved in the particular RFID applications or their oversight. Employees with knowledge of technical, marketing and other disciplines may also be needed participants in the process, depending upon the nature of the RFID application and their relation to it. RFID operators may have consultation mechanisms by which external stakeholders, whether individuals, organizations or authorities, can interact with them and provide feedback. As far as is appropriate, the RFID operator should use consultation mechanisms to gain input from the groups representing the individuals whose privacy will be directly impacted by the proposals, e.g. employees and customers of the RFID operator.

THE PIA PROCESS

The purpose of the framework is to provide guidance to RFID application operators for conducting PIAs on specific RFID applications, as called for in the recommendation, and to define the common organizational structure and content categories of the PIA reports in which the results from such PIAs have to be documented.

In addition, because many RFID application operators within particular sectors may be considering the same or similar RFID applications, the framework provides a basis for the development of PIA templates for particular applications or industry sectors. PIA templates can assist these sectors to conduct PIAs and produce the resulting PIA reports for these similar RFID applications more efficiently[2].

Because common RFID applications may be offered in a number of Member States, the framework is designed to harmonize requirements for RFID application operators consistent with local laws, regulations, best practices and other binding agreements. The framework addresses the process for conducting PIAs of RFID applications before deployment and specifies the scope of resulting PIA reports.[3]

RFID application operators must develop a PIA for each RFID application they operate. If they deploy several related RFID applications (potentially in the same context or at the same premises) they may create one PIA report if the boundaries and differences of the Applications are explicitly described in the PIA report. If RFID application operators reuse one RFID application in the same way for multiple products, services or processes, they may create one PIA report for all products, services or processes that are similar (for example, a car manufacturer deploying the same anti-theft mechanisms in all cars and under the same service conditions). The execution and reporting, where appropriate, of PIAs are in addition to other obligations that the RFID application operators may have under specific applicable laws, regulations, and other binding agreements.

The PIA process has two phases:

1. *Initial Analysis Phase:* the RFID application operator will follow the steps outlined in this Section to determine:
    a. Whether a PIA of its RFID application is required or not; and
    b. If a Full or Small Scale PIA is warranted.
2. *Risk Assessment Phase:* it outlines the criteria and elements of Full and Small Scale PIAs.

Phase I: Initial Analysis

As a prerequisite to conducting a PIA for a specific application, each organization must understand how to implement such a process based on the nature and sensitivity of the data it deals with, the nature and type of processing or stewardship of information it engages in, and the type of RFID application in question. For those organizations that may already have privacy risk assessment processes in place for other applications, the classification criteria and process steps should help them map their existing PIA processes to this framework.

In order to conduct the initial assessment, an RFID application operator has to go through the decision tree depicted in Figure 2. This will help the RFID application operator to determine whether and to what extent a PIA is needed for the RFID application at hand.

The resulting level in the initial analysis phase helps determine the level of detail necessary in the risk assessment (e.g., either a Full Scale or Small Scale PIA). This initial analysis must be documented and made available to data protection authorities upon request. For documentation guidelines, please see *Appendix 1.*

A *Full Scale PIA* is required for applications that are determined to be Level 2 or Level 3 by the initial analysis phase. Examples of applications requiring a Full Scale PIA include applications that process personal information (Level 2) or where the RFID tag contains personal data (Level 3). While both Level 2 and Level 3 result in a Full Scale PIA, they identify different risk environments and as such will have different mitigation strategies. For example, Level 2 applications may have controls to protect back-end data while Level 3 applications may have controls to protect both back-end data and tag data. Industry may further refine these levels and how they impact the PIA process with further experience.

Since the application processes personal data, a highly detailed risk assessment (Full Scale)

Figure 2. Decision tree on whether and at what level of detail to conduct a PIA

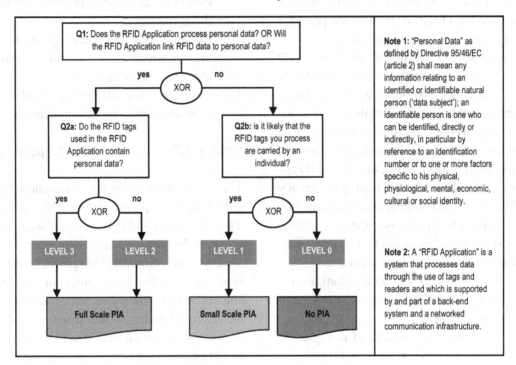

is necessary to ensure that mitigations are well elaborated. This will help the RFID application operator to identify relevant risks and develop appropriate controls. In this context, operators should also consider whether the RFID tag's information is likely to be used beyond the initial purpose or context understood by the individual, particularly if it could be used to process or link to personal data, and whether a new PIA analysis is warranted or other mitigating controls should be employed.

Small Scale PIAs follow the same process as Full Scale PIAs, but given the lower risk profile a Small Scale PIA is more restricted in scope and level of detail in both the inquiry and the report than a Full Scale PIA. Small Scale PIAs are relevant for Level 1 Applications. While a Small Scale PIA follows a similar process to the Full Scale PIA, since the relevant risks of a Level 1 application are lower than Level 2 or Level 3, the required controls and corresponding documentation in the PIA report are simplified.

Risk Assessment Phase

The objective of a risk assessment is to identify the privacy risks caused by an RFID application –ideally at an early stage of system development– and to document how these risks are *pro-actively* mitigated through technical and organizational controls. In this way a PIA plays an important role in compliance and the legal requirements of privacy (Directive 95/46) and is a measure by which we judge the effectiveness of the mitigation procedures. To save time and cost, it is recommended to run through this risk assessment phase well before final decisions on an RFID application's architecture are taken so that technical privacy mitigation strategies can be embedded into the system's design, and do not need to be 'bolted on' later.

A risk assessment process typically considers in the first instance the risks of an RFID application in terms of their likelihood of occurrence and magnitude of their consequences. RFID application Operators are advised to use the privacy targets of the EU Directive as a starting point for their risk assessment (see *Appendix 2*).

Privacy risks may be high, because the RFID application implementation could be susceptible to malicious attacks or because organizational or environmental privacy controls do not exist. Privacy risks may also be small, simply because their occurrence is unlikely in the environment or organization at hand, or because the RFID application is already configured in a highly privacy friendly way. The PIA process aims to consider all potential risks and then reflects on their magnitude, likelihood and potential mitigation. The result of this reflection is the identification of those privacy risks that are really relevant for the organization's RFID deployment and that need to be mitigated through effective controls.

The PIA Process (as visualized in Figure 3) requires any RFID application operator to:

1. Describe the RFID application;
2. Identify and list how the RFID application under review could threaten privacy and estimate the magnitude and likelihood of those risks;
3. Document current and proposed technical and organizational controls to mitigate identified risks; and
4. Document the resolution (results of the analysis) regarding the application.

Step 1: Characterization of Application

The Application characterization should give a comprehensive and full picture of the application, its environment and system boundaries. The application design, its adjacent interfaces with other systems, and information flows are described. Data flow diagrams that show processing of primary and secondary data are recommended to visualize information flows. Data structures need to be documented, too, so that potential links can be analyzed. *Appendix 1* summarizes the elements that characterize an RFID application for the purposes of conducting a PIA.

In addition, information related to the application's operational and strategic environment is recommended. This may include the immediate

Figure 3. The PIA process

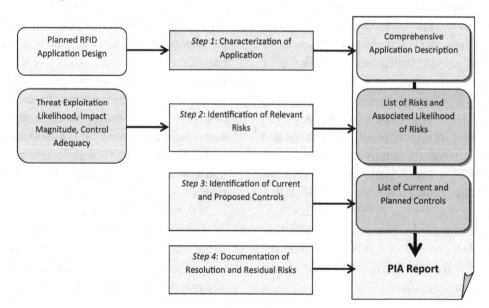

and longer-term system mission, stakeholders in information collection, functional requirements, all potential users and a description of the RFID application's architecture and data flows (in particular, interfaces to external systems that may process personal data).

Step 2: Identification of Risks

The goal of this step is to identify conditions that may threaten or compromise personal data using the EU Directive as a guide for important hallmarks of privacy targets to protect. Risks may be related to the RFID application components, its operations (collection, storage and processing infrastructure) and the data sharing and processing environment in which it is embedded.

A list of potential privacy risks may be found in *Appendix 3*. They serve as a guide for a systematic identification of potential risks that threaten the EU Directive targets (*Appendix 2*).

In addition to the identification of risks, a PIA requires a relative quantification of these risks. An RFID application operator should consider, as informed by the principles of proportionality and under reasonable terms, the *likelihood* of privacy risks occurring. Risks can occur from inside and, where warranted, outside of the particular RFID application under consideration. These risks may be derived from both the likely uses and possible misuses of the information, and in particular if the RFID tags used within the RFID application remain operational once in possession of individuals.

The risk assessment requires evaluating the applicable risks from a privacy perspective; the RFID operator should consider:

1. The significance of a risk and the likelihood of its occurrence.
2. The magnitude of the impact should the risk occur.

The resulting risk level can then be classified as low, medium or high.

A risk that has caused a prime subject of debate is that RFID tags could be used for the profiling and/or tracking of individuals. In this case the RFID tag's information – in particular its identifier(s) – would be used to re-identify a particular individual. Retailers who pass RFID tags on to customers without automatically deactivating or removing them at the checkout *may* unintentionally enable this risk.

A key question, though, is whether this risk is likely and actually materializes into a dismissible risk or not. According to point 11 of the EU RFID recommendation, retailers should deactivate or remove at the point of sale tags used in their application unless consumers, after being informed of the policy in accordance with this Framework, give their consent to keep the tags operational. Retailers are not required to deactivate or remove tags if the PIA report concludes that tags that are used in a retail application and would remain operational after the point of sale do not represent a likely threat to privacy or the protection of personal data as stated in point 12 of the same Recommendation. Deactivation of the tags should be understood as any process that stops those interactions of a tag with its environment which do not require the active involvement of the consumer.

Sector specific templates that shall be developed over time on the basis of this Framework and for use in different industries may inform risk identification in greater detail.

Step 3: Identification and Recommendation of Controls

The goal of this step is to analyze the controls that have been implemented or are planned for implementation, to minimize, mitigate or eliminate the identified privacy risks.

Controls are either of a technical or nontechnical nature. Technical controls are incorporated into the Application through architectural choices or technically enforceable policies, e.g. default settings, authentication mechanisms, and encryption methods. Nontechnical controls on the other hand are management and operational controls, e.g. operational procedures. Controls can be categorized as being preventive or detective. The former ones inhibit violation attempts and the latter ones warn of violations or attempted violations.

There can also be 'natural' controls created by the environment. For example, if there are no readers installed that could conduct a tracking of items or individuals (i.e. because there is no business case for it), then naturally there is also no (likely) risk.

The identified risks and their associated risk levels should guide the decision on which of the identified controls are relevant and thus need to be implemented. The PIA documentation should explain how the controls relate to specific risks, and should elaborate on how this mitigation will result in an acceptable level of risk.

Examples of controls are provided in *Appendix 1V.*

Step 4: Documentation of Resolution and Residual Risks

Once the risk assessment has been completed, the final resolution about the application should be documented in the PIA report, along with any further remarks concerning risks, controls and residual risks.

- An RFID application is approved for operations once the PIA process has been completed with relevant risks identified and appropriately mitigated to assure no significant residual risks remain in order to meet the requirements of compliance, with appropriate internal reviews and approvals.
- Where an RFID application is not approved for operations in its current state, further consideration will require a specific corrective action plan to be developed, and a new privacy impact assessment to be

completed in order to determine if the application has reached an approvable state.

The resolution should be associated with the following information:

- Name of the person signing the resolution.
- Title of the person.
- Date of the resolution.

The PIA Report

PIAs are internal processes containing sensitive information that can have security implications as well as potentially confidential and proprietary information of the company related to products and processes. A PIA report should typically include:

1. The Description of the RFID application as outlined in *Appendix 1.*
2. Documentation of the four steps outlined above.

The signed PIA report that contains an approved resolution should be given to the assigned company's data privacy/security official in accordance with the RFID application operator's internal procedures. This report is provided without prejudice to the obligations set forth in the EU Directive 95/46/EC for data controllers, most notably the independent obligation to notify the competent authority as described in section IX of EU Directive 95/46/EC.

F2F: PILOT

In the following subsections we describe the information model and hardware requirements for physical allocation of the information system planned by a wine pilot of the F2F project.

The pilot in the wine sector was developed in the frame of the present project. The pilot takes place in Ribadavia, Spain, with the company

"Vitivinícola del Ribeiro". The pilot is led by University of Vigo. The basic scope of the pilot involves one farm (vineyard) and the winery. The information model and collection of information are following described.

Information Model

The identified security threats related to RFID tags are tag cloning, tag snatching, tag data manipulation, and disruption. They are commented in the following paragraphs.

Regarding the processes taking place along the supply chain of the Vitivinícola del Ribeiro winery, we have identified their different steps, and the information to be managed in each of them, as follows:

- Plants monitoring and caring:
- Meteorological conditions: ambient temperature and humidity, solar radiation.
- Growing conditions of vines and grapes (botanic information): soil moisture and temperature, soil water content and leaf wetness.
- Treatments applied:
  ◦ Fertilizer, sulfates: date, type, quantity.
  ◦ Pruning, purged: date.
- Sensor location and ID.

Regarding the meteorological and growing conditions data, they have been collected using wireless sensor networks (WSNs). In addition to these environmental data, not less important additional information has been recorded for traceability purposes, namely:

- Grape collection: Collection date/time (beginning, end), GPS location, delivery note number (to link the box with the supplier's data if necessary), field's name, vines' plantation date, type of grape, box/container ID, reader ID.

- Winery entrance:
- Entrance date/time.
- Selection parameters measure: weight, quality, and type.
- Quantity of grapes collected.
- Reader ID.
- Chute dumping: Containers ID, chute ID, reader ID.
- Press: Type of press fractions (light, hard), entrance date/time (for each press fractions if necessary), volume, chute ID, press ID, reader ID.
- Decantation (débourbage): entrance/exit date, input/output volume, chemical data (analysis results and date), press ID, decantation tank ID, reader ID.
- Fermentation: entrance date, exit date, temperature, density, chemical data (analysis results and date, treatments applied, quantity and date), input/output volume, decantation tank ID, fermentation tank ID, holding vat data (location, material...), reader ID.
- Conservation: entrance / exit date, racking data (number, data, input/output volume, tanks ID), temperature, chemical data, holding vat location, fermentation tank ID, conservation tanks ID, reader ID.
- Treatments: type (filtering, spin, refrigeration), date, employee ID, equipment ID, last conservation tank ID, reader ID.
- Blend: types of wine, date, last conservation tank ID, blending tank ID, reader ID.
- Bottles: bottling date, extra material data (label, cork ...), total number of bottles, bottling filters, blending tank ID, packing box ID, reader ID.
- Storage: entrance date, location, temperature, humidity, number of bottles per box, packing box ID, pallet ID, reader ID.
- Delivery: delivery date, retail/supplier ID, truck registration number, packing box ID, reader ID.

From the above description we can observe that the pilot has deployed both WSNs for environmental data monitoring and RFID systems for traceability purposes. An information system was planned based on three different databases, namely:

- One database for the data coming from the sensors. At the moment a database in Excel is used but it might be migrated to MySQL. The meteorological measures obtained from the sensors (temperature, humidity, solar radiation, soil moisture ...) are periodically stored in the excel database, together with the data capture time.
- One local database with the traceability data needed for internal management operations. The structure of this database is not defined yet.
- EPCIS repository to store business events and master data.

The following software was used for the development of the pilot:

- Taylor made software in java for sensor network data management.
- MySQL Community Server databases.
- Aspire/Fosstrak RFID middleware.
- GlassFish 3.x web application server to run the MySQL server and Aspire/Fosstrak middleware (or Tomcat).
- Web Applications to send sensors and RFID data to the DB.

In order to store the relevant information it has been identified the following hardware requirements for physical allocation of the information system:

- One local PC to run the software controlling and processing the data coming from the sensors and to send it to a server that stores the filtered information received in an Excel database.
- It is also necessary to use a gateway connecting the sensor network with the PC, and a GPRS modem linking the local PC and the server.
- One server to store the internal database for traceability and to run the RFID middleware software.
- Network infrastructure to connect readers and computers.
- One PC as a terminal in the winery packing portal.
- Two RFID printers (one in the vineyard and another in the winery).

The data provided by the coordinator of the WSN has been registered in a local database. These data from the vineyards are related to each wine bottle by means of the GPS location and the grapes selection. Mean values of different parameters could be added to the European database, to be accessible by the final consumers of each bottle. Internal databases also store the different data provided by the RFID system. Once the wine is bottle, the technicians of the winery will decide the information added to the European database, to be accessible by the distribution partners and the final consumers. A more detailed scheme of the information model in the winery company is shown in Figure 4.

Sensor data gathered with the aid of the WSN are locally stored in a PC. Both the computer and the gateway have been installed in a hut to ensure power supply for the field equipment used during the pilot duration. The data stored in the local PC are transmitted to a remote server at the University of Vigo. Thus, all the sensor data are real-time available outside the vineyard. To achieve this data transmission, a GPRS modem is needed, since there is no line of sight between the hut location and the "Vitivinícola del Ribeiro" central building. Figure 5 details the hardware architecture of the whole system.

Figure 4. Information model design for the wine pilot in Vigo, Spain

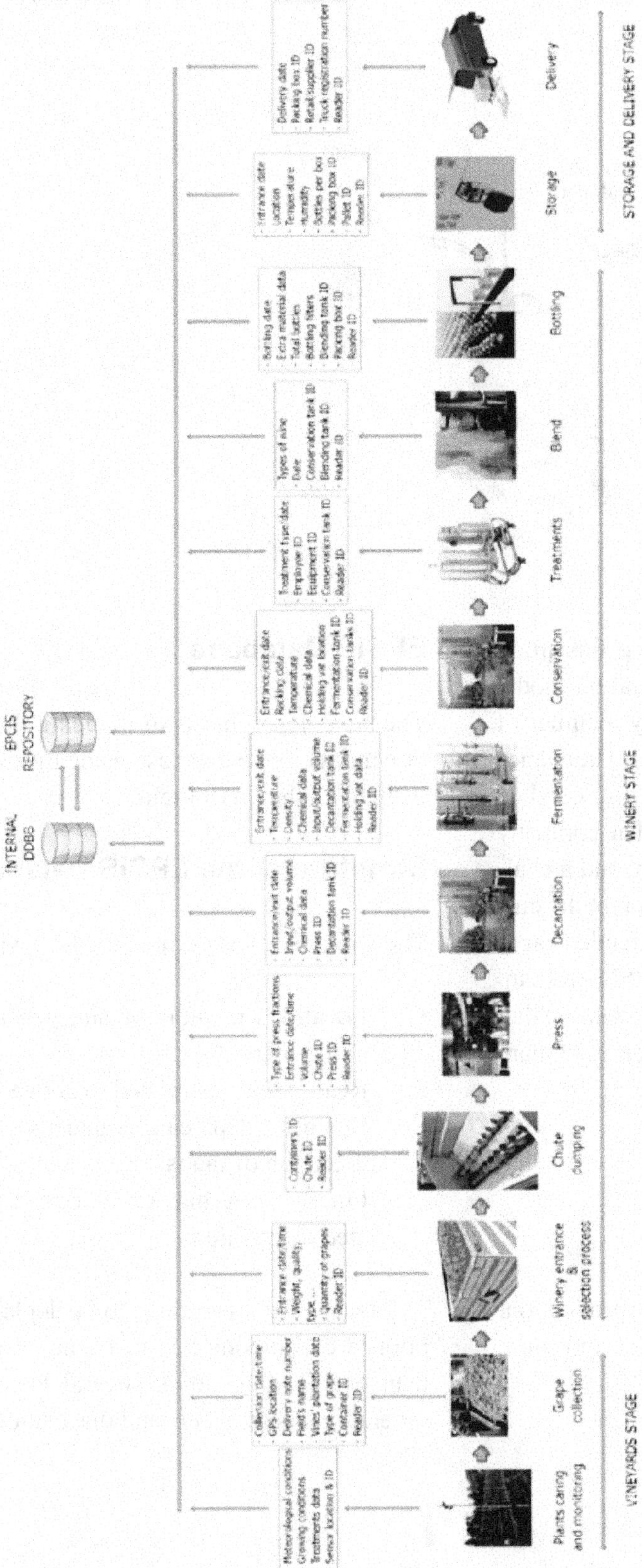

Figure 5. Hardware requirements for physical allocation of the information system planned for the wine pilot in Vigo, Spain

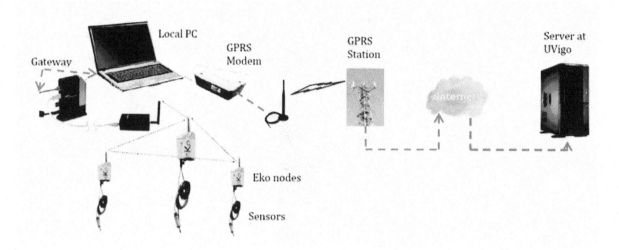

Other relevant issues concerning this pilot, and that will affect the final information model is that the company is not currently member of the Global Standard organization GS1. The chance to become a member need to be discussed in future meetings with the company. The company is currently using a bar code system, and a challenge of the pilot will be to complement the currently used bar code system with extended capabilities provided by the RFID and WSNs system. Finally, it has been identified the information that could be placed in international databases to share among pan-European end users:

- Origin of the grapes.
- Type of grape.
- Quality of grape.
- Type of wine.
- Treatments applied during wine production.
- Storage conditions (i.e. max. and min. temperature).

EPCIS Database

The next step consists of mapping the information model into a database using the middleware EPICS, as shown in Figure 4.

Structure of the EPCIS Database

The wine pilot has defined the following items:

- Locations to monitor and generate traceability data
- Read points, as related to above locations
- Business steps can be generated upon the detection of items
- Finally, items may be in one of the enumerated dispositions

Traceability operations to be deployed in the pilots are also complex, involving always more than one reading point, several locations, and several business events and dispositions.

Table 1. EPCIS database structure for F2F wine pilot in Spain

LOCATIONS	READ POINTS	BUSINESS STEPS	DISPOSITIONS
Vineyard	Vineyard handheld	Monitoring	Monitored
Winery-receiving room	Winery handheld	Collection	Collected
Winery-making room	Winery packing portal	Selection	Classified
Winery-bottling room	Fork lift fixed reader	Chute-damping	Dumping
Winery-warehouse		Press-entry	Pressing
		Decantation	Decanting
		Fermentation	Fermenting
		Conservation	In conservation
		Treatment	In treatment
		Blend	Blended
		Bottling	Bottled
		Storage	Stored
		Delivery	Sold

EPCIS: RFID MIDDLEWARE

The middleware is the software component between the readers and the backend applications. This module is often responsible for processing streams of tag or sensor data coming from reader devices.

By nature, RFID tags are dumb devices. Upon query from a reader, they reply with an identifier, usually a number or short string that is used to uniquely identify the tag and the item it is attached to. The real brain of any RFID deployment is in the middleware and backend systems. In most given deployments, the backend is usually a database that provides an interface for users to obtain meaningful data.

The system will not work without middleware, and the database application will not be functional if it cannot place data into it. A reader spits out numbers or strings with no real form; therefore, a database needs a piece of middleware to translate between the reader and the database, which is usually done through an application that interacts with the tag. The middleware application then plays "fill in the blank" when talking to the database, creating SQL statements and inserting the relevant information into the right place.

Many of the unique challenges come from the vastly larger quantity of fine-grained data that originates from radio frequency (RF) tag readers, as compared to the granularity of data that traditional enterprise applications are accustomed to. Hence, a lot of processing performed by RFID middleware concerns data reduction operations such as filtering, aggregation, and counting of tag data, reducing the volume of data prior to sending it to Enterprise Applications.

EPCglobal Architecture Framework Version 1.0 defines how RFID middleware works, and how it defines the interface to enterprise applications. In particular, RFID middleware role is responsible for the following tasks (Fummi 2007):

- Reader coordination
- Access coordination for concurrent access
- Data filtering and aggregation
- Logical readers (readers grouping)

RFID middleware must be able to transform data from the variety of existent RFID readers and redirect it to applications or networks of any type. A schematic of the EPCglobal Architecture Framework is shown in Figure 6.

Specific requirements for EPC processing vary greatly from application to application. Therefore, the emphasis in the RFID middleware specification is on extensibility rather than specific processing features. As this layer is rich in computational resources, data gathered from reader layer auditing may be processed for security vio-

Figure 6. The EPCglobal architecture framework (www.epcglobalinc.org)

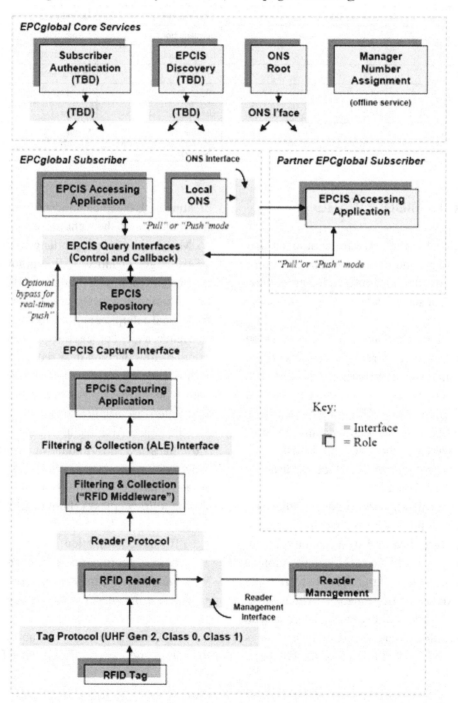

lations using a detection module (Thamilarasu 2008, Sanghera 2007). More details are available at www.epcglobalinc.org.

Middleware Security Threats and Solutions

RFID Middleware can be considered as an application server; therefore the application server security threats are also applicable to RFID middleware (Koindala 2006). Summarizing this section, Table 2-7 contains the identified security threats and the proposed solutions in the middleware level

Table 2. Security threats and solutions in middleware

Security threat	Security solution
Malicious code injection Viruses DoS	System authentication and authorization Check input data Disable backend scripting languages Isolate the RFID middleware server Security audit Antivirus SW, firewall
Buffer overflow	Radio-shielded environment Bounds checking

Table 3. Security threats and solutions in EPCIS capturing and accessing application

Security threat	Security solution
Flooding	Staging area
Purposeful tag duplication Spurious events Readability rates	Events checking and correlation
Virus attacks	System authentication and authorization Data checking Antivirus SW, firewall
Buffer overflow	Data checking

Table 4. Security threats and solutions in EPCIS repository

Security threat	Security solution
Privilege abuse	Query-level access control
Privilege elevation	Query-level access control Intrusion prevention systems (IPS) Audition
SQL injection	Query-level access control Intrusion prevention systems (IPS) Event correlation
Platform vulnerabilities	Software Updates
Denial of Service	Timing control Connection rate control Intrusion prevention systems (IPS) Query access control
Backup data exposure	Encryption

Table 5. Security threats and solutions in EPCIS interfaces

Security threat	Security solution
MIM attack	Secure gateway (SSL-TLS / EAP-TLS)
TCP replay attack	Authentication

Table 6. Security threats and solutions in the ONS

Security threat	Security solution
Packet interception and manipulation ID guessing attacks	VPN or SSL Tunneling
Name chaining or cache poisoning Betrayal by trusted server Denial of Service (DoS) Authenticated denial of domain names Unauthorized updates	Limit Usage DNS Security Extensions (DNSSEC)

Table 7. Security threats and solutions on web user interfaces

Security threat	Security solution
Cross site scripting (XSS)	Content filtering Escaping Browser collaboration Use proxy devices and firewalls
Cross site request forgery (XSRF)	Use proxy devices and firewalls Session expiration date Security tokens View state Challenge-response options Checking Referrer header Custom HTTP headers Double authentication Double submitting cookies

THE PIA QUESTIONNAIRE FOR F2F PROJECT PILOTS

As aforementioned, this questionnaire is intended to identify whether the planned activity impacts privacy concerns, in every pilot undertaken in the F2F project. The first step to follow when conducting a PIA for an RFID application is to determine whether and what extent the PIA should be. The first part of the questionnaire for the PIA of the F2F pilots is focused on it, involving these questions:

- *Will the RFID application process personal data? OR Will the RFID application link RFID data to personal data?*
  - *If so, do the RFID tags used in the RFID application contain personal data?*
  - *If not, is it likely that the RFID tags you process are carried by an individual?*

The answers given to these questions allow the classification of the RFID application in one of four possible levels following the decision tree shown in Figure 2, and determine whether the PIA is needed and its level of detail necessary on risk assessment.

Applications with level 0 do not require a PIA. In the other levels, the higher the level is, the more detailed the analysis of the RFID application should be. Applications that process personal information (Level 2) or where the RFID tag contains personal data (Level 3) need a full scale PIA. Applications of level 1 have lower risk profile and so require less level of detail.

The rest of questions are focused on the performance of a risk assessment, in order to identify the privacy risks caused by the RFID application.

The questions are divided into several blocks relating to different aspects involved in the protection of personal data:

A. *Privacy management in the organization:* This section includes questions regarding the current privacy policies implemented in the pilot and the people responsible for it and their level of awareness.

- *Does the pilot have a privacy policy in place?*
- *Is there an appointed privacy or information governance contact person?*
- *How often does your organization perform a risk assessment procedure to assess security-related risks from internal and external threats to your facility, its assets, and/or personnel? Who performs the risk assessment?*
- *Are employees with access to personal data information in the organization provided with training related to privacy protection?*
- *Who is responsible for assuring proper use of the data?*

B. *Information Collected and Maintained:* Questions in this section are about what information is intended to be collected from individuals, how it is going to be done and the proposed used of the information.

- *What information is going to be collected? Generally describe the information to be used in the system in each of the following categories: Customer, Employee, and Other.*
- *How will the information be collected from individuals?*
- *Are the data subjects aware of the collection of their personal information and of the uses it will have? Do individuals have an opportunity and/or right to decline to provide information?*
- *What are the proposed uses of the personal information? How do these uses relate to the purpose for which the information was collected? That means if the use of data is both relevant and necessary to the purpose for which the information was collected and for which the system was designed.*
- *Have the data subjects consented to their personal information being used in this manner? Describe the consent process.*

C. *Technical Access and Security:* This time the questions are focused on the access control to personal data.

- *Who will have access to the data in the system (Users, Managers, System Administrators, Developers, Other)?*
- *Does the system use "roles" to assign privileges to users of the system?*
- *What procedures are in place to determine which users may access the system?*
- *What controls are in place to protect the data from unauthorized access or to prevent the misuse of data by those having access? Describe briefly the administrative, technical and physical safeguards in place to protect personal data information against theft, loss, unauthorized use or disclosure and unauthorized copying, modification or disposal. For instance: electronic access control, electronic gate*

access control system, intrusion detection system, information security products...

- *Are criteria, procedures, controls, and responsibilities regarding access documented?*
- *Do other systems share data or have access to data in this system? If yes, explain.*

D. *Sharing:* The questions are related to the sharing of information with third parts and the protection mechanisms used to protect personal data in this case.

- *Is the information shared with third parts?*
- *How the data subject has been informed of this?*
- *How is the information transmitted or disclosed from sites? What security measures will be taken to protect the shared information from loss, unauthorized access, use, modification, disclosure or other misuse?*

E. *Individual Access, Redress and Correction:* Questions in this section are about the procedures for individuals to access and modify their data.

- *What are the procedures which allow individuals to gain access to their own information?*
- *What are the procedures for individuals to correct erroneous information or erase their data and how are they notified of their existence?*

F. *Retention:* This section includes questions regarding the time the data remain in the system and its final deletion.

- *What is the retention period for the data in the system?*
- *Is the data retained only the time necessary to fulfill the specified purpose?*

- *What are the procedures for eliminating the data at the end of the retention period? Are the procedures documented?*

G. *Privacy protection and RFID tags:* The last sections are common to all the systems handling personal data from individuals, but this one is specifically for data protection in RFID applications in which the tags will contain personal data (or will be a possible link to them), or that contain no personal data but will be transported by individuals.

- *Is personal information stored on the tag?*
- *Is there any possibility to link data on the tag with data in the backend, which would result in personal data?*
- *Do tags include the possibility to disable them after they are not used anymore? Yes, we are planning to use tags with the kill capability.*
- *What controls will be used to prevent unauthorized monitoring of the users by reading the tags they carry (e.g. passwords or other authentication mechanisms)?*

To evaluate the RFID systems where any personal data could be linked with the system or will be stored on the RFID tag (Level 1) is enough to answer this section of the questionnaire.

PIA Questionnaire Results for Wine Pilot in Vigo

According to the decision tree of Figure 2 on whether and at what level of detail to conduct a PIA, for the initial phase analysis of the wine pilot deployed in Vigo, we have only the following two questions:

- *Will the RFID application process personal data? OR Will the RFID application link RFID data to personal data?*

No. Consumer personal data, if so, will be stored in another independent database.

- *Is it likely that the RFID tags you process are carried by an individual?*

Yes, as we are planning to put RFID tags on bottles.

The assessment would accomplish a Level 1 application that is considered of lower risk and it does not require a full scale PIA. For the phase of risk assessment, the questions corresponding to block *G* must be answered:

- *Privacy protection and RFID tags*
- *Is personal information stored on the tag?*

No. It´s going to store the EPC-URI only and it is only related with the product not with the consumer.

- *Is there any possibility to link data on the tag with data in the backend, which would result in personal data?*

No, consumer personal data, if so, will be stored in another independent database.

- *Do tags include the possibility to disable them after they are not used anymore?*

Yes, we are planning to use tags with the kill capability.

- *What controls will be used to prevent unauthorized monitoring of the users by reading the tags they carry (e.g. passwords or other authentication mechanisms)?*

We intend to use tags with access password.

The PIA questionnaire for the F2F wine pilot is relatively short due to the Level 1 of the application which means that any personal data could be linked with the system (Level 2) or will be stored on the RFID tag (Level 3). This fact implies that only the questions included in the block *G* must be answered.

RFID TAGS, PIA, AND WISP

One of the plausible points of the RFID application susceptible of security and privacy threats are the tags. As aforementioned explained the tags basically consist of thumb devices queried by a reader usually under the EPC Gen1 or Gen2 protocol (EPCGlobal 2005) so resulting largely exposed to security threads. The radiofrequency link established between the RFID reader and the tags is also susceptible of risks.

Table 8 contains the main identified security threats and the proposed solutions in RFID tags

Table 8. Security threats and solutions in RFID tags and in the link between RFID tags and readers

	Security threat	Security solution
TAG	Tag snatching	Anti - tamper tag and packaging
	Tag cloning Tag data manipulation	Limit the amount of data stored Tag / reader authentication Shielded enclosure
	Disruption	Shielded enclosure Reader authentication or tag´s access password (to avoid illegal deactivations)
TAG-READER	Eavesdropping Man in the middle attack (MIM attack) Replay attack	Limit the amount of data stored Data encryption Authentication
	Denial of service (DoS)	Shielded enclosure

and in the link between RFID tags and readers. Many of the solutions require involving the middleware given that the tags are passive devices not provided of computing capacity to assess data privacy risks. However, the Wireless Identification and Sensing Platform (WISP) can ensure this capacity for applications of Level 2 and Level 3 that can involve a link between sensing and personal data on the tag.

WISP technology merges the features of RFID tags, with sensing and computing capabilities (Chae 2007; Sample 2008; Yeager 2008; Yeager-Sample 2008). WISPs tags are powered by harvested energy from off-the-shelf UHF RFID readers. To a RFID reader, a WISP is just a normal EPC gen1 or gen2 tag; but inside the WISP, the harvested energy is operating a 16-bit general purpose microcontroller. The microcontroller can perform a variety of computing tasks, including sampling sensors, and reporting that sensor data back to the RFID reader. WISPs have been built with light sensors, temperature sensors, and strain gauges.

Furthermore, WISPs can write to flash memory and perform cryptographic computations. For instance, the RC5 encryption algorithm was implemented on the Intel WISP tags. This provides the tag with the possibility to protect itself of some attacks indicated in Table 8 by following the indications given in *Appendix 1* to *IV*. However, the PIA questionnaire is not affected by the technology used in the tag.

CONCLUSION

Security and privacy threats are present in each layer of the RFID systems: physical, network, application. This chapter presents the objectives and initial work of the project "RFID from Farm to Fork", highlighting the radio propagation concerns of a RFID system within an industrial environment. This chapter shows that the information security risks can also be found in the F2F RFID system, since it consists of traditional network elements

(database servers, application servers…) in addition to specific ones (RFID tags and readers).

Different countermeasure mechanisms must be designed and implemented in order to protect each part of the system with the desired level. It should be ensured that RFID operators, notwithstanding their other obligations pursuant to Directive 95/46/EC:

- Conduct an assessment of the implications of the application implementation for the protection of personal data and privacy, including whether the application could be used to monitor an individual. The level of detail of the assessment should be appropriate to the privacy risks possibly associated with the application;
- Take appropriate technical and organizational measures to ensure the protection of personal data and privacy;
- Make available the assessment to the competent authority at least six weeks before the deployment of the application;

The specific security and privacy risk depend strongly on the application, but all RFID information systems require continuous monitoring, assessment, guidance and regulation in order to avoid security and privacy risks. Raising awareness among the public and small and medium-sized enterprises about the features and capabilities of RFID will help allow this technology to fulfill its economic promise, mitigating at the same time the risks of being used to the detriment of the public interests, thus enhancing its acceptability.

The tags provided of WISP technology present the advantage of embedded computing and encryption capabilities offering a more robust option than simple passive tags. Many of the threads summarized in Table 8 can be solved by WISP. However, the PIA questionnaire is not affected by the technology used in the tag and it must be completed according to the Level 0-3 of the RFID application.

REFERENCES

Chae, H. J., Yeager, D. J., Smith, J. R., & Fu, K. (2007). Maximalist cryptography and computation on the wisp UHF RFID tag. In *Proc. Conference on RFID Security*.

Commission of the European Communities. (2009, May 12). *Commission recommendation on the implementation of privacy and data protection principles in applications supported by radio-frequency identification*. Retrieved from http://ec.europa.eu/information_society/policy/rfid/documents/recommendationonrfid2009.pdf

Commission of the European Communities. (2009, May 12). *Commission staff working document accompanying the commission recommendation on the implementation of privacy and data protection principles in applications supported by radio frequency identification: Summary of the impact assessment*. Retrieved from http://ec.europa.eu/information_society/policy/rfid/documents/recommendationonrfid200i9impact.pdf

EPCglobal. (2005). *EPC radio-frequency identity protocols class-1 generation-2 UHF RFID protocol for communications at 860MHz- 960MHz*, version 1.0.9.

European Parliament. (1995, November 23). *Directive 95/46/EC of the European Parliament and of the Council of 24 October 1995 on the protection of individuals with regard to the processing of personal data and on the free movement of such data*. Retrieved from http://ec.europa.eu/justice_home/fsj/privacy/docs/95-46-ce/dir1995-46_part1_en.pdf

European Parliament. (2002, July 12). *Directive 2002/58/EC of the European Parliament and of the Council of 12 July 2002 concerning the processing of personal data and the protection of privacy in the electronic communications sector (Directive on privacy and electronic communications)*. Retrieved from http://eur-lex.europa.eu/LexUriServ/LexUriServ.do?uri=OJ:L:2002:201:0037:0047:EN:PDF

European Parliament. (2005, January 19). *Working document on data protection issues related to RFID technology*. Article 29 Data Protection Working Party. Retrieved from http://ec.europa.eu/justice_home/fsj/privacy/docs/wpdocs/2005/wp105_en.pdf

European Parliament. (2007). *Opinion 4/2007 on the concept of personal data*. Article 29 Data Protection Working Party. Retrieved from http://ec.europa.eu/justice_home/fsj/privacy/docs/wpdocs/2007/wp136_en.pdf

European Parliament. (2009, November 25). *Directive 2009/136/EC of the European Parliament and of the Council of 25 November 2009 amending Directive 2002/22/EC on universal service and users' rights relating to electronic communications networks and services, Directive 2002/58/EC concerning the processing of personal data and the protection of privacy in the electronic communications sector and Regulation (EC) No 2006/2004 on cooperation between national authorities responsible for the enforcement of consumer protection laws*. Retrieved from http://eur-lex.europa.eu/LexUriServ/LexUriServ.do?uri=OJ:L:2009:337:0011:0036:EN:PDF

European Parliament. (2012). *Status of implementation of Directive 95/46 on the protection of Individuals in regards to the processing of personal data*. Retrieved from http://ec.europa.eu/justice_home/fsj/privacy/law/implementation_en.htm

Fummi, F., & Perbellini, G. (2007). eEPC: An EPCglobal-compliant embedded architecture for RFID-based solutions. *The Journal of Communication, 2*(7).

Information Commissioner's Office. (n.d.). *Privacy impact assessment handbook*. Retrieved from http://www.ico.gov.uk/upload/documents/pia_handbook_html_v2/files/PIAhandbookV2.pdf

Koindala, D. M., Kim, W., & Ki, K. (2006). *Security assessment of EPCglobal architecture framework.* Auto-ID Labs White Paper. Retrieved on June 12, 2011, from http://www.autoidlabs.org/uploads/media/AUTOIDLABS-WP-SWNET-017.pdf

Sample, A. P., Yeager, D. J., Powledge, P. S., & Smith, J. R. (2008). Design of an RFID-based battery-free programmable sensing platform. *IEEE Transactions on Instrumentation and Measurement, 57*(11), 2608–2615. doi:10.1109/TIM.2008.925019

Sanghera, P., Thornton, F., Haines, B., Kung Man, F., Fung, F., Kleinschmidt, J., … Campbell, A. (2007). *How to cheat at deploying and securing RFID.* Syngress, 2007.

Thamilarasu, G., & Sridhar, R. (2008). Intrusion detection in RFID systems. In *IEEE Military Communications Conference,* (pp. 1-7).

Thorton, F. (2010). RFID security part 3: Threat and target identification. *EE Times Design,* September 2010.

Yeager, D. J., Sample, A. P., & Smith, J. R. (2008). Wisp: A passively powered UHF RFID tag with sensing and computation. In Syed, M. I., & Ahson, A. (Eds.), *RFID handbook: Applications, technology, security, and privacy.* CRC Press.

ADDITIONAL READING

Ahson, S., & Ilyas, M. (2008). *RFID handbook: Applications, technology, security and privacy.* CRC Press. doi:10.1201/9781420055009

Barth, A., Collin, J., & Mitchel, J. C. (2008). Robust defenses for cross-site request forgery. *15th ACM Conference on Computer and Communications Security,* (pp. 75-88).

Chandra, P., Bensky, A., et al. (2008). *Wireless security: Know it all.*

Cole, E. (2009). *Network security bible.* Wiley.

El-Bakry, H. M., Riad, A. M., Abu-Elsoud, M., Mohamed, S., Hassan, A. E., Kandel, M. S., & Mastorakis, N. (2010). Adaptive user interface for web applications. In *4th WSEAS International Conference on Business Administration,* (pp. 190-211).

Granström, M. (2009). *What is wrong with DNS?* TKK Technical Reports in Computer Science and Engineering, Helsinki University of Technology. Retrieved on June 12, 2011, from http://cse.tkk.fi/en/publications/B/5/papers/Granstrom_final.pdf

Iha, G., & Doi, H. (2009). An implementation of the binding mechanism in the Web browser for preventing XSS attacks: Introducing the bind-value headers. In *International Conference on Availability, Reliability and Security,* (pp. 966- 971).

Ismail, O., Etoh, M., Kadobayashi, Y., & Yamaguchi, S. (2004). A proposal and implementation of automatic detection/collection system for cross-site scripting vulnerability. In *18th International Conference on Advanced Information Networking and Applications,* Vol. 1, (pp. 145-151).

Jovanovic, N., Kirda, E., & Kruegel, C. (2006). Preventing cross site request forgery attacks. In *Second International Conference on Security and Privacy in Communications Networks and the Workshops,* (pp. 1-10).

Karygiannis, M., Eydt, B., Barber, G., Bunn, L., & Phillips, T. (2007). *Guidance for securing radio frequency identification (RFID) systems.* Recommendations of the National Institute of Standards and Technology. Retrieved on June 12, 2011, from http://csrc.nist.gov/publications/nistpubs/800-98/SP800-98_RFID-2007.pdf

Kim, T., & Kim, H. (2006). Authorization policy for middleware in RFID system. In *IEEE Tenth International Symposium on Consumer Electronics,* (pp. 1-4).

Lehtonen, M., Ostojic, D., Ilic, A., & Michahelles, F. (2009). Securing RFID systems by detecting tag cloning. In Seventh International Conference on Pervasive Computing, Pervasive.

Lo, N. W., & Yeh, K. (2010). A secure communication protocol for EPCglobal class 1 generation 2 RFID systems. In *IEEE 24th International Conference on Advanced Information Networking and Applications Workshops,* (pp. 562-566).

Martin, T. (2008). *RFID authentication resistant to compromised readers.* Master Thesis, Université Catholique de Louvain and Université Bordeaux I. Retrieved on June 12, 2011, from http://www.labri.fr/perso/fleury/courses/MasterCSI/Stages/archives/2008/martin_tania-rapport_stage.pdf

Mitrokotsa, A., Rieback, M. R., & Tanenbaum, A. S. (2008). Classification of RFID attacks. In *Proceedings of the 2nd International Workshop on RFID Technology - Concepts, Applications, Challenges* (pp. 73-86).

Park, J., Kang, S., & Lee, I. (2007). A study on secure RFID authentication protocol in insecure communication. In *Third International Conference on Security and Privacy in Communications Networks and the Workshops,* (pp. 133-143).

Piramuthu, S. (2007). Protocols for RFID tag/reader authentication. *Decision Support Systems Journal, 43*(3), 897–914. doi:10.1016/j.dss.2007.01.003

RFID-F2F From Farm to Fork. (n.d.). Retrieved on October, 2011, from http://www.rfid-f2f.eu

RFID Gazette. (n.d.). Retrieved on October, 2011, from http://www.rfidgazette.org

RFID Journal. (n.d.). Retrieved on October, 2011, from http://www.rfidjournal.com

Rieback, M. R., Crispo, B., & Tanenbaum, A. S. (2006). Is your cat infected with a computer virus? In *Fourth Annual IEEE International Conference on Pervasive Computing and Communications,* (pp. 10-179).

Rieback, M. R., Simpson, P. N. D., Crispo, B., & Tanenbaum, A. S. (2006). RFID malware: Design principles and examples. *Pervasive and Mobile Computing, 2*(4), 405–426. doi:10.1016/j.pmcj.2006.07.008

Shulman, A. (2006). *Top ten database security threats.* Imperva, Inc., White Paper, Retrieved on June 12, 2011, from http://www.schell.com/Top_Ten_Database_Threats.pdf

Suliman, A., Shankarapani, M. K., Mukkamala, S., & Sung, A. H. (2008). RFID malware fragmentation attacks. In *International Symposium on Collaborative Technologies and Systems,* (pp. 533-539).

Ter Louw, M., & Venkatakrishnan, V. N. (2009). Blueprint: Robust prevention of cross-site scripting attacks for existing browsers. In *30th IEEE Symposium on Security and Privacy,* (pp. 331-346).

Xiaoli, L., Zavarsky, P., Ruhl, R., & Lindskog, D. (2009). Threat modeling for CSRF attacks. In *International Conference on Computational Science and Engineering,* Vol. 3, (pp. 486-491).

KEY TERMS AND DEFINITIONS

Individual: A natural person who interacts with or is otherwise involved with one or more components of an RFID application (e.g., back-end system, communications infrastructure, RFID tag), but who does not operate an RFID application or exercise one of its functions. In this respect, an Individual is different from a User. An individual may not be directly involved with the functionality of the RFID application, but rather, for example, may merely possess an item that has an RFID tag.

Information Security: Preservation of the confidentiality, integrity and availability of information.

Monitor: Carrying out an activity for the purpose of detecting, observing, copying or recording the location, movement, activities, or state of an individual.

Personal Data: Any information relating to an identified or identifiable natural person (data subject); an identifiable person is one who can be identified, directly or indirectly, in particular by reference to an identification number or to one or more factors specific to his physical, physiological, mental, economic, cultural or social identity.

Privacy and Data Protection Impact Assessment (PIA): Guidance and questionnaire for the privacy impact assessment on RFID applications prepared in accordance with the "Privacy and Data Protection Impact Assessment (PIA) Framework for RFID Applications", which has been adopted in 2011 by the European Data Protection Commission, to help to uncover the privacy risks associated with an RFID application and to address them.

RFID: Technology that uses electromagnetic waves to communicate with RFID tags, with the possibility of reading the unique identification numbers of the RFID tags or perhaps other information stored in them.

RFID Tag: Electronic memory that is readable and perhaps writable, and antenna. An RFID device having the ability to produce a radio signal or an RFID device which re-couples, back-scatters or reflects (depending on the type of device) and modulates a carrier signal received from a reader or writer.

RFID Application: Process information developed through the interaction of RFID tags and RFID readers.

RFID Operators: The natural or legal person, public authority, agency, or any other body, which, alone or jointly with others, determines the purposes and means of operating an application, including controllers of personal data using an RFID application supported by back-end systems and networked communication infrastructures.

RFID Reader: A fixed or mobile data capture and identification device using a radio frequency electromagnetic wave or reactive field coupling to stimulate and effect a modulated data response from a tag or group of tags.

RFID Tag Information: The information contained in an RFID tag and transmitted when the RFID tag is queried by an RFID reader.

User: Specifically, an RFID application user, i.e., a person (or other entity, such as a legal entity) who directly interacts with one or more components of an RFID application (e.g., back-end system, communications infrastructure, RFID tag) for the purposes of operating an RFID application or exercising one or more of its functions.

ENDNOTES

[1]   In this case the applicable definition shall be the one provided in the directive 2009/136/EC amending directive 2002/58 see page 29 http://eur-lex.europa.eu/LexUriServ/LexUriServ.do?uri=OJ:L:2009:337:0011:0036:EN:PDF

[2]   The concept of mutual or multiple recognition across entities and sectors for the deployment of previously vetted RFID Applications should be explored.

[3]   Point 5 (a) of the European Commission Recommendation of May 2009 on the implementation of privacy and data protection principles in Applications supported by radiofrequency identification C(2009) 3200 final.

[4]   Point 12/13 of the EC Recommendation of 12 May 2009. {SEC (2009) 585}: *Any deactivation or removal method should be made available free of charge, either immediately or at a later stage, without any reduction or termination of the legal obligations of the retailer or manufacturer towards the consumer.*

APPENDIX 1: CHARACTERISATION OF THE RFID APPLICATION DESCRIPTION

The RFID application operator should include, where applicable, the below information in the PIA Report, as per Table 9.

Table 9. PIA report information

Information	Description
RFID Application Operator	• Legal entity name and location • Person or office responsible for PIA timeliness • Point(s) of contact and inquiry method to reach the Operator
RFID Application Overview	• RFID Application name • Purpose(s) of RFID Application(s) • Basic use case scenarios of the RFID Application • RFID Application components and technology used (i.e. frequencies, etc) • Geographical scope of the RFID Application • Types of users/individuals impacted by the RFID Application • Individual access and control
PIA Report Number	• Version Number of PIA Report (distinguishing new PIA or just minor changes) • Date of last change made to PIA Report
RFID Data Processing	• List of types of data elements processed • Presence of Sensitive information in the data being processed, e.g., health
RFID Data Storage	• List of types of data elements stored • Storage duration
Internal RFID Data Transfer (if applicable)	• Description or diagrams of data flows of internal operations involving RFID data • Purpose(s) of transferring the personal data
External RFID Data Transfer (if applicable)	• Type of data recipient(s) • Purpose(s) for transfer or access in general • Identified and/or identifiable (level of) personal data involved in transfer • Transfers outside the European Economic Area (EEA)

APPENDIX 2: CHARACTERISATION OF THE RFID APPLICATION DESCRIPTION

There are today 9 privacy targets embedded in the Directive 95/46/EC, as shown in Table 10. The PIA process was developed by considering these targets and the associated risks of RFID. This Annex summarizes these privacy targets. While all targets are essential elements of organizational compliance, in many cases only a subset of these requirements will be at issue in the RFID application under consideration. Thus the role of these targets is to inform the creation and development of the PIA process more than the operation of any specific PIA.

Table 10. Privacy targets

Privacy target	Description and example
Safeguarding quality of personal data	Data avoidance and minimization, purpose specification and limitation, quality of data and transparency are the key targets that need to be ensured.
Legitimacy of processing personal data	Legitimacy of processing personal data must be ensured either by basing data processing on consent, contract, legal obligation, etc.
Legitimacy of processing sensitive personal data	Legitimacy of processing sensitive personal data must be ensured either by basing data processing on explicit consent, a special legal basis, etc.
Compliance with the data subject's right to be informed	It must be ensured that the data subject is informed about the collection of his data in a timely manner.
Compliance with the data subject's right of access to data, correct and erase data	It must be ensured that the data subject's wish to access, correct, erase and block his data is fulfilled in a timely manner.
Compliance with the data subject's right to object	It must be ensured that the data subject's data is no longer processed if he or she objects. Transparency of automated decisions vis-à-vis individuals must be ensured especially.
Safeguarding confidentiality and security of processing	Preventing unauthorized access, logging of data processing, network and transport security and preventing accidental loss of data are the key targets that need to be ensured.
Compliance with notification requirements	Notification about data processing, prior compliance checking and documentation are the key targets that need to be ensured.
Compliance with data retention requirements	Retention of data should be for the minimum period of time consistent with the purpose of the retention or other legal requirements.

APPENDIX 3: PRIVACY RISKS

This Annex provides a list of possible privacy risks related to the use of the RFID application under review. It is recommended that – in particular for Full Scale PIAs - risks are systematically identified with the help of standard risk assessment procedures that would include threats and vulnerabilities to an RFID application.

Table 11 and Table 12 provide examples of risks that may affect an entity's ability to meet the privacy targets described in *Appendix 2*. RFID application operators can use this list as a starting point; however, not all of these risks may apply to all RFID applications. RFID operators should make sure each identified risk is appropriately mitigated by one or more controls in light of the likelihood of risk occurrence and magnitude of impact. RFID application operators may need to combine controls or augment existing controls based on factors including the technology in use, nature of their implementation, type of information, and applicable policies, among others.

Table 11. Privacy risk examples (part 1)

Privacy risk	Description and example
Unspecified and unlimited purpose	The purpose of data collection has not been specified and documented or more data is used than is required for the specified purpose. Example: No documentation of purposes for which RFID data is used and/or use of RFID data for all kinds of feasible analysis.
Collection exceeding purpose	Data is collected in identifiable form that goes beyond the extent that has been specified in the purpose. Example: RFID payment card information is not only used for the purpose of processing transactions but also to build individual profiles.
Incomplete information or lack of transparency	The information provided to the data subject on the purpose and use of data is not complete, data processing is not made transparent, or information is not provided in a timely manner. Example: RFID Information available to consumers that lacks clear information on how RFID data is processed and used, the identity of the Operator, or the user's rights.
Combination exceeding purpose	Personal data is combined to an extent that is not necessary to fulfill the specified purpose. Example: RFID payment card information is combined with personal data obtained from a third party.
Missing erasure policies or mechanisms	Data is retained longer than necessary to fulfill the specified purpose. Example: Personal data is collected as part of the Application and is saved for longer than legally allowed.
Invalidation of explicit consent	Consent has been obtained under threat of disadvantage. Example: Cannot return/exchange/use legal warranties for products when RFID Tag is deactivated or removed.
Secret data collection by RFID Operator	Some data is secretly recorded and thus unknown to the data subject, e.g. movement profiles. Example: Consumer information is read while walking in front of stores or in mall and no Logo or Emblem is warning him or her about RFID readouts.

APPENDIX 4: CHARACTERISATION OF THE RFID APPLICATION DESCRIPTION

This Annex provides a list of examples of potential controls that can help an RFID application operator to identify appropriate mitigating strategies. Risks identified as relevant for an RFID application operator in Step 2 of the PIA risk process can be mitigated through one or several mitigation strategies, some of which are outlined in this *Appendix 1V*. The goal is that by running through a PIA process, the RFID application operator identifies and implements the controls necessary to mitigate the relevant privacy risks.

Potential control mechanisms include: RFID Application Governing Practices; Individual access and control; System Protection Measures (including Security Controls); Tag Protection; Accountability Measures.

These practices are ancillary to the existing European Union data protection regulatory framework and are not intended to replace it or modify its scope. They are following described.

Table 12. Privacy risk examples (part II)

Privacy risk	Description and example
Inability to grant access	There is no way for the data subject to initiate a correction or erasure of his data. Example: Employer cannot give employee a full picture of what is saved about him or her on the basis of RFID access and manufacturing data.
Prevention of objections	There are no technical or operational means to allow complying with a data subject's objection. Example: Hospital visitor cannot opt out of reading out sensitive personal information on tags (i.e. medications).
A lack of transparency of automated individual decisions	Automated individual decisions based on personal aspects are used but the data subjects are not informed about the logic of the decision-making. Example: Without notice to consumers, an RFID Operator reads all tags carried by an individual, including tags provided by another entity, and determines what type of marketing message the individual should receive based on the tags.
Insufficient access right management	Access rights are not revoked when they are no longer necessary. Example: Through an RFID card, an ex-trainee gets access to parts of an enterprise where he or she should not.
Insufficient authentication mechanism	A suspicious number of attempts to identify and authenticate are not prevented. Example: Personal data contained on tags is not protected by default with a password or another authentication mechanism.
Illegitimate data processing	Processing of personal data is not based on consent, a contract, legal obligation, etc. Example: An RFID Operator shares collected information with a third party without notice or consent as otherwise legally allowed.
Insufficient logging mechanism	The implemented logging mechanism is insufficient. It does not log administrative processes. Example: It is not logged who has accessed the RFID employee card data.
Uncontrollable data gathering from RFID Tags	The risk that RFID Tags could be used for regular profiling and/or tracking of individuals. Example: Retailer reads all tags that they can see.

RFID Application Governing Practices

Governing practices may include:

- Management practices by the RFID application operator.
- Disposal of and erasure policies for RFID data.
- Policies related to lawful processing of personal information.
- Provisions in place for data minimization in handling RFID data, where feasible.
- Processing or storing of information from tags that do not belong to the RFID operator.
- Security governance practices.

Providing Individual Access and Control

- Providing information about the purposes of the processing and the categories of personal data involved.
- Description of how to object to the processing of personal data or withdraw consent.
- Identification of process to request rectification or erasure of incomplete or inaccurate personal data.

System Protection

System protection concepts apply to back-end systems and communication infrastructure in so far as they are relevant to the RFID application. Where they do apply, it should be recognized that backend systems are often complex and may have been the subject of their own PIA. That analysis may need to be reviewed to assure that it considered information of the nature used by the RFID application. Where no such PIA exists, the following components of the backend system should be considered:

- Access controls related to the type of personal data and functionality of the systems are in place.
- Controls and policies put in place to ensure the Operator does not link personal data in the RFID application in a manner inconsistent with the PIA report.
- Whether appropriate measures are in place to protect the confidentiality, integrity, and availability of the personal data in the systems and in the communication infrastructure.
- Policies on the retention and disposal of the personal data.
- Existence and implementation of information security controls, such as:
  - Measures that address the security of networks and transport of RFID data.
  - Measures that facilitate the availability of RFID data through appropriate back-ups and recovery.

RFID Tag Protection

RFID Tag Protection controls related to privacy and personal data should be indicated. They are particularly relevant to RFID applications that use RFID tags containing personal data.
  These protection controls include the following:

- Access control to functionality and information, including authentication of readers, writers, and underlying processes, and authorization to act upon the RFID tag.
- Methods to assure/address the confidentiality of the information (e.g., through encryption of the full RFID Tag or of selective fields).
- Methods to assure/address the integrity of the information.
- Retention of the information after the initial collection (e.g., duration of retention, procedures for eliminating the data at the end of the retention period or for erasing the information in the RFID tag, procedures for selective field retention or deletion).
- Tamper resistance of the RFID tag itself.
- Deactivation or removal, if required or otherwise provided.

Mitigation can include user based controls that address situations where different needs or sensitivities related to privacy may be at issue. Deactivation or removal is currently the two most common form of end-user/consumer mitigation. These may either be required as part of a PIA analysis, in certain circumstances by law or as a customer option after the point of sale to enhance confidence. In addition, the EC recommendation on RFID privacy and data protection for RFID applications suggests certain methodologies and best practices associated with implementation of deactivation or removal in retail.[4]

Accountability Measures

These measures are designed to address procedural data protection, in the area of accountability. Through these measures external awareness regarding RFID applications is raised.

- Ensuring the easy availability of a comprehensive *information policy* that includes:
  - Identity and address of the RFID application operator.
  - Purpose of the RFID Application
  - Types of data processed by the RFID application, in particular if personal data are processed.
  - Whether the locations of RFID tags will be monitored when possessed by an individual.
  - Likely privacy and data protection impacts, if any, relating to the use of RFID tags in the RFID application and the measures available to mitigate these impacts.
- Ensuring concise, accurate and easy to understand *notices* of the presence of RFID readers that include:
  - The identity of the RFID application operator.
  - A point of contact for Individuals to obtain the information policy.
- Noting if and how *redress* mechanisms are made available:
  - RFID application operator accountable legal entity (-ies) (may be one for each jurisdiction or operating area).
  - Point(s) of contact of the designated person or office responsible for reviewing the assessments and the continued appropriateness of the technical and organizational measures related to the protection of personal data and privacy.
  - Inquiry methods (e.g., methods through which the RFID application operator may be reached to ask a question, make a request, file a complaint, or exercise a right).
  - Methods to object to processing, to exercise access rights to personal data (including deleting and correcting personal data), to revoke consent, or to change controls and other choices regarding the processing of personal data, if required or otherwise provided.
  - Other redress methods, if required or otherwise provided.

Compilation of References

1-800-Wheelchair. (n.d.). *Bed alarm sensor pad*. Retrieved June 16, 2011, from

Abramson, S. (1970). The ALOHA system-Another alternative for computer communications. *Proceedings of Fall Joint Computer Conference, AFIPS Conference 1970*, Vol. 37.

Ahson, S., & Ilyas, M. (2008). *RFID handbook: Applications, technology, security and privacy*. CRC Press. doi:10.1201/9781420055009

Alejos, A., García Sánchez, M., Cuiñas, I., & Garcia Valladares, J. C. (2010). Sensor area network for active RTLS for RFID tracking applications at 2.4GHz. *Progress in Electromagnetics Research, 110*, 43–58. doi:10.2528/PIER10100204

Al-Medhwahi, M., Alkholidi, A., & Hamam, H. (2010). A new hybrid frame ALOHA and binary splitting algorithm for anti-collision in RFID systems. *International Conference on Software, Telecommunications and Computer Networks* (SoftCOM), (pp. 219 –224).

Alotaibi, M., Bialkowski, K. S., & Postula, A. (2010). Improving the time efficiency of QTA anti-collision algorithm. *IEEE International Conference on RFID-Technology and Applications* (RFID-TA), (pp. 205–210).

Altus, D. E., Mathews, R. M., Xaverius, P. K., Engelman, K. K., & Nolan, B. A. D. (2000). Evaluating an electronic monitoring system for people who wander. *American Journal of Alzheimer's Disease and Other Dementias, 15*(2), 121–125. doi:10.1177/153331750001500201

Alwan, M., Dalal, S., Mack, D., Kell, S., Turner, B., & Leachtenauer, J. (2006). Impact of monitoring technology in assisted living: Outcome pilot. *IEEE Transactions on Information Technology in Biomedicine, 10*(1), 192–198. doi:10.1109/TITB.2005.855552

Angluin, D., & Laird, P. (1988). Learning from noisy examples. *Machine Learning, 2*(4), 343–370. doi:10.1007/BF00116829

Anon., (1997). Editor's choice. Compliance - A broken concept. *British Medical Journal, 314*(7082), 1.

Asokan, N., Saxena, N., Uddin, M., & Voris, J. (2011). *Vibrate-to-unlock: Mobile phone assisted user authentication to multiple personal RFID tags*. In International Conference on Pervasive Computing and Communications, 2011.

Aumasson, J., Henzen, L., Meier, W., Naya-Plasencia, M., Mangard, S., & Standaert, F. (2010). Quark: A lightweight hash. *Cryptographic Hardware and Embedded Systems, CHES, 2010*, 1–15.

Auto-ID Center. (March 2003). *900 MHz class 0 radio frequency (RF) identification tag specication*.

Avoine, G., & Oechslin, P. (2005). RFID traceability: A multilayer problem. *Lecture Notes in Computer Science, Security and Cryptology, 3570*, 125-140.

Avoine, G., Carpent, X., & Martin, B. (2010). Strong authentication and strong integrity (SASI) is not that strong. *Proceedings of the 6th International Conference on Radio Frequency Identification: Security and Privacy Issues*, (pp. 50-64).

Avoine, G., & Oechslin, P. (2005). RFID traceability: A multilayer problem. In . *Proceedings of Financial Cryptography, 2005*, 125–140.

Ayoade, J. (2007). Roadmap to solving security and privacy concerns in RFID systems. *Computer Law & Security Report, 23*(6), 555–561. doi:10.1016/j.clsr.2007.09.005

Bagnato, G., Maselli, G., Petrioli, C., & Vicari, C. (2009). *Performance analysis of anti-collision protocols for RFID systems*. IEEE 69th Vehicular Technology Conference, VTC Spring 2009, April 2009.

Barlow, J., Singh, D., Bayer, S., & Curry, R. (2007). A systematic review of the benefits of home telecare for frail elderly people and those with long-term conditions. *Journal of Telemedicine and Telecare, 13*, 172–179. doi:10.1258/135763307780908058

Barreto, P. S. L. M. (2001b, 2008.11.19). *The Khazad legacy-level block cipher*. Retrieved from http://www.larc.usp.br/~pbarreto/khazad-tweak.zip

Barreto, P. S. L. M. Rijmen. (2001a, 2008.11.19). *The Anubis block cipher*. Retrieved from http://www.larc.usp.br/~pbarreto/anubis.zip

Barth, A., Collin, J., & Mitchel, J. C. (2008). Robust defenses for cross-site request forgery. In *15th ACM Conference on Computer and Communications Security*, (pp. 75-88).

Basrur, P., & Parry, D. (2006). "Where are my glasses?": An object location system within the home. *Bulletin of Applied Computing and Information Technology, 4*(2).

Bellare, M., Pointcheval, D., & Rogaway, P. (2000). Authenticated key exchange secure against dictionary attacks. *Eurocrypt '00 . LNCS, 1807*, 139–155.

Benelli, G., & Pozzebon, A. (2009). NFcare - Possible applications of NFC technology in sanitary. *Proceedings of the Second International Conference on Health* (pp. 58-65). Porto, Portugal: INSTICC Press.

Berners-Lee, T., Hendler, J., & Lassila, O. (2001, May). The Semantic Web. *Scientific American*, 29–37.

Biryukov, A. (2003). Analysis of involutional ciphers: Khazad and Anubis *Fast Software Encryption – 2003. Lecture Notes in Computer Science*, 4–5.

Blum, A., Kalai, A., & Wasserman, H. (2003). Noise-tolerant learning, the parity problem, and the statistical query model. *Journal of the ACM, 50*(4), 506–519. doi:10.1145/792538.792543

Bogdanov, A., Knežević, M., Leander, G., Toz, D., Varici, K., & Verbauwhede, I. (2011). SPONGENT: A lightweight hash function. *In Proceedings of the 13th International Conference on Cryptographic Hardware and Embedded Systems (CHES'11)*, (pp. 312-325).

Bogdanov, A., Knudsen, A., Leander, L., Paar, G., Poschmann, C., & Robshaw, A. (2007). PRESENT: An ultra-lightweight block cipher. In *Proceedings of the 9th international workshop on Cryptographic Hardware and Embedded Systems (CHES '07)*, (pp. 450-466).

Bogdanov, A., Knudsen, L. R., Le, G., Paar, C., Poschmann, A., Robshaw, M. J. B., et al. (2007). PRESENT: An ultra-lightweight block cipher. *The Proceedings of CHES 2007*. Springer.

Bogdanov, A., Leander, G., Paar, C., Poschmann, A., Robshaw, M., & Seurin, Y. (2008). Hash functions and RFID tags: Mind the gap. In *Proceeding sof the 10th international workshop on Cryptographic Hardware and Embedded Systems (CHES '08)*, (pp. 283-299).

Bogdanov, A. (2008). Linear slide attacks on the KeeLoq block cipher . In Dingyi, P., Moti, Y., Dongdai, L., & Chuankun, W. (Eds.), *Information security and cryptology* (pp. 66–80). Springer-Verlag. doi:10.1007/978-3-540-79499-8_7

Bolic, M., Simplot-Ryl, D., & Stojmenovic, I. (2010). *RFID systems: Research trends and challenges*. Wiley Series in Wireless Communications and Mobile Computing, 2010.

Bolotnyy, L., & Robins, G. (2006). *Generalized yoking-proof for a group of radio frequency identification tags*. International Conference on Mobile and Ubiquitous Systems (MOBIQUITOUS). San Jose, CA.

Bolotnyy, L., & Robins, G. (2009). Generalized "yoking-proofs" and inter-tag communication . In Turcu, C. (Ed.), *Development and implementation of RFID technology* (pp. 447–462). doi:10.5772/6538

Bonuccelli, M. A., Lonetti, F., & Martelli, F. (2006). Tree slotted aloha: A new protocol for tag identification in RFID networks. *Proceedings of International Symposium on World of Wireless, Mobile and Multimedia Networks, WOWMOM*, (pp. 603–608).

Bonuccelli, M. A., Lonetti, F., & Martelli, F. (2007a). Exploiting ID knowledge for tag identification in RFID networks. *Proceedings of the 4th ACM Workshop on Performance Evaluation of Wireless Ad Hoc, Sensor, and Ubiquitous Networks*, PE-WASUN, (pp. 70–77).

Bonuccelli, M., Lonetti, F., & Martelli, F. (2009). Exploiting signal strength detection and collision cancellation for tag identification in RFID systems. *IEEE Symposium on Computers and Communications* (ISCC), (pp. 500–506).

Bonuccelli, M. A., Lonetti, F., & Martelli, F. (2007b). Instant collision resolution for tag identification in RFID networks. *Ad Hoc Networking*, *5*, 1220–1232. doi:10.1016/j.adhoc.2007.02.016

Brands, S., & Chaum, D. (1993). *Distance-bounding protocols.* In Advances in Cryptology - EUROCRYPT, International Conference on the Theory and Applications of Cryptographic Techniques, 1993.

Brent, R. P. (2004). Note on Marsaglia's Xorshift random number generators. *Journal of Statistical Software*, *11*(4), 1–5.

Bringer, J., & Chabanne, H. (2008). *Trusted-HB: A low-cost version of HB+ secure against man-in-the-middle attacks.* Cryptology ePrint Archive, Report 2008/042.

Bringer, J., Chabanne, H., & Dottax, E. (2006). HB++: A lightweight authentication protocol secure against some attacks. *Second International Workshop on Security, Privacy and Trust in Pervasive and Ubiquitous Computing (SecPerU'06)*, (pp. 28-33).

Bringer, J., Chabanne, H., & Dottax, E. (2006). HB++: A lightweight authentication protocol secure against some attacks . In *Security*. Privacy and Trust in Pervasive and Ubiquitous Computing. doi:10.1109/SECPERU.2006.10

Brumley, B. B. (2010). *Secure and fast implementations of two involution ciphers.* IACR Eprint archive.

Buckland, M., Frost, B., & Reeves, A. (2006). Liverpool Telecare Pilot: Telecare as an information tool. *Informatics in Primary Care*, *14*(3), 191–196.

Buettner, M., Greenstein, B., Sample, A., & Smith, J. R. (2008). *Demonstration: RFID sensor networks with the Intel WISP*. In 6th ACM Conference on Embedded Networked Sensor Systems.

Buettner, M., Prasad, R., Philipose, M., & Wetherall, D. (2009). *Recognizing daily activities with RFID-based sensors.* In International Conference on Ubiquitous Computing, 2009.

Burmester, M., Tri van Le, de Medeiros, B., & Tsudik, G. (2009). Provably secure ubiquitous systems: Universally composable RFID authentication protocols. *ACM Transactions on Information and System Security (TISSEC)*, *12*(4).

Burmester, M., & de Medeiros, B. (2008). The security of EPC Gen2 compliant RFID protocols . In Bellovin, S. M., Gennaro, R., Keromytis, A. D., & Yung, M. (Eds.), *LNCS 5037* (pp. 490–506). Springer. doi:10.1007/978-3-540-68914-0_30

Burmester, M., de Medeiros, B., & Mota, R. (2008). Provably secure grouping-proofs for RFID tags . In Grimaud, G., & Standaert, F. X. (Eds.), *CARDIS* (*Vol. 5189*, pp. 176–190). Lecture Notes in Computer Science Springer.

Burmester, M., & Medeiros, B. (2009). On the security of route discovery in MANETs. *IEEE Transactions on Mobile Computing*, *8*(9), 1180–1188. doi:10.1109/TMC.2009.13

Burmester, M., & Munilla, J. (2011). Lightweight RFID authentication with forward and backward security. *ACM Transactions on Information and System Security*, *14*(1). doi:10.1145/1952982.1952993

Canetti, R. (1995). *Studies in secure multiparty computation and application.* Ph.D. thesis, Weizmann Institute of Science.

Canetti, R. (2001). Universally composable security: A new paradigm for cryptographic protocols. In *Proceedings of the IEEE Symposium on Foundations of Computer Science (FOCS'01)* (pp. 136—145).

Canetti, R. (2000). Security and composition of multiparty cryptographic protocols. *Journal of Cryptology*, *13*(1), 143–202. doi:10.1007/s001459910006

Cannière, C., Dunkelman, O., & Knežević, M. (2009). KATAN and KTANTAN -- A family of small and efficient hardware-oriented block ciphers. *Proceedings of the 11th International Workshop on Cryptographic Hardware and Embedded (CHES '09)*, (pp. 272-288).

Capetanakis, J. I. (1979, September). Tree algorithms for packet broadcast channels. *IEEE Transactions on Information Theory, IT-25*, 505–515. doi:10.1109/TIT.1979.1056093

Carskadon, M., & Dement, W. (1989). Normal human sleep: An overview. In Kryger, M. H., Roth, T., & Dement, W. C. (Eds.), *Principles and practice of sleep medicine*. Philadelphia, PA: W B Saunders.

Catarinucci, L., Colella, R., & Tarricone, L. (2010). Sensor data transmission through passive RFID tags to feed wireless sensor networks. In *Proceedings of IEEE MTT International Microwave Symposium*.

Catarinucci, L., Cappelli, M., Colella, R., & Tarricone, L. (2008). A novel low-cost multisensor-tag for RFID applications in healthcare. *Microwave and Optical Technology Letters, 50*, 2877–2880. doi:10.1002/mop.23837

Catarinucci, L., Colella, R., & Tarricone, L. (2009). *A cost-effective UHF RFID tag for transmission of generic sensor data in wireless sensor networks*. IEEE Transaction on Microwave Theory and Techniques – Special Issue on RFID Technology.

Center, M. A.-I. (2003). *Draft protocol specification for a 900 mhz class 0 radio frequency identification tag*. Retrieved from http://www.epcglobalinc.org

Cha, J., & Kim, J. (2005). Novel anti-collision algorithms for fast object identification in RFID system. *International Conference on Parallel and Distributed Systems*, Vol. 2, (pp. 63–67).

Cha, J., & Kim, J. (2006). Dynamic framed slotted ALOHA algorithms using fast tag estimation method for RFID system. *Proceedings of IEEE Consumer Communications and Networking Conference*, Vol. 2, (pp. 768–772).

Chae, H. J., Yeager, D. J., Smith, J. R., & Fu, K. (2007). Maximalist cryptography and computation on the wisp UHF RFID tag. In *Proc. Conference on RFID Security*.

Chae, M.-J., Yeager, D. J., Smith, J. R., & Fu, K. (2007). *Maximalist cryptography and computation on the WISP UHF RFID tag*. Paper presented at the Conference on RFID Security, Malaga, Spain.

Chari, S., Rao, J. R., & Rohatgi, P. (2002). Template attacks. In B. S. Kaliski Jr., Ç.-K. Koç, & C. Paar (Eds.), *4th International Workshop on Cryptographic Hardware and Embedded Systems - CHES 2002* (pp. 13-28). Berlin, Germany: Springer.

Chaudhri, R., Lester, J., & Borriello, G. (2008). An RFID based system for monitoring free weight exercises. In *Proceedings of the 6th ACM Conference on Embedded Network Sensor Systems* (SenSys '08), (pp. 431-432). New York, NY: ACM.

Chaudhry, N., Thompson, D., & Thompson, C. (2005). *RFID technical tutorial and threat modeling*. Arkansas.

Che, W., Deng, H., Tan, W., & Wang, J. (2008). A random number generator for application in RFID tags . In Ranasinghe, D. C., & Cole, P. C. (Eds.), *Networked RFID Systems and Lightweight Cryptography, 2008 (* (pp. 279–287). doi:10.1007/978-3-540-71641-9_16

Chiang, K. W., Hua, C., & Yum, T.-S. P. (2006). Prefix-randomized query-tree protocol for RFID systems. *Proceedings of IEEE International Conference on Communications*, (pp. 1653–1657).

Chien, H. (2007). SASI: A new ultralightweight RFID authentication protocol providing strong authentication and strong integrity. *IEEE Transactions on Dependable and Secure Computing, 4*(4), 337–340. doi:10.1109/TDSC.2007.70226

Chien, H. Y., Yang, C. C., Wu, T. C., & Lee, C. F. (2011). Two RFID-based solutions to enhance inpatient medication safety. *Journal of Medical Systems, 35*(3), 369–375. doi:10.1007/s10916-009-9373-7

Chien, H., & Chen, C. (2007). Mutual authentication protocol for RFID conforming to EPC Class 1 Generation 2 standards. *Computer Standards & Interfaces, 29*(2), 254–259. doi:10.1016/j.csi.2006.04.004

Choi, E. Y., Lee, S. M., & Lee, D. H. (2005). Efficient RFID authentication protocol for ubiquitous computing environment. In *Proceedings of SECUBIQ '05*.

Choi, H., Cha, J., & Kim, J. (2005). Improved bit-by-bit binary tree algorithm in ubiquitous Id System. *Advances in Multimedia Information Processing - PCM 2004, Vol. 3332 of Lecture Notes in Computer Science*, (pp. 696–703).

Choi, J. M., Choi, B. H., Seo, J. W., Sohn, R. H., Ryu, M. S., Yi, W., et al. (2004, 1-5 September). *A system for ubiquitous health monitoring in the bedroom via a Bluetooth network and wireless LAN.* Paper presented at the 26th Annual International Conference of the IEEE Engineering in Medicine and Biology Society, IEMBS '04.

Choi, J., & Lee, W. (2007). Comparative evaluation of probabilistic and deterministic tag anti-collision protocols for RFID networks. *Proceeding of 2007 Conference on Emerging Direction in Embedded and Ubiquitous Computing (EUC'07), Lecture Notes in Computer Science,* (pp. 538-549).

Cohen, H., Miyaji, A., & Ono, T. (1998). Efficient elliptic curve exponentiation using mixed coordinates . In Ohta, K., & Pei, D. (Eds.), *Advances in Cryptogoly – Asiacrypt'99* (pp. 51–65). Berlin, Germany: Springer. doi:10.1007/3-540-49649-1_6

Cole, R. J., Kripke, D. F., Gruen, W., Mullaney, D. J., & Gillin, J. C. (1992). Automatic sleep/wake identification from wrist activity. *Sleep, 15*(5), 461–469.

Comba, P. (1990). Exponentiation cryptosystems on the IBM PC. *IBM Systems Journal, 29*(4), 526–538. doi:10.1147/sj.294.0526

Commission of the European Communities. (2009, May 12). *Commission recommendation on the implementation of privacy and data protection principles in applications supported by radio-frequency identification.* Retrieved from http://ec.europa.eu/information_society/policy/rfid/documents/recommendationonrfid2009.pdf

Coron, S. (1999). Resistance against differential power analysis for elliptic curve cryptosystems. In Ç.-K. Koç, & C. Paar (Eds.), *First International Workshop on Cryptographic Hardware and Embedded Systems - CHES 1999* (pp. 292-302). Berlin, Germany: Springer.

Crossbow Technology Inc. (2009). *Eko PRO series user's manual, rev. C.*

Cuiñas, I., Catarinucci, L., & Trebar, M. (2010). RFID from farm to fork: Traceability along the complete food chain. In *Proceedings of Progress in Electromagnetic Research Symposium.*

Cuinas, I., Gay-Fernandez, J. A., Alejos, A., & Sanchez, M. (2010). A comparison of radioelectric propagation in mature forests at wireless network frequency bands. In *European Conference on Antennas and Propagation* (pp. 1-5).

Czeskis, A., Koscher, K., Smith, J., & Kohno, T. (2008). *RFIDs and secret handshakes: Defending against Ghost-and-Leech attacks and unauthorized reads with context-aware communications.* In ACM Conference on Computer and Communications Security, 2008.

Daemen, J., Knudsen, L. R., & Rijmen, V. (1997). *The block cipher square.* Paper presented at the 4th International Workshop on Fast Software Encryption.

Datasheet Helion Technology. (2005). *MD5, SHA-1, SHA-256 hash core for Asic.* Retrieved from http://www.heliontech.com

David, M., Ranasinghe, D., & Larsen, T. (2011). A2U2: A stream cipher for printed electronics RFID tags. *2011 IEEE International Conference on RFID (RFID),* (pp. 176 -183).

Demsar, J., Zupan, B., Leban, G., & Curk, T. (2004). Orange: From experimental machine learning to interactive data mining . In *ECML/PKDD (Vol. 3202,* pp. 537–539). Berlin, Germany: Springer. doi:10.1007/978-3-540-30116-5_58

der Loos, H. V., Kobayashi, H., Liu, G., Tai, Y., Ford, J., & Norman, J..... Osada, T. (2001). *Unobtrusive vital signs monitoring from a mulitsensor bed sheet.* In RESNA, 2001, Arlington, VA.

Deshpande, A. M., Deshpande, M. S., & Kayatanavar, D. N. (2009, 4-6 June 2009). *FPGA implementation of AES encryption and decryption.* Paper presented at the International Conference on Control, Automation, Communication and Energy Conservation, INCACEC 2009.

Dimitriou, T. (2005). A lightweight RFID protocol to protect against traceability and cloning attacks. In *IEEE Conference on Security and Privacy in Communication Networks, 2005.*

Dobkin, D. M., & Wandinger, T. (2005). A radio-oriented introduction to RFID protocols, tags and applications. *High Frequency Electronics,* August, (pp. 32–46).

Dobkin, D. (2007). *The RF in RFID: Passive UHF RFID in practice*. Newton, MA: Newnes.

Dolin, R. H., Alschuler, L., Boyer, S., Beebe, C., Behlen, F. M., & Biron, P. V. (2006). HL7 clinical document architecture, release 2. *Journal of the American Medical Informatics Association, 13*(1), 30–39. doi:10.1197/jamia.M1888

Doughty, K., Lewis, R., & McIntosh, A. (2000). The design of a practical and reliable fall detector for community and institutional telecare. *Journal of Telemedicine and Telecare, 6*(Supp. 1), S150–S154. doi:10.1258/1357633001934483

Drimer, S., & Murdoch, S. (2007). *Keep your enemies close: Distance bounding against smartcard relay attacks*. In 16th USENIX Security Symposium, 2007.

Duc, D., & Kim, K. (January de 2007). Securing HB+ against grs man-in-the-middle attack. *Insitute of Electronics, Information and Comunication Engineers, Syposium on Crytography and Information Security*, (pp. 23-26).

Duc, D., Park, J., Lee, H., & Kim, K. (2006). Enhancing security of EPC global Gen-2 RFID tag against traceability and cloning. In *The Proceedings of the 2006 Symposium on Cryptography and Information Security (SCIS'06)*, (pp. 17–20).

Egan, D. (2005). The emergence of ZigBee in building automation and industrial controls. *Computing & Control Engineering Journal, 16*(2), 14–19. doi:10.1049/cce:20050203

Engels, D. W., & Sarma, S. E. (2002). The reader collision problem. In *IEEE International Conference on Systems, Man and Cybernetics*, Vol. 3, October 2002.

Engels, D., Fan, X., Gong, G., Hu, H., & Smith, E. M. (2010). *Hummingbird: ultra-lightweight cryptography for resource-constrained devices*. Paper presented at the 14th International Conference on Financial Cryptograpy and Data Security.

EPCGlobal Standard. (n.d.). *Web page*. Retrieved October 15, 2011, from http://www.epcglobalinc.org

EPCglobal. (2005). *EPC radio-frequency identity protocols class-1 generation-2 UHF RFID protocol for communications at 860MHz- 960MHz version 1.0.9.*

EPCglobal. (2007). *Tag class definitions*. Retrieved 2011, from http://www.gs1.org/docs/epcglobal/TagClassDefinitions_1_0-whitepaper-20071101.pdf

EPCglobal. (2008). *EPC radio-frecuency protocols class-1 Generation-2 UHF RFID Version 1.2.0*. Retrieved 2011, from http://www.gs1.org/gsmp/kc/epcglobal/uhfc1g2/uhfc1g2_1_2_0-standard-20080511.pdf

EPCglobal. (2008). *EPCTM radio-frequency identity protocols class-1 generation-2 UHF RFID protocol for communications at 860MHz-960MHz*, Version 1.2.0. Retrieved October, 2008, from http://www.epcglobalinc.org

EPCGlobal. (2011). *EPC™ radio-frequency identity protocols EPC class-1 HF RFID air interface protocol for communications at 13.56 MHz Version 2.0.3.*

Ester, M., Kriegel, H., Sander, J., & Xu, X. (1996). A density-based algorithm for discovering clusters in large spatial databases with noise. In E. Simoudis, J. Han, & U. Fayyad (Eds.), *The Second International Conference on Knowledge Discovery and Data Mining (KDD-96)*.

EuroFIR (European Food Information Resource). (n.d.). *Website*. Retrieved October 15, 2011, http://www.eurofir.net

European Digital Rights (EDRI-gram). (2006). *Cloning an electronic passport* (pp. 4–16).

European Parliament. (1995, November 23). *Directive 95/46/EC of the European Parliament and of the Council of 24 October 1995 on the protection of individuals with regard to the processing of personal data and on the free movement of such data*. Retrieved from http://ec.europa.eu/justice_home/fsj/privacy/docs/95-46-ce/dir1995-46_part1_en.pdf

European Parliament. (2002, July 12). *Directive 2002/58/EC of the European Parliament and of the Council of 12 July 2002 concerning the processing of personal data and the protection of privacy in the electronic communications sector (Directive on privacy and electronic communications)*. Retrieved from http://eur-lex.europa.eu/LexUriServ/LexUriServ.do?uri=OJ:L:2002:201:0037:0047:EN:PDF

European Parliament. (2005, January 19). *Working document on data protection issues related to RFID technology.* Article 29 Data Protection Working Party. Retrieved from http://ec.europa.eu/justice_home/fsj/privacy/docs/wpdocs/2005/wp105_en.pdf

European Parliament. (2007). *Opinion 4/2007 on the concept of personal data.* Article 29 Data Protection Working Party. Retrieved from http://ec.europa.eu/justice_home/fsj/privacy/docs/wpdocs/2007/wp136_en.pdf

European Parliament. (2009, November 25). *Directive 2009/136/EC of the European Parliament and of the Council of 25 November 2009 amending Directive 2002/22/EC on universal service and users' rights relating to electronic communications networks and services, Directive 2002/58/EC concerning the processing of personal data and the protection of privacy in the electronic communications sector and Regulation (EC) No 2006/2004 on cooperation between national authorities responsible for the enforcement of consumer protection laws.* Retrieved from http://eur-lex.europa.eu/LexUriServ/LexUriServ.do?uri=OJ:L:2009:337:0011:0036:EN:PDF

European Parliament. (2012). *Status of implementation of Directive 95/46 on the protection of Individuals in regards to the processing of personal data.* Retrieved from http://ec.europa.eu/justice_home/fsj/privacy/law/implementation_en.htm

Fabregas, S., Johnstone, J., & Shambroom, J. (2009). Performance of a wireless dry sensor system in automatically monitoring sleep and wakefullness. *Sleep, 32,* A388.

Feldhofer, M., & Rechberger, C. (2006). A case against currently used hash functions in RFID protocols. *First International Workshop on Information Security (IS'06),* (pp. 372-381).

Feldhofer, M., & Wolkerstorfer, J. (2007, 27-30 May 2007). *Strong crypto for RFID tags - A comparison of low-power hardware implementations.* Paper presented at the Circuits and Systems, 2007. ISCAS 2007. IEEE International Symposium on.

Feldhofer, M., Dominikus, S., & Wolkerstorfer, J. (2004). Strong authentication for RFID systems using the AES algorithm . In Joye, M., & Quisquater, J.-J. (Eds.), *CHES 2004* (pp. 357–370). doi:10.1007/978-3-540-28632-5_26

Fosstrak (Free and Open Source Software for Track and Trace). (n.d.). *Web page.* Retrieved October 15, 2011, http://www.fosstrak.org

Francillon, A., Danev, B., & Capkun, S. (2011). *Relay attacks on passive keyless entry and start systems in modern cars.* In 18th Annual Network and Distributed System Security Symposium, 2011.

Frenger, P. (2000). The ultimate RISC: A zero-instruction computer. *SIGPLAN Notice, 35*(2), 17–24. doi:10.1145/345105.345111

Fummi, F., & Perbellini, G. (2007). eEPC: An EPCglobal-compliant embedded architecture for RFID-based solutions. *The Journal of Communication, 2*(7).

Garcia-Alfaro, J., Barbeau, M., & Kranakis, E. (2008). Security threats on EPC based RFID systems. *Fifth International Conference on Information Technology: New Generations, ITNG 2008,* (pp. 1242 -1244).

Garfinkel, S., Juels, A., & Pappu, R. (2005). *RFID privacy: An overview of problems and proposed solutions. Security Privacy* (pp. 34–43). IEEE.

Garsden, H., Basilakis, J., Celler, B. G., Huynh, K., & Lovell, N. H. (2004). *A home health monitoring system including intelligent reporting and alerts.* Paper presented at the 26th Annual International Conference of the IEEE Engineering in Medicine and Biology Society, IEMBS '04.

Gay-Fernandez, J. A., Garcia Sanchez, M., Cuiñas, I., Alejos, A. V., Sánchez, J. G., & Miranda-Sierra, J. L. (2010). Propagation analysis and deployment of a wireless sensor network in a forest. *Progress in Electromagnetics Research, 106,* 121–145. doi:10.2528/PIER10040806

Gebotys, C., & Gebotys, R. (2003). Secure elliptic curve implementations: An analysis of resistance to power-attacks in a DSP processor. In B. S. Kaliski Jr., Ç.-K. Koç, & C. Paar (Eds.), *4th International Workshop on Cryptographic Hardware and Embedded Systems - CHES 2002* (pp. 114-128). Berlin, Germany: Springer.

Giganti, F., Ficca, G., Gori, S., & Salzarulo, P. (2008). Body movements during night sleep and their relationship with sleep stages are further modified in very old subjects. *Brain Research Bulletin, 75*(1), 66–69. doi:10.1016/j.brainresbull.2007.07.022

Gilbert, H., Robshaw, M., & Seurin, Y. (2008). Good variants of HB+ are hard to find (G. Tsudik, Ed.) *LNCS, 5143*, (pp. 156-170).

Gilbert, H., Robshaw, M., & Seurin, Y. (2008). *HB#: Increasing the security and efficiency of HB+*. In Advances in Cryptology - EUROCRYPT, International Conference on the Theory and Applications of Cryptographic Techniques, 2008.

Gilbert, H., Robshaw, M., & Seurin, Y. (2008). *HB#: Increasing the security and efficiency of HB+. En Advances in Cryptology – EUROCRYPT 2008* (Vol. 4965, pp. 361–378). Berlin, Germany: Springer.

Gilbert, H., Robshaw, M., & Sibert, H. (2005). An active attack against HB+ - A provably secure lightweight protocol. *Electronics Letters, 41*, 1169–1170. doi:10.1049/el:20052622

Gillespie, L. G., Robertson, M. C., Lamb, S. E., Cumming, R. G., & Rowe, B. H. (2006). *Interventions for preventing falls in elderly people*. Cochrane Database of Systematic Reviews.

Gilreath, W. F., & Laplante, P. A. (2003). *Computer architecture: A minimalist perspective*. Kluwer Academic Publishers.

Global, E. P. C. (n.d.). *EPC tag data standards, vs. 1.3*. Retrieved from http://www.epcglobalinc.org/standards/EPCglobal_Tag Data Standard TDS Version 1.3.pdf.

Glover, B., & Bhatt, H. (2006). *RFID essentials*. O'Reilly Media, Inc.

Goldreich, O. (2001). *The foundations of cryptography*. Cambridge, UK: Cambridge University Press. doi:10.1017/CBO9780511546891

Good, T., & Benaissa, M. (2006). Very small FPGA application-specific instruction processor for AES. *IEEE Transactions on Circuits and Systems. I, Regular Papers, 53*(7), 1477–1486. doi:10.1109/TCSI.2006.875179

Goundar, R., Joye, M., & Miyaji, A. (2010). Co-Z addition formulae and binary ladders on elliptic curves. In Mangard, S., & Standaert, F.-X. (Eds.), *Cryptographic Hardware and Embedded Systems -CHES 2010* (pp. 65–79). Berlin, Germany: Springer. doi:10.1007/978-3-642-15031-9_5

Gouvêa, C., & Lopez, J. (2009). Software implementation of pairing-based cryptography on sensor networks using the MSP430 microcontroller . In Roy, B., & Sendrier, N. (Eds.), *Progress in Cryptology - INDOCRYPT 2009, LNCS 5922* (pp. 248–262). Berlin, Germany: Springer. doi:10.1007/978-3-642-10628-6_17

Guajardo, J., Blümel, R., Krieger, U., & Paar, C. (2001). Efficient implementation of elliptic curve cryptosystems on the TI MSP430x33x family of microcontrollers. In K. Kim (Ed.), *4th International Workshop on Practice and Theory in Public Key Cryptosystems – PKC 2001, LNCS 1992* (pp. 365-382). Berlin, Germany: Springer.

Guo, J., Peyrin, T., Poschmann, A., & Robshaw, M. (2011). The LED block cipher. In *Proceedings of the 13th International Conference on Cryptographic Hardware and Embedded Systems (CHES'11)*, (pp. 326-341).

Halevi, T., & Saxena, N. (2010). *On pairing constrained wireless devices based on secrecy of auxiliary channels: The case of acoustic eavesdropping*. In ACM Conference on Computer and Communications Security, 2010.

Halevi, T., Saxena, N., & Halevi, S. (2009). *Using HB family of protocols for privacy-preserving authentication of RFID tags in a population*. In Workshop on RFID Security, 2009.

Halperin, D., Heydt-Benjamin, T., Ransford, B., Clark, S., Defend, B., & Morgan, W. ... Maisel, W. (2008). *Pacemakers and implantable cardiac defibrillators: Software radio attacks and zeropower defenses*. In IEEE Symposium on Security and Privacy, 2008.

Halperin, D., Kohno, T., Heydt-Benjamin, T., Fu, K., & Maisel, W. (2008, January-March). Security and privacy for implantable medical devices. *IEEE Pervasive Computing / IEEE Computer Society and IEEE Communications Society, 7*(1), 30–39. doi:10.1109/MPRV.2008.16

Hammouri, G., & Sunar, B. (2008). PUF-HB: A tamper-resilient HB based authentication protocol. *Proceedings of the 6th International Conference on Applied Cryptography and Network Security*, (pp. 346-365).

Hancke, G., & Kuhn, M. (2005). *An RFID distance bounding protocol*. In the 1st International Conference on Security and Privacy for Emerging Areas in Communications Networks, 2005.

Hankerson, D., Menezes, A. J., & Vanstone, S. (2003). *Guide to elliptic curve cryptography.* New York, NY: Springer.

Hashemi, H. (2008). Propagation channel modeling for ad hoc networks. In *European Microwave Week,* (pp. 39-43).

Helal, S., Mann, W. H., King, J., Kaddoura, Y., & Jansen, E. (2005). The Gator Tech smart house: A programmable pervasive space. *Computer, 38*(3), 50–60. doi:10.1109/MC.2005.107

Hell, M., Johansson, T., & Meier, W. (2007). Grain - A stream cipher for constrained environments. *International Journal of Wireless and Mobile Computing, 2*(1), 86–93. doi:10.1504/IJWMC.2007.013798

Henrici, D., & Muller, P. (2004). Hash-based enhancement of location privacy for radio radiofrequency identification devices using varying identifiers. *Proceedings of PERCOMW'04, Second IEEE Annual Conference on Pervasive Computing and Communications Workshops,* (pp. 149–153).

Holcomb, D., Burleson, W., & Fu, K. (2007). Initial SRAM state as a fingerprint and source of true random numbers for RFID tags. In *Conference on RFID Security, 2007.*

Holleman, J., Yeager, D., Prasad, R., Smith, J., & Otis, B. (2008). NeuralWISP: An energy-harvesting wireless neural interface with 1-m range. In *Biomedical Circuits and Systems Conference, 2008.* Benoit Huyghe, B., & Doutreloigne, J. (2009). *3D orientation tracking based on unscented Kalman filtering of accelerometer and magnetometer data.* In IEEE Sensors Application Symposium, 2009.

Homewood, A., Hall, A., Singh, A., Croft, F., & Parry, D. (2010). *Are personal health records secure? A security Audit of "Google health."* (Poster). Paper presented at the Health Informatics New Zealand, Wellington.

Hoppe, N. J., & Blum, M. (2001). Secure human identification protocols. *Conference on the Theory and Application of Cryptology and Information Security International* (pp. 52-66). ASIACRYPT 2001.

Hoque, E., Dickerson, R., & Stankovic, J. (2010). *Monitoring body positions and movements during sleep using WISPs.* In Wireless Health' 10. San Diego, USA.

http://www.1800wheelchair.com/asp/view-product.asp?product_id=3060

Huang, H. H., & Ku, C. Y. (2009). A RFID grouping-proof protocol for medication safety of inpatient. *Journal of Medical Systems, 35*(3), 369–375.

Hush, D. R., & Wood, C. (1998). Analysis of tree algorithms for RFID arbitration. *Proceedings of IEEE International Symposium on Information Theory,* (p. 107).

Hutter, M., & Wenger, E. (2011). Fast multi-precision multiplication for public-key cryptography on embedded microprocessors. In B. Preneel & T. Takagi (Eds.), *13th International Workshop on Cryptographic Hardware and Embedded Systems - CHES 2011,* (pp. 459-474). Berlin, Germany: Springer.

Hutter, M., Joye, M., & Sierra, Y. (2011). Memory-constrained implementations of elliptic curve cryptography in Co-Z coordinate representation. In A. Nitaj & D. Pointcheval (Eds.), *Progress in Cryptology - AFRICACRYPT 2011 Fourth International Conference on Cryptology,* LNCS 6737, (pp. 170-187). Berlin, Germany: Springer.

Hutter, M., Medwed, M., Hein, D., & Wolkerstorfer, J. (2009). Attacking ECDSA-enabled RFID devices. In M. Abdalla, D. Pointcheval, P.-A. Fouque, & D. Vergnaud (Eds.), *7th International Conference on Applied Cryptography and Network Security - ACNS 2009* (pp. 519-534). Berlin, Germany: Springer.

Impinj. (2012). *Speedway revolution - Superior performance made easy.* Speedway Revolution Reader Specifications. Retrieved January 2012, from www.impinj.com

Information Commissioner's Office. (n.d.). *Privacy impact assessment handbook.* Retrieved from http://www.ico.gov.uk/upload/documents/pia_handbook_html_v2/files/PIAhandbookV2.pdf

Inoue, M., Uemura, K., Minagawa, Y., Esaki, M., & Honda, Y. (1985). A home automation system. *IEEE Transactions on Consumer Electronics, CE-31*(3), 516–527. doi:10.1109/TCE.1985.289966

Intel Research Seattle. (2010). *WISP: Wireless identification and sensing platform.* Retrieved January 2012, from http://seattle.intel-research.net/wisp

International Telecommunication Union (2007). Attenuation in Vegetation. *ITU-R Recommendation 833-6.*

ISO/IEC 18000. (2010). *Information technology AIDC techniques – RFID for item management – Air interface.*

Israsena, P. (2006, 16-18 January). *Securing ubiquitous and low-cost RFID using tiny encryption algorithm.* Paper presented at the Wireless Pervasive Computing, 2006 1st International Symposium on.

Jiang, T., Sidiropoulos, N. D., & Giannakis, G. B. (2003). Kalman filtering for power estimation in mobile communications. *IEEE Transactions on Wireless Communications, 2*(1), 151–161. doi:10.1109/TWC.2002.806386

Joye, M., & Yen, S.-M. (2003). The Montgomery powering ladder. In G. Goos, J. Hartmanis, & J. van Leeuwen (Eds.), *4th International Workshop on Cryptographic Hardware and Embedded Systems* (pp. 291-302). Berlin, Germany: Springer.

Juels, A. (2004). Yoking-proofs for RFID tags. *Proceedings of the First International Workshop on Pervasive Computing and Communication Security* (pp. 138-143).

Juels, A., & Weis, S. (2005-3). Authenticating pervasive devices with human protocols. In *International Cryptology Conference, 2005.*

Juels, A., Rivest, R., & Szydlo, M. (2003). *The blocker tag: Selective blocking of RFID tags for consumer privacy.* In ACM Conference on Computer and Communications Security, 2003.

Juels, A., Syverson, P., & Bailey, D. (2005-1). *High-power proxies for enhancing RFID privacy and utility.* In Privacy Enhancing Technologies, 2005.

Juels, A. (2005). Minimalist cryptography for low-cost RFID tags. *Security in Communication Networks, LNCS, 3352,* 149–164. doi:10.1007/978-3-540-30598-9_11

Juels, A. (2006). RFID security and privacy: A research survey. *IEEE Journal on Selected Areas in Communications, 24*(2), 381–394. doi:10.1109/JSAC.2005.861395

Juels, A., & Weis, S. (2005). Authenticating pervasive devices with human protocols . In Shoup, V. (Ed.), *CRYPTO 2005, LNCS 3126* (pp. 293–198). doi:10.1007/11535218_18

Juels, A., & Weiss, S. A. (2007). Defining strong privacy for RFID. *Proceedings of PerCom, 07,* 342–347.

Kaelber, D. C., Jha, A. K., Johnston, D., Middleton, B., & Bates, D. W. (2008). A research agenda for personal health records (PHRs). *Journal of the American Medical Informatics Association, 15*(6), 729–736. doi:10.1197/jamia.M2547

Karabuk, T., Hefeeda, M., & Bagheri, M. (2007). Wireless sensor networks for early detection of forest fires. In *IEEE International Conference on Mobile Adhoc and Sensor Systems,* (pp. 1-6).

Karlen, W. (2009). *Adaptive wake and sleep detection for wearable systems.* PhD thesis, Ecole Polytechnique Federale de Lausanne (EPFL).

Karthikeyan, S., & Nesterenko, M. (2005). RFID security without extensive cryptography. *Proceedings of the 3rd ACM Workshop on Security of Ad Hoc and Sensor Networks,* (pp. 63-67).

Karygicmnis, A., Phillips, T., & Tsibertzopoulos, A. (2006). RFID security: A taxonomy of risk. *First International Conference on Communications and Networking in China. ChinaCom '06,* (pp. 1-8).

Katz, J., & Shin, J. (2006). Parallel and concurrent security of the HB and HB+ protocols. In *Advances in Cryptology - EUROCRYPT, International Conference on the Theory and Applications of Cryptographic Techniques, 2006.*

Kavun, E., & Yalcin, T. (2010). A lightweight implementation of Keccak hash function for radio-frequency identification applications. In *Proceedings of the 6th International Conference on Radio Frequency Identification: Security and Privacy Issues (RFIDSec '10),* (pp. 258-269).

Kfir, Z., & Wool, A. (2005). Picking virtual pockets using relay attacks on contactless smartcard. In *Security and Privacy for Emerging Areas in Communications Networks, 2005.*

Kiltz, E., Pietrzak, K., Cash, D., Jain, A., & Venturi, D. (2011). Efficient authentication from hard learning problems. In *Proceedings of the 30th Annual international conference on Theory and Applications of Cryptographic Techniques: Advances in Cryptology* (pp. 7-26). Tallinn, Estonia: Springer-Verlag.

Klair, D., Chin, K.-W., & Raad, R. (2010). A survey and tutorial of RFID anti-collision protocols. *IEEE Communications Surveys Tutorials*, *12*(3), 400–421. doi:10.1109/SURV.2010.031810.00037

Knudsen, L., Leander, G., Poschmann, A., & Robshaw, M. (2010). PRINTcipher: A block cipher for IC-printing. *Proceedings of the 12th International Conference on Cryptographic Hardware and Embedded Systems (CHES '10)*, (pp. 16-32).

Kocer, F., & Flynn, M. P. (2006). A new transponder architecture with on-chip ADC for long-range telemetry applications. *IEEE Journal of Solid-state Circuits*, *41*(5), 1142–1148. doi:10.1109/JSSC.2006.872741

Kocher, P. (1996). Timing attacks on implementations of Diffie-Hellman, RSA, DSS, and other systems. In N. Koblitz (Ed.), *Proceedings of the 16th Annual International Cryptology Conference on Advances in Cryptology, CRYPTO 1996* (pp. 104-113). Berlin, Germany: Springer.

Kocher, P., Jaffe, J., & Jun, B. (1999). Differential power analysis. In M. Wiener (Ed.), *Proceedings of the 19th Annual International Cryptology Conference on Advances in Cryptology, CRYPTO 1999* (pp. 388-397). Berlin, Germany: Springer.

Kodialam, M., & Nandagopal, T. (2006). Fast and reliable estimation schemes in RFID systems. *Proceedings of the 12th Annual International Conference on Mobile Computing and Networking, MobiCom*, (pp. 322–333).

Kogure, T., Shirakawa, S., Shimokawa, M., & Hosokawa, Y. (2011). Automatic sleep/wake scoring from body motion in bed: Validation of a newly developed sensor placed under a mattress. *Journal of Physiological Anthropology*, *30*(3), 103–109. doi:10.2114/jpa2.30.103

Koindala, D. M., Kim, W., & Ki, K. (2006). *Security assessment of EPCglobal architecture framework*. Auto-ID Labs White Paper. Retrieved on June 12, 2011, from http://www.autoidlabs.org/uploads/media/AUTOIDLABS-WP-SWNET-017.pdf

Kong, J. H., Ang, L. M., Seng, K. P., & Adejo, A. O. (2010, 5-8 December). *Minimal instruction set FPGA AES processor using Handel — C*. Paper presented at the 2010 International Conference on Computer Applications and Industrial Electronics (ICCAIE).

Koscher, K., Juels, A., Brajkovic, V., & Kohno, T. (2009). EPC RFID tag security weaknesses and defenses: Passport cards, enhanced drivers licenses, and beyond. In *ACM Conference on Computer and Communications Security, 2009*.

Krauchi, K., Cajochen, C., & Wirz-Justice, A. (2004). Waking up properly: Is there a role of thermoregulation in sleep inertia? *Journal of Sleep Research*, *13*(2), 121–127. doi:10.1111/j.1365-2869.2004.00398.x

La Porta, T., Maselli, G., & Petrioli, C. (2011). Anticollision protocols for single-reader RFID systems: Temporal analysis and optimization. *IEEE Transactions on Mobile Computing*, *10*(2), 267–279. doi:10.1109/TMC.2010.58

LaGrone, A., & Chapman, C. (1961). Some propagation characteristics of high UHF signals in the immediate vicinity of trees. *IRE Transactions on Antennas and Propagation*, *9*(5), 487–491. doi:10.1109/TAP.1961.1145049

Lamport, L. (1979). *Constructing digital signatures from a one way function*. Technical Report CSL-98, SRI International, October 1979.

Landt, J. (2005). The history of RFID. *IEEE Potentials*, *24*(4), 8–11. doi:10.1109/MP.2005.1549751

Law, C., Lee, K., & Siu, K. Y. (2000). Efficient memoryless protocol for tag identification (extended abstract). *Proceedings of the 4th International Workshop on Discrete Algorithms and Methods for Mobile Computing and Communications, DIALM*, (pp. 75–84).

Law, Y. W., Doumen, J., & Hartel, P. (2006). Survey and benchmark of block ciphers for wireless sensor networks. *ACM Transactions in Sensor Networks*, *2*(1), 65–93. doi:10.1145/1138127.1138130

Lee, S., Joo, S., & Lee, C. (2005). An enhanced dynamic framed slotted aloha algorithm for RFID tag identification. *Proceedings of Mobiquitous*, (pp. 166–172).

Lee, Y. K., & Verbauwhede, I. (2007). A compact architecture for Montgomery elliptic curve scalar multiplication processor. In S. Kim, M. Yung, & H.-W. Lee (Eds.), *8th International Workshop on Information Security Applications - WISA 2007* (pp. 115-127). Berlin, Germany: Springer.

Lei, H., Xin-Me, L., Song-He, Y., & Zeng-Yu, C. (2010). A one-way hash based low-cost authentication protocol with forward security in RFID system. In *Informatics in Control, Automation and Robotics (CAR), 2010 2nd International Asia Conference on* (Vol. 2, pp. 269 -272).

Leng, X., Lien, Y., Mayes, K., & Markantonakis, K. (2010). An RFID grouping proof protocol exploiting anti-collision algorithm for subgroup dividing. In *International Journal of Security and Networks (IJSN) Special Issue on "Security and Privacy in RFID Systems*, 2010.

Leng, X., Mayes, K., & Markantonakis, K. (2008). HB-M+ protocol: An improvement on the HB-MP protocol. *IEEE International Conference on RFID*, (pp. 118-124).

Lexware Labs. (n.d.). *Sleep cycle*. Retrieved June 16, 2011 from http://www.lexwarelabs.com/sleepcycle/

Lien, Y., Leng, X., Mayes, K., & Chiu, J.-H. (2008). *Reading order independent grouping proof for RFID tags*. In 2008 IEEE International Conference on Intelligence and Security Informatics, June 17-20, 2008, Taipei, Taiwan.

Lim, C., & Lee, P. (1994). More flexible exponentiation with precomputation. In Y. Desmedt (Ed.), *Proceedings of the 14th Annual International Cryptology Conference on Advances in Cryptology - CRYPTO 1994*, (pp. 95-107). Berlin, Germany: Springer.

Lin, C.-C., Lai, Y.-C., Tygar, J. D., Yang, C.-K., & Chiang, C.-L. (2007). Coexistence proof using chain of timestamps for multiple RFID tags. In *The Proceeding of APWeb/WAIM International Workshop, LNCS 5189*, (pp. 634-643). Springer-Verlag.

Liu, A., & Ning, P. (2008). TinyECC: A configurable library for elliptic curve cryptography in wireless sensor networks. *7th International Conference on Information Processing in Sensor Networks - IPSN 2008* (pp. 245-256).

López, J., & Dahab, R. (1999). Fast multiplication on elliptic curves over GF(2^m) without precomputation . In Koç, C. K., & Paar, C. (Eds.), *Cryptographic Hardware and Embedded Systems – CHES 1999* (pp. 316–327). Berlin, Germany: Springer. doi:10.1007/3-540-48059-5_27

Mack, D., Kell, S., Alwan, M., Turner, B., & Felder, R. (2003). *Non-invasive analysis of physiological signals (naps): A vibration sensor that passively detects heart and respiration rates as part of a sensor suite for medical monitoring*. In 2003 Summer Bioengineering Conference.

Mack, D., Alwan, M., Turner, B., Suratt, P., & Felder, R. (2006). *A passive and portable system for monitoring heart rate and detecting sleep apnea and arousals: Preliminary validation* (pp. 51–54). Arlington, VA: Distributed Diagnosis and Home Healthcare.

Mangard, S., Oswald, M. E., & Popp, T. (2007). *Power analysis attacks- Revealing the secrets of smart cards.* Springer 2007.

Mangard, S., Oswald, M. E., & Popp, T. (2007). *Power analysis attacks - Revealing the secrets of smart cards.* Berlin, Germany: Springer.

Marsaglia, G. (2003). Xorshift RNGs. *Journal of Statistical Software, 8*(14), 1–6.

Martin, H., San Millan, E., Entrena, L., Hernandez-Castro, J., & Peris-Lopez, P. (2011). AKARI-X: A pseudorandom number generator for secure lightweight systems. *2011 IEEE 17th International On-Line Testing Symposium (IOLTS),* (pp. 228 -233).

Martin, S., Kelly, G., Kernohan, W. G., McCreight, B., & Nugent, C. (2008). Smart home technologies for health and social care support. *Cochrane Database of Systematic Reviews, 2008*(4). doi:10.1002/14651858. CD006412.pub2

Maselli, G., Petrioli, C., & Vicari, C. (2008). Dynamic tag estimation for optimizing tree slotted aloha in RFID networks. *Proceedings of the 11th ACM Symposium on Modeling, Analysis and Simulation of Wireless and Mobile Systems,* (pp. 315–322).

McPherson, K., Parry, D., & Symonds, J. (2007, December 4). *Radio frequency identification (RFID) for assisted living: Testing the aura object location (AOL) model.* Paper presented at the 18th Australasian Conference on Information Systems, Toowoomba, Queensland, Australia.

Medwed, M., & Oswald, E. (2008). Template attacks on ECDSA. In K. Chung, M. Yung, & K. Sohn (Eds.), *9th International Workshop on Information Security Applications - WISA 2008,* (pp. 14-27), Berlin, Germany: Springer.

Melia-Segui, J., Garcia-Alfaro, J., & Herrera-Joancomarti, J. (2010). Analysis and improvement of a pseudorandom number generator for EPC Gen2 tags. *Proceedings of the 14th International Conference on Financial Cryptograpy and Data Security (FC '10)*, (pp. 34-46).

Meloni, N. (2006). *Fast and secure elliptic curve scalar multiplication over prime fields using special addition chains*. Cryptology ePrint Archive, Report 2006/216.

Menezes, A., Van Oorschot, P., Vanstone, S., & Rivest, R. (1996). *Handbook of applied cryptography*. Boca Raton, FL: CRC Press, Inc.

Micic, A., Nayac, A., Simplot-Ryl, D., & Stojmenovic, I. (2005). A hybrid randomized protocol for RFID tag identification. *Proceedings of 1st IEEE International Workshop on Next Generation Wireless Networks (WoNGeN)*.

Milenkovic, A., Otto, C., & Jovanov, E. (2006). Wireless sensor networks for personal health monitoring: Issues and an implementation. *Computer Communications*, *29*(13-14), 2521–2533. doi:10.1016/j.comcom.2006.02.011

Mitrokotsa, A., Rieback, M., & Tanenbaum, A. (2010). Classifying RFID attacks and defenses. *Information Systems Frontiers*, *12*(5), 491–505. doi:10.1007/s10796-009-9210-z

Molnar, D., & Wagner, D. (2004). *Privacy and security in library RFID: Issues, practices, and architectures*. In ACM Computer and Communications Security, 2004.

Montefiore, A., Parry, D., & Philpott, A. (2010). *A radio frequency identification (RFID)-based wireless sensor device for drug compliance measurement*. Paper presented at the Health Informatics New Zealand, Wellington.

Montgomery, P. L. (1987). Speeding the Pollard and elliptic curve methods of factorization. *Mathematics of Computation*, *48*(177), 243–264. doi:10.1090/S0025-5718-1987-0866113-7

Mowry, M. (2008). *A survey of RFID in the medical industry*. Retrieved from http://faculty.ist.psu.edu/xu/papers/jips.pdf

Munilla, J., & Peinado, A. (2007). HB-MP: A further step in the hb-family of lightweight authentication protocols. *Computer Networks*, *51*, 2262–2267. doi:10.1016/j.comnet.2007.01.011

Munilla, J., & Peinado, A. (2009). Distance bounding protocols for RFID . In Kitsos, P. (Ed.), *Security in RFID and sensor networks* (pp. 151–169). CRC Press. doi:10.1201/9781420068405.ch7

Myung, J., & Lee, W. (2006). Adaptive binary splitting: A RFID tag collision arbitration protocol for tag identification. *Mobile Networks and Applications*, *11*, 711–722. doi:10.1007/s11036-006-7797-6

Myung, J., Lee, W., & Shih, T. (2006). An adaptive memoryless protocol for RFID tag collision arbitration. *IEEE Transactions on Multimedia*, *8*, 1096–1101. doi:10.1109/TMM.2006.879817

Myung, J., Lee, W., Srivastava, J., & Shih, T. K. (2007). Tag-splitting: Adaptive collision arbitration protocols for RFID tag identification. *IEEE Transactions on Parallel and Distributed Systems*, *18*, 763–775. doi:10.1109/TPDS.2007.1098

Nakahara, J. Jr. (2010). *WG2 - Lightweight cryptographic algorithms*. European Network of Excellence in Cryptology II.

Namjun, C., Seong-Jun, S., Sunyoung, K., Shiho, K., & Hoi-Jun, Y. (2005, 12-16 September). A 5.1-μW UHF RFID tag chip integrated with sensors for wireless environmental monitoring. *Proceedings of the 31st European Solid-State Circuits Conference*, ESSCIRC 2005.

National Institute of Standards and Technology (NIST). (2009). *FIPS-186-3: Digital signature standard* (DSS). Retrieved June, 2011, from http://www.itl.nist.gov/fipspubs/

New Zealand Privacy Commissioner. (2008, 2008). *The New Zealand health information privacy code*. Retrieved 1st May 2011, from http://privacy.org.nz/health-information-privacy-code/

Ng, N. G. (2009). *Programmable medication alert & dispenser*. Singapore: SIM.

Nithyanand, R., Tsudik, G., & Uzun, E. (2010). *Readers behaving badly: Reader revocation in PKI-based RFID systems*. In European Symposium on Research in Computer Security, 2010.

Nükhet, S., & Haldun, A. (2009). The importance of using wireless sensor networks for forest fire sensing and detection in Turkey. In *5th International Advanced Technologies Symposium*, (pp. 1323-1327).

Ohkubo, M., Suzuki, K., & Kinoshita, S. (2003). *Cryptographic approach to "privacy friendly" tags*. RFID Privacy Workshop.

Okeya, K., & Sakurai, K. (2000). Power analysis breaks elliptic curve cryptosystems even secure against the timing attack. In B. K. Roy & E. Okamoto (Eds.), *First International Conference in Cryptology in India Progress in Cryptology - INDOCRYPT 2000*, (pp. 178-190). Springer.

Oksenberg, A., & Silverberg, D. (1998). The effect of body posture on sleep-related breathing disorders: Facts and therapeutic implications. *Sleep Medicine Reviews*, 2(3), 139–162. doi:10.1016/S1087-0792(98)90018-1

Oren, Y., & Feldhofer, M. (2009). *A low-resource public-key identification scheme for RFID tags and sensor nodes*. Paper presented at Second ACM Conference on Wireless Network Security.

Oren, Y., & Wool, A. (2009). *Relay attacks on RFID-based electronic voting systems*. Cryptology ePrint Archive, Report 2009/422. Retrieved from http://eprint.iacr.org/2009/422

Oren, Y., & Shamir, A. (2007). Remote password extraction from RFID tags. *IEEE Transactions on Computers*, 56(9), 1292–1296. doi:10.1109/TC.2007.1050

Örs, B., Batina, L., & Prenel, B. (2003). Hardware implementation of elliptic curve processor over GF(p). *International Journal of Embedded Systems*, 3(4), 229–240. doi:10.1504/IJES.2008.022394

Ouafi, K., Overbeck, R., & Vaudenay, S. (2008). On the security of HB# against a man-in-the-middle attack . In Pieprzyk, J. (Ed.), *Advances in Cryptology -- Asiacrypt 2008* (pp. 108–124). doi:10.1007/978-3-540-89255-7_8

Ouafi, K., & Phan, R. C.-W. (2008). Privacy of recent RFID authentication protocols. *Proceedings of ISPEC, LNCS, 4991*, 263–277.

Papadimitratos, P., & Jovanovic, A. (2008). *GNSS-based positioning: Attacks and countermeasures*. In IEEE Military Communications Conference, 2008.

Pare, G., Jaana, M., & Sicotte, C. (2007). Systematic review of home telemonitoring for chronic diseases: The evidence base. *Journal of the American Medical Informatics Association*, 14(3), 269–277. doi:10.1197/jamia.M2270

Parhami, F. M. A. B. (1987). *URISC: The ultimate reduced instruction set computer*. Department of Computer Science, University of Waterloo.

Pasquet, M., Reynaud, J., & Rosenberger, C. (2008). Secure payment with NFC mobile phone in the SmartTouch project. *International Symposium on Collaborative Technologies and Systems, CTS 2008*, (pp. 121-126).

Patterson, D. A., & Ditzel, D. R. (1980). The case for the reduced instruction set computer. *SIGARCH Computer Architecture News, 8*(6), 25–33. doi:10.1145/641914.641917

Peng, Q., Zhang, M., & Wu, W. (2007). Variant enhanced dynamic frame slotted aloha algorithm for fast object identification in RFID system. *Proceedings of IEEE International Workshop on Anti-counterfeiting, Security, Identification*, (pp. 88–91).

Peng, Y.-T., Lin, C.-Y., & Sun, M.-T. (2006). Multimodality sensors for sleep quality monitoring and logging. *22nd International Conference on Data Engineering Workshops (ICDEW'06)* (p. 108).

Peris-Lopez, P., Hernandez-Castro, J. C., Estevez-Tapiador, J. M., & Ribagorda, A. (2007). Solving the simultaneous scanning problem anonymously: Clumpling proofs for RFID tags. In *Third International Workshop on Security, Privacy and Trust in Pervasive and Ubiquitous Computing*.

Peris-Lopez, P., Hernandez-Castro, J., Estevez-Tapiador, J., & Ribagorda, A. (2006). *LMAP: A real lightweight mutual authentication protocol for low-cost RFID tags*. Workshop on RFID Security (RFIDSec'06).

Peris-Lopez, P., Hernandez-Castro, J., Estevez-Tapiador, J., & Ribagorda, A. (2006). M2AP: A Minimalist mutual authentication protocol for low-cost RFID tags. *International Conference on Ubiquitous Intelligence and Computing (UIC'06)*, (pp. 912-923).

Peris-Lopez, P., Tong-Lee, L., & Tieyan, L. (2008). Providing stronger authentication at a low cost to RFID tags operating under the EPCglobal framework. *IEEE/IFIP International Conference on Embedded and Ubiquitous Computing '08* (pp. 159 -166).

Peris-Lopez, P., Hernandez-Castro, J. C., Tapiador, J. M., & Ribagorda, A. (2009). Advances in ultralightweight cryptography for low-cost RFID tags: Gossamer protocol . In Kyo-Il, C., Kiwook, S., & Moti, Y. (Eds.), *Information security applications* (pp. 56–68). Springer-Verlag. doi:10.1007/978-3-642-00306-6_5

Peris-Lopez, P., Hernández-Castro, J., Estévez-Tapiador, J., & Ribagorda, A. (2006). RFID systems: A survey on security threats and proposed solutions. *Personal Wireless Communicaitons, 4217*, 159–170. doi:10.1007/11872153_14

Peris-Lopez, P., Hernandez-Castro, J., Estevez-Tapiador, J., & Ribagorda, A. (2009). LAMED - A PRNG for EPC class-1 generation-2 RFID specification. *Computer Standards & Interfaces, 31*(1), 88–97. doi:10.1016/j.csi.2007.11.013

Peris-Lopez, P., Orfila, A., Hernandez-Castro, J. C., & van der Lubbe, J. C. A. (2011). Flaws on RFID grouping-proofs. Guidelines for future sound protocols. *Journal of Network and Computer Applications, 34*(3). doi:10.1016/j.jnca.2010.04.008

Peris-Lopez, P., Orfila, A., Palomar, E., & Hernandez-Castro, J. (2011). A secure dustance-based RFID identification protocol with an off-line back-end database. *Personal and Ubiquitous Computing, 16*(3), 1–15.

Philipose, M., Smith, J. R., Jiang, B., Sundara-Rajan, K., Mamishev, A., & Roy, S. (2005). Battery-free wireless identification and sensing. *IEEE Pervasive Computing / IEEE Computer Society and IEEE Communications Society, 4*(1), 37–45. doi:10.1109/MPRV.2005.7

Philips actiwatch. (n.d.). *Website*. Retrieved June 16, 2011, from http://www.actiwatch.respironics.com/

Piramuthu, S. (2006). *On existence proofs for multiple RFID tags*. In IEEE International Conference on Pervasive Services, Workshop on Security, Privacy and Trust in Pervasive and Ubiquitous Computing.

Plos, T. (2008). Susceptibility of UHF RFID tags to electromagnetic analysis. In T. Malkin (Ed.), *The Cryptographers' Track at the RSA Conference - CT-RSA 2008*, (pp. 288-300). Springer.

Population Division United Nations. (2002). *World population ageing: 1950-2050*. New York, NY: Author. Retrieved from http://www.un.org/esa/population/publications/worldageing19502050/

Product, W. I. S. P. (n.d.). *MCRF200IWQ23: Passive RFID IC device*. Retrieved from http://www.datasheetcatalog.com/datasheets_pdf/M/C/R/F/MCRF202.shtml

Qiu, R. G. (2005). *An Internet computing model for ensuring continuity of healthcare*. Paper presented at the Systems, Man and Cybernetics, 2005 IEEE International Conference on.

Raddum, H. (2002). *The statistical evaluation of the NESSIE submission Anubis*.

Rasmussen, K., & Capkun, S. (2010). *Realization of RF distance bounding*. In the USENIX Security Symposium, 2010.

Razaq, A., Luk, W., & Cheng, L. (2007). Privacy and security problems in RFID. *IEEE International Workshop on Anti-counterfeiting, Security, Identification* (pp. 402 -405).

Redmond, S., & Heneghan, C. (2006). Cardio respiratory-based sleep staging in subjects with obstructive sleep apnea. *IEEE Transactions on Bio-Medical Engineering, 53*(3), 485–496. doi:10.1109/TBME.2005.869773

RFID-F2F (RFID From Farm to Fork). (n.d.). *Website*. Retrieved October 15, 2011, http://www.rfidf2f.eu/

Richter, J., Caldeirinha, R. F. S., Al-Nuaimi, M. O., Seville, A., Rogers, N. C., & Savage, N. (2005). A generic narrowband model for radiowave propagation through vegetation. In *Vehicular Technology Conference*, Vol. 1, (pp. 39- 43).

Rieback, M., Crispo, B., & Tanembaum, A. (2006). *Is your cat infected with a computer virus?* In Pervasive Computing and Communications, Pisa, Italy, March 2006. IEEE Computer Society Press.

Roberts, L. G. (1975). ALOHA packet system with and without slots and capture. *SIGCOMM Computer Communications Review*, *5*(2), 28–42. doi:10.1145/1024916.1024920

Rolfes, C., Poschmann, A., Leander, G., & Paar, C. (2008). *Ultra-lightweight implementations for smart devices - Security for 1000 gate equivalents*. Paper presented at the 8th IFIP WG 8.8/11.2 International Conference on Smart Card Research and Advanced Applications.

Römer, K., Schoch, T., Mattern, F., & Dübendorfer, T. (2004). Smart identification frameworks for ubiquitous computing applications. *Wireless Networks*, *10*(6), 689–700. doi:10.1023/B:WINE.0000044028.20424.85

Rotter, P. (2008, April-June). A framework for assessing RFID system security and privacy risks. *IEEE Pervasive Computing / IEEE Computer Society and IEEE Communications Society*, *7*(2), 70–77. doi:10.1109/MPRV.2008.22

Rouvroy, G., Standaert, F. X., Quisquater, J. J., & Legat, J. D. (2004, 5-7 April 2004). *Compact and efficient encryption/decryption module for FPGA implementation of the AES Rijndael very well suited for small embedded applications*. Paper presented at the International Conference on Information Technology: Coding and Computing, ITCC 2004.

Ruhanen, A., et al. (2008). *Sensor-enabled RFID tag handbook*. Retrieved January 2008 from http://www.bridge-project.eu/data/File/BRIDGE_WP01_RFID_tag_handbook.pdf

Ryu, J., Lee, H., Seok, Y., Kwon, T., & Choi, Y. (2007). A hybrid query tree protocol for tag collision arbitration in RFID systems. *Proceedings of IEEE International Conference on Communications*, (pp. 5981–5986).

Sadeh, A., & Acebo, C. (2002). The role of actigraphy in sleep medicine. *Sleep Medicine Reviews*, *6*(2), 113–124. doi:10.1053/smrv.2001.0182

Saito, J., & Sakurai, K. (2005). Grouping-proof for RFID tags. *19th International Conference on Advanced Information Networking and Applications* (pp. 621-624).

Salajegheh, M., Clark, S., Ransford, B., Fu, K., & Juels, A. (2009). *CCCP: Secure remote storage for computational RFIDs*. In 18th USENIX Security Symposium, August, 2009.

Sample, A., Yeager, D., & Smith, J. (2009). *A capacitive touch interface for passive RFID tags*. In IEEE International Conference on RFID, 2009.

Sample, A., Yeager, D., Powledge, P., & Smith, J. (2007). *Design of a passively-powered, programmable sensing platform for UHF RFID systems*. In IEEE International Conference on RFID, 2007.

Sample, A. P., Yeager, D. J., Powledge, P. S., Mamishev, A. V., & Smith, J. R. (2008). Design of an RFID-based battery-free programmable sensing platform. *IEEE Transactions on Instrumentation and Measurement*, *57*(11), 2608–2615. doi:10.1109/TIM.2008.925019

Sample, A. P., Yeager, D. J., Powledge, P. S., & Smith, J. R. (2008). Design of an RFID-based battery-free programmable sensing platform. *IEEE Transactions on Instrumentation and Measurement*, *57*(11), 2608–2615. doi:10.1109/TIM.2008.925019

Sanghera, P., Thornton, F., Haines, B., Kung Man, F., Fung, F., Kleinschmidt, J., … Campbell, A. (2007). *How to cheat at deploying and securing RFID*. Syngress, 2007.

Sarma, S. E., Weis, S. A., & Engels, D. W. (2002). RFID systems and security and privacy implications. In *Proceedings of CHES'02, 2523*, (pp. 454–470).

Saxena, N., & Voris, J. (2009). *We can remember it for you wholesale: Implications of data remanence on the use of RAM for true random number generation on RFID tags*. In Conference on RFID Security, 2009.

Saxena, N., & Voris, J. (2009, November). *Accelerometer based random number generation on RFID tags*. Paper presented at the 1st Workshop on Wirelessly Powered Sensor Networks and Computational RFID, Berkely, California.

Saxena, N., & Voris, J. (2010). *Still and silent: Motion detection for enhanced RFID security and privacy without changing the usage model*. In Workshop on RFID Security, 2010.

Saxena, N., Ekberg, J., Kostiainen, K., & Asokan, N. (2006). *Secure device pairing based on a visual channel* (short paper). In IEEE Symposium on Security and Privacy, 2006.

Schoute, F. C. (1983). Dynamic frame length aloha. *IEEE Transactions on Communications*, *31*, 565–568. doi:10.1109/TCOM.1983.1095854

Scott, M., & Szczechowiak, P. (2007). *Optimizing multiprecision multiplication for public key cryptography.* Cryptology ePrint Archive, Report 2007/299. Retrieved January, 2012, from http://eprint.iacr.org

Shambroom, J., Johnstone, J., & Fabregas, J. (2009). Evaluation of portable monitor for sleep staging. *Sleep*, A386.

Shinar, Z., Akselrod, S., Dagan, Y., & Baharav, A. (2006). Autonomic changes during wake-sleep transition: A heart rate variability based approach. *Autonomic Neuroscience*, *130*(1-2), 17–27. doi:10.1016/j.autneu.2006.04.006

Shoup, V. (1999). *On formal models for secure key exchange* (version 4), 15 November 1999. Revision of IBM Research Report RZ 3120 (April 1999).

Simon, D. (2006). *Optimal state estimation: Kalman, H-infinity, and nonlinear approaches*. John Wiley & Sons. doi:10.1002/0470045345

Sleep Foundation. (n.d.). *Sleep related problems.* Retrieved June 16, 2011 from http://www.sleepfoundation.org/articles/sleep-related-problems

Smith, J., Sample, A., Powledge, P., Roy, S., & Mamishev, A. (2006). A wirelessly-powered platform for sensing and computation. *Proceedings of Ubicomp 2006: 8th International Conference on Ubiquitous Computing*, Orange Country, CA, USA, September 17-21 2006, (pp. 495-506).

Smith, J. R., Fishkin, K. P., Jiang, B., Mamishev, A., Philipose, M., & Rea, A. D. (2005). RFID-based techniques for human-activity detection. *Communications of the ACM*, *48*(9), 39–44. doi:10.1145/1081992.1082018

Song, B. A. (2008). RFID authentication protocol for low-cost tags. In *Proceedings of the First ACM Conference on Wireless Network Security* (pp. 140-147). Alexandria, VA, USA.

Standard, E. P. C. (2008). *EPC radio-frequency identity protocols class-1 generation-2 UHF RFID protocol for communications at 860 MHz – 960*, MHz Version 1.2.0. Retrieved from http://www.epcglobalinc.org/standards/.

Stauffer, H. B. (1991). Smart enabling system for home automation. *IEEE Transactions on Consumer Electronics*, *37*(2), xxix–xxxv. doi:10.1109/30.79314

Stefanov, D. H., Bien, Z., & Bang, W.-C. (2004). The smart house for older persons and persons with physical disabilities: Structure, technology arrangements, and perspectives. *IEEE Transactions on Neural Systems and Rehabilitation Engineering*, *12*(2), 228–250. doi:10.1109/TNSRE.2004.828423

Story, M. (1998). Maximizing usability: The principles of universal design. *Assistive Technology*, *10*(1), 4–12. doi:10.1080/10400435.1998.10131955

Surhone, L. M., Tennoe, M. T., & Henssonow, S. F. (2010). *Anubis (Cipher)*. VDM Verlag Dr. Mueller AG & Co. Kg.

Suzuki, R., Ogawa, M., Otake, S., Izutsu, T., Tobimatsu, Y., & Iwaya, T. (2006). Rhythm of daily living and detection of atypical days for elderly people living alone as determined with a monitoring system. *Journal of Telemedicine and Telecare*, *12*(4), 208–214. doi:10.1258/135763306777488780

Szczechowiak, P., Oliveira, L., Scott, M., Collier, M., & Dahab, R. (2008). NanoECC: Testing the limits of elliptic curve cryptography in sensor networks. In R. Verdone (Ed.), *5th European Conference on Wireless Sensor Networks - EWSN 2008*, (pp. 305-320). Berlin, Germany: Springer.

Szekely, A., Höfler, M., Stögbuchner, R., & Aigner, M. (2013). Security enhanced WISPs: Implementation challenges . In Smith, J. R. (Ed.), *Wirelessly Powered Sensor Networks and Computational RFID, Springer (to appear in 2013)*.

Tae Youn, W., Ji Young, C., & Dong Hoon, L. (2008, 17-20 December). *Strong authentication protocol for secure RFID tag search without help of central database.* Paper presented at the IEEE/IFIP International Conference on Embedded and Ubiquitous Computing, EUC '08.

Technology, N. I. o. S. a. (2001). *Advanced encryption standard*.

Thamilarasu, G., & Sridhar, R. (2008). Intrusion detection in RFID systems. In *IEEE Military Communications Conference*, (pp. 1-7).

Thorton, F. (2010). RFID security part 3: Threat and target identification. *EE Times Design*, September 2010.

Timmons, N. F., & Scanlon, W. G. (2004). Analysis of the performance of IEEE 802.15.4 for medical sensor body area networking. In *First Annual IEEE Communications Society Conference on Sensor and Ad Hoc Communications and Networks*, (pp. 16-24).

Tippenhauer, N. O., Popper, C., Rasmussen, K. B., & Capkun, S. (2011). *On the requirements for successful GPS spoofing attacks.* In ACM Conference on Computer and Communication Security (CCS'11), October 2011.

Tormanen, T. (2010). *Analog IC design 2010: Lecture 9 - Noise.* Retrieved from http://framtiden.eit.lth. se/fileadmin/eit/courses/eti063/lectures2010/AnalogIC_F9.pdf

Turner, C. (2003). *Deterministic or probabilistic – Which is best? Why would you even care?* Retrieved from http://www.rfip.eu/downloads/Deterministic_or_Probabilistic.pdf

UK Audit Commission. (2004, 12th February). *Assistive technology independence and well-being 4.* Retrieved 1 November 2006, from http://www.audit-commission.gov.uk/reports/national-report.asp?CategoryID=&ProdID=BB070AC2-A23A-4478-BD69-4C19BE942722

UK Department of Health. (2005, 2nd March 2007). *Our health, our care, our say.* Retrieved 1st July 2007, from http://www.dh.gov.uk/en/Publicationsandstatistics/Publications/PublicationsPolicyAndGuidance/DH_4127453

Vajda, I., & Buttyán, L. (2003). Lightweight authentication protocols for low-cost RFID tags. In *Proceedings of UBICOMP '03.*

Vaudenay, S. (2006). *RFID privacy based on public-key cryptography.* ICISC 2006, LNCS (pp. 1–6). Springer.

Vermeire, E., Hearnshaw, H., Van Royen, P., & Denekens, J. (2001). Patient adherence to treatment: three decades of research. A comprehensive review. *Journal of Clinical Pharmacy and Therapeutics*, *26*, 331–342. doi:10.1046/j.1365-2710.2001.00363.x

Vogt, H. (2002). Efficient object identification with passive RFID. *Proceedings of The First International Conference on Pervasive Computing,* (pp. 98–113).

Voris, J., Saxena, N., & Halevi, T. (2011). *Accelerometers and randomness: Perfect together.* In ACM Conference on Wireless Network Security, 2011. Smith, J., Powledge, P., Roy, S., & Mamishev, A. (2006). *A wirelessly-powered platform for sensing and computation.* In 8th International Conference on Ubiquitous Computing, 2006.

Wagner, D. (2005). *Privacy in pervasive computing: What can technologists do?* Invited talk at the 1st International Conference on Security and Privacy in Communication Networks, 2005. Retrieved from http://www.cs.berkeley.edu/~daw/talks/SECCOM05.ppt

Waldrop, J., Engels, D. W., & Sarma, S. E. (2003). Colorwave: A MAC for RFID reader networks. In *IEEE Wireless Communications and Networking, WCNC 2003*, Vol. 3, (pp. 1701–1704). New Orleans, Louisiana, March 2003.

Want, R. (2006). An introduction to RFID technology. *Pervasive Computing*, *5*(1), 25–33. doi:10.1109/MPRV.2006.2

Wieselthier, J. E., Ephremides, A., & Michaels, L. A. (1989). An exact analysis and performance evaluation of framed aloha with capture. *IEEE Transactions on Communications*, *37*, 125–137. doi:10.1109/26.20080

Wiki, W. I. S. P. (n.d.). Retrieved January, 2012, from http://wisp.wikispaces.com

WISP. (n.d.). *Wiki.* Retrieved from http://wisp.wikispaces.com/

Wisp: Wireless identification and sensing platform. (n.d.). Retrieved June 16, 2011, from http://www.seattle.intel-research.net/wisp/

Woo Kwon, K., Hwaseong, L., Yong Ho, K., & Dong Hoon, L. (2008, 24-26 April 2008). *Implementation and analysis of new lightweight cryptographic algorithm suitable for wireless sensor networks.* Paper presented at the International Conference on Information Security and Assurance, ISA 2008.

Wood, A., Stankovic, J., Virone, G., Selavo, L., He, Z., & Cao, Q.,…. Stoleru, R. (2008). Context-aware wireless sensor networks for assisted-living and residential monitoring. *IEEE Network*, *22*(4), 26–33. doi:10.1109/MNET.2008.4579768

Wright, K., Johnstone, J., Fabregas, S., & Shambroom, J. (2008). Evaluation of a portable, dry sensor-based automatic sleep monitoring system. *Sleep*, *31*, A337.

Xiao, Y., Hu, F., & Kumar, S. (2007, Jan. 2007). *Towards a secure, RFID/sensor based telecardiology system.* Paper presented at the 4th IEEE Consumer Communications and Networking Conference, CCNC 2007.

Yang, J., Park, J., Lee, H., Ren, K., & Kim, K. (2005). *Mutual authentication protocol for low-cost RFID.* Ecrypt Workshop on RFID and Lightweight Crypto.

Yang, Y., Shin, J., Jang, S., Lee, H., & Yoon, Y. (2006). Research of body movement during sleep with an infrared triangulation distance sensor, wavelets and neuro-fuzzy reasoning. *Proceedings of International Federation for Medical and Biological Engineering, 14*(7), 760–763.

Yeager, D., Prasad, R., Wetherall, D., Powledge, P., & Smith, J. (2008). Wirelessly-charged uhf tags for sensor data collection. In *Proceedings of IEEE RFID*.

Yeager, D. J., Sample, A. P., & Smith, J. R. (2008). WISP: A passively powered UHF RFID tag with sensing and computation . In Ahson, S., & Ilyas, M. (Eds.), *RFID handbook: Applications, technology, security, and privacy*. CRC Press.

Yeager, D. J., Sample, A. P., & Smith, J. R. (2008). Wisp: A passively powered UHF RFID tag with sensing and computation . In Syed, M. I., & Ahson, A. (Eds.), *RFID handbook: Applications, technology, security, and privacy*. CRC Press.

Yeager, D., Holleman, J., Prasad, R., Smith, J., & Otis, B. (2009). NeuralWISP: A wirelessly powered neural interface with 1-m Range. *IEEE Transactions on Biomedical Circuits and Systems, 3*(6), 379–387. doi:10.1109/TBCAS.2009.2031628

Yeager, D., Holleman, J., Prasad, R., Smith, J., & Otis, B. (in press). Neuralwisp: A wirelessly powered neural interface with 1-m range. *IEEE Transactions on Biomedical Circuits and Systems*.

Yeh, T., Wang, Y., Kuo, T., & Wang, S. (2010). Securing RFID systems conforming to EPC class 1 generation 2 standard. *Expert Systems with Applications, 37*(12), 7678–7683. doi:10.1016/j.eswa.2010.04.074

Yoon, B. (2009). *HB-MP++ protocol: An ultra lightweight authentication protocol for RFID system.*

Yüksel, K. (2004). *Universal hashing for ultra-low-power cryptographic hardware applications.*

Yun, X., Bachmann, E., & McGhee, R. (2008). A simplified Quaternion-based algorithm for orientation estimation from earth gravity and magnetic field measurements. *IEEE Transactions on Instrumentation and Measurement, 57*(3).

Zeo. (n.d.). *Personal sleep coach.* Retrieved June 16, 2011 from http://www.myzeo.com/

Zhang, H., Gummeson, J., Ransford, B., & Fu, K. (2011). Moo: A batteryless computational RFID and sensing platform. Tech report UM-CS-2011-020, Department of Computer Science, University of Massachusetts Amherst.

Zhang, N., & Vojcic, B. (2005). Binary search algorithms with interference cancellation RFID systems. *Proceedings of Military Communications Conference, MILCOM,* Vol. 2, (pp. 950–955).

Zimmermann, H. (1980). OSI reference model--The ISO model of architecture for open systems interconnection. *IEEE Transactions on Communications, 28*(4), 425–432. doi:10.1109/TCOM.1980.1094702

About the Contributors

Pedro Peris-Lopez has a M.Sc. in Telecommunications Engineering and Ph.D. in Computer Science. His research interests are in the field of protocols design, primitives design, lightweight cryptography, cryptanalysis, et cetera. Nowadays, his research is focused on Radio Frequency Identification Systems (RFID) and Implantable Medical Devices (IMD). In these fields, he has published a great number of papers in specialized journals and conference proceedings. He has cited over 800 times and his h-index is 12.

Julio C. Hernandez-Castro has a degree in Maths and an MSc in Coding Theory and Network Security. He has worked heavily in the past on the applications of artificial intelligence techniques (notably evolutionary computation) to Cryptography and Cryptanalysis. He got his Ph.D. in Computer Science from Carlos III University in Madrid in 2003 and was there till February 2009 when he was appointed Senior Lecturer in the School of Computing of Portsmouth University, UK. He is currently Visiting Professor at the Department of Computer Science, Universidad Carlos III de Madrid. He has published more than 30 papers in international journals and more than 50 in international conferences. He is a keen chess player.

Tieyan Li He is a Senior Researcher at Institute for Infocomm Research (I2R, Singapore) from Oct. 2001. He obtained his Ph.D. degree in 2003 at School of Computing, National University of Singapore. Dr. Li is experienced in practical system developments such as networking, system integration, and software programming. He is also active in academic security research fields with tens of journal and conference publications and several patents. Currently his areas of research are in applied cryptography and network security, as well as security issues in RFID, sensor, multimedia and tamper resistant hardware/software, et cetera. Dr. Li has served as the PC member and reviewer for a number of security conferences and journals.

* * *

Ana V. Alejos received the M.S. and PhD. degree from University of Vigo, Spain, in 2000 and 2006. Her Master's Thesis obtained in 2002 the Ericcson Award by the Spanish Association of Electrical Engineers, as the best Multimedia Wireless Project. She is granted with the Marie Curie International Outgoing Fellowship, carrying out the outgoing phase in the New Mexico State University (NM, USA), at the Klipsch School of ECE, with a research focused on measurement and modeling of propagation through dispersive media, and radar waveform generation. Her research work includes radio propagation, communication electronics, wideband radio channel modeling, multimedia wireless systems, waveform and noise code design, and radar. Dr. Alejos is an IEEE member and also a reviewer of several IEEE and IET journals.

Li-Minn Ang received his PhD and Bachelor degrees from Edith Cowan University, Australia in 2001 and 1996, respectively. He was a Lecturer at Monash University (Malaysia Campus) and associate professor at Nottingham University (Malaysia Campus). He is currently with the Centre for Communications Engineering Research at Edith Cowan University. His research interests are in the fields of visual information processing, embedded systems, and wireless sensor networks.

Mike Burmester is a Professor of Computer Science at Florida State University, USA, and Director of the Center for Security and Assurance in IT (C-SAIT). He received his Ph.D. from the University of Rome (La Sapienza), Italy. He joined the faculty at FSU after more than 30 years of research and teaching at leading institutions around the world. Dr. Burmester is Editor of four journals in information security, has organized several workshops and conferences, has two books, five book chapters, and over 120 journal and refereed conference publications covering a wide range of security topics including: privacy/anonymity, pervasive/ubiquitous network systems, lightweight cryptographic applications, RFIDs and sensor applications, trust management, and group key exchange.

Iñigo Cuiñas was born in Vigo, Spain, in 1971. He received his degree in Telecommunication Engineering in 1996, and his Ph.D. degree in 2000, both from the University of Vigo, Spain. His major field of study is the radio channel characterization, including measurements, modelling, and simulation, at microwave and millimetre wave frequencies, both in narrow and wide band conditions. Nowadays, his main research interest is focused on environmental aspects of radiofrequency systems, as the development of techniques to reduce the electromagnetic pollution in populated areas, and on rural applications of wireless technologies, as WSN and RFID. He is currently a Profesor Titular de Universidad in the Departamento de Teoría do Sinal e Comunicacións at the Universidade of Vigo, Spain, where he teaches courses in Remote Sensing, and Satellite Communication Systems, at the School of Telecommunication Engineering. Currently, he is the Vice-Dean of the School. He was a Visiting Researcher at the TELEMIC division, ESAT, Katholieke Universiteit Leuven (Belgium), and at the Department of Electronics and Information Technology, University of Glamorgan (Wales, United Kingdom). Dr. Cuiñas is a reviewer of *IEEE Transactions on Vehicular Technology, IEEE Communications Magazine, IEE Proceedings – Communications, IET Sonar Radar and Navigation, Wireless Personal Communications*, and several international conferences. He received the 2001 Best Doctoral Thesis on Access Networks Award of the Colegio Oficial de Ingenieros de Telecomunicación (COIT), the Spanish Association of Electrical Engineers. He is a member of IEEE.

Robert F. Dickerson completed his Bachelor of Science in Computer Engineering from the University of Florida and his Master of Computer Science (MCS) Degree from the University of Virginia. He is now a Ph.D. candidate working with the UVa Center for Wireless Health. His current research interest is behavioral and psychological monitoring for healthcare applications.

Isabel Expósito was born in Vigo, Spain. She received the Telecommunication Engineering degree from the Universidade de Vigo, Spain, in 2009. She has been working at the Departamento de Teoría do Sinal e Comunicacións, Universidade de Vigo, as a Ph.D. researcher since 2010. Her major research interests include RFID and wireless sensor networks.

Jose Antonio Gay Fernandez was born in Baiona, Spain, in 1981. He received the Telecommunication Engineering degree and Signal Theory and Communications M.S. degree from the Universidade de Vigo, Spain, in 2007 and 2011 respectively. He has been working at the Departamento de Teoría do Sinal e Comunicacións, Universidade de Vigo, as a Ph.D. researcher since 2008. His major research interests include wireless sensor networks, location algorithms, and radio propagation models for outdoor environments.

Kong Jia Hao received his Bachelor of Engineering degree (with honours) in the field of Electrical and Electronics from the University of Nottingham Malaysia Campus in 2009. He is currently a PhD Research Student at The University of Nottingham Malaysia Campus. His research interest is in the field of reconfigurable crypto-coprocessors, computer architecture, cryptography, and wireless sensor network.

Enamul Hoque completed his Bachelor of Science degree in Computer Science from Bangladesh University of Engineering and Technology in 2007 and obtained a Master in Computer Science (MCS) Degree from University of Virginia in 2010. He is now a Ph.D. candidate in Computer Science Department of University of Virginia supervised by Professor John A. Stankovic. His current research directions are on middleware for wireless sensor networks and applications of wireless sensor networks in behavioral monitoring for healthcare and energy conservation.

Michael Hutter is a Research Assistant at Graz University of Technology in Austria. Since 2007, he has been working in the security group of the Institute for Applied Information Processing and Communications (IAIK). His main research interests include applied cryptography, RFID security and privacy, side-channel attacks, and fault analyses. He holds a PhD and MSc in Computer Science and participated in various European Commission and Austrian government funded projects. Currently, he gives two courses in Computer Organization and Networks and System on Chip Architectures and Modeling.

Francesca Lonetti got the PhD degree in Computer Science from University of Pisa in 2007. She had one year post-doctoral position at Scuola Superiore Sant'Anna, in Pisa, working on modelling solutions for pervasive and heterogeneous infrastructures for the urban mobility control. She is a researcher at Istituto di Scienza e Tecnologie dell'Informazione "A. Faedo" of the National Research Council of Pisa. Her research interests span from RFID systems and video coding and transcoding techniques for wireless networks to monitoring and testing approaches for web-based systems. She is member of the Editorial Board of *International Journal of Communication Networks and Information Security* (IJC-NIS) and *Communication Technologies Journal* (NCT). She serves the research community as program committee member and reviewer of relevant conferences in wireless networks, internet services, image processing, and software testing topics. She is author of papers, published on international journals and conferences, related to these research topics. She also did teaching activity and was tutor of many Master's degree theses at University of Pisa.

Di Ma is an Assistant Professor in the Computer and Information Science Department at the University of Michigan - Dearborn. She obtained her PhD degree from the University of California, Irvine in 2009. She also received a B.Eng. degree from Xi'an Jiaotong University, China and a M.Eng. degree from Nanyang Technological University, Singapore. She was with the Institute for Infocomm Research, Singapore (2000-05). Her research interests include computer and network security, data and storage security, and applied cryptography. She won the Tan Kah Kee Young Inventor Award in 2004.

Francesca Martelli is a post-doc researcher at the Istituto di Informatica e Telematica (IIT) of the Italian National Researcher Council (CNR). She got her Laurea and PhD degrees in Computer Science at University of Pisa in 2000 and 2005, respectively. She has been a research associate at the Istituto di Scienza e Tecnologie dell'Informazione (CNR) in Pisa, from 2001 till 2006. From 2006 till 2010 she has been a research associate at the Department of Computer Science of the University of Pisa. She joined the IIT institute in July 2010. Her research interests focus on distributed systems and algorithms, in particular on medium access control problems in wireless networks, such as packet scheduling algorithms. She is interested also in RFID systems, focusing on the tag identification problem. Other research topics are vehicular networks and MIMO networks. She has published several articles on international journals and conferences related to her research topics.

Lara Ortiz-Martin received her B.S. degree in Computer Science from the Technical University of Madrid, Spain in 2010, and the M.S degree in Computer Science and Technology at the Carlos III University of Madrid, Spain in 2011. She is currently a Ph.D. student in Department of Applied Mathematics at the Technical University of Madrid. Her research interested includes RFID applications, security, privacy, authentication and design, and cryptanalysis of lightweight protocols.

Alan Montefiore is an Honours student in Computer Science AUT, Auckland New Zealand working in the area of smart mobile devices. He has wide experience in Software development and RFID technology. He currently runs his own software consulting firm, mostly focussing on Android App development. Alan is a keen BMX rider and adventurer.

Jorge Munilla was born in Málaga (Spain). He is a Telecommunication Engineer and he has worked in the IT industry in roles including analysis, design, and technical support. Now, he works as an Associate Professor for the Communication Engineering Department of the University of Málaga. His research interests include cryptography, security in RFID, security in VANETs, and mobile communications. He completed his Ph. D. at the University of Málaga in December of 2010 with his thesis "Advances in RFID Authentication Protocols." His dissertation addresses the problems of reliability, security, and privacy in RFID systems. For the time being, he collaborates with a project which involves the investigation of the security of NFC (Near Field Communication) technology, and the potential problems which arise with its convergence with the mobile phone technology.

David Parry is a Senior Lecturer and director of the AUT Radio frequency Identification (RFID) laboratory (AURA) in the AUT School of Computer and Information Sciences. He was awarded a degree in Physics from Imperial College London, followed by a Masters in Medical Physics from St. Bartholomewís Medical College. He gained a research MSc. in Computer Science from the University of Otago, New Zealand. His research interests include Health Informatics, ontology based information retrieval, and RFID applications for pervasive computing. Current research projects include the use of RFID in healthcare, fuzzy ontology development and implementation, and information systems to support health care in developing nations. He has published over 50 refereed articles and conference papers.

Markus Pelnar was born in June 1986. He raised up near Linz where he later attended a technical college (2000-2005) with orientation on mechanical engineering combined with electronic engineering, in short mechatronics. After a year filled with military service and a four month long internship, he started studying computer science ("Telematik") at Graz University of Technology. In 2010 he finished his BSc and continued with the MSc. One year later he went to the Swiss Federal Institute of Technology in Zurich, Switzerland, for writing his Master thesis (Integrated Systems Laboratory, supervisor: Norbert Felber) focusing on the combination of the well known cipher AES with the new hash function Groestl, which is one of the SHA-3 finalists.

Christian Pendl was born in July 1985 and grew up in Graz where he also attended grammar school. After one year of military service he enrolled in Computer Science ("Telematik") at Graz University of Technology. He finished the BSc in 2010 and proceeded with the MSc. In late 2011 he moved to the Swiss Federal Institute of Technology in Zurich in order to work on his Master's thesis focusing on high-speed authenticated encryption in hardware (Integrated Systems Laboratory, supervisor: Norbert Felber). He is especially interested in applied cryptography, integrated circuit design, and embedded systems.

Anne Philpott is a Senior Lecturer in the Auckland University of Technology's School of Computer and Information Sciences. She gained an MSc(Hons) in Computer Science from the University of Auckland in 1995. Anne worked in software engineering research and development roles in industry in the areas of industrial control and in the development of graphical software engineering tools. Anne's research interests and the areas in which she has published include software development practices, processes and tools, and computer science education. Anne currently teaches in the area of software development practice and supervises many undergraduate and post-graduate students of software development. Anne is also the Academic Director of the Programming Challenge for Girls, an international organisation which provides activities and resources to encourage the participation of female high school students in Computer Science.

Pablo Picazo-Sanchez is a PhD student at the Department of Applied Mathematics in the University School of Computer Science (UPM) of Madrid. He has a B.Sc and a M.Sc. in Computer Science and Technology. His research interests are in the field of protocols design, authentication, privacy, lightweight cryptography, cryptanalysis, et cetera. Nowadays, his research is focused on Radio Frequency Identification Systems (RFID).

Manuel García Sánchez (S'88-M'93) received the Ingeniero de Telecomunicación degree from the Universidad de Santiago de Compostela, Spain, in 1990 and the Doctor Ingeniero de Telecomunicación (Ph.D.) degree from the Universidad de Vigo, Spain, in 1996. Professor Sánchez is currently with the Teoría de la Señal y Comunicaciones Department at Universidad de Vigo, where he teaches Broadcasting and Radiodetermination to undergraduate students. He is the head of the Radio Systems research group.

His research interests include studies of indoor and outdoor radio channel sounding and modeling for narrow and wide-band applications at microwave and millimetre wave frequencies and radio interference detection and mitigation, radio network deployment, radio location, and radio system development.

Nitesh Saxena is an Assistant Professor in the Department of Computer and Information Sciences at the University of Alabama, Birmingham (UAB). He works in the broad areas of computer and network security and applied cryptography, and has a strong interest in the emerging field of usable security. Nitesh obtained his Ph.D. in Information and Computer Science from UC Irvine, an M.S. in Computer Science from UC Santa Barbara, and a Bachelor's degree in Mathematics and Computing from the Indian Institute of Technology, Kharagpur, India. Before joining UAB, he was an Assistant Professor in the Department of Computer Science and Engineering at the Polytechnic Institute of New York University (NYU-Poly).

Kah Phooi Seng received the B.Eng (1st class) and PhD from the University of Tasmania, Australia in 1997 and 2001, respectively. She is currently a Professor in School of Computer Technology, Sunway University Malaysia. Before joining Sunway University, she was an Associate Professor in Faculty of Engineering, University of Nottingham Malaysia Campus. Her current research interests include the fields of intelligent visual processing, multi-biometrics, artificial intelligence, and signal processing.

John A. Stankovic is the BP America Professor in the Computer Science Department at the University of Virginia. He served as Chair of the department for 8 years. He is a Fellow of both the IEEE and the ACM. He won the IEEE Real-Time Systems Technical Committee's Award for Outstanding Technical Contributions and Leadership. He also won the IEEE Technical Committee on Distributed Processing's Distinguished Achievement Award (inaugural winner). He has won four Best Paper awards, including one for ACM SenSys 2006. He has given more than 25 keynote talks at conferences and many distinguished lectures at major universities. He was the Editor-in-Chief for the *IEEE Transactions on Distributed and Parallel Systems* and was founder and co-editor-in-chief for the *Real-Time Systems Journal*. His research interests are in real-time systems, distributed computing, wireless sensor networks, and cyber physical systems. Prof. Stankovic received his PhD from Brown University.

Erich Wenger studied Computer Science at Graz University of Technology. He finished the BSc in 2007 and the MSc in 2010. During his Master's studies he went on exchange to the McMaster University in Hamilton, Ontario, Canada (8 months, Joint Study) and to the Swiss Federal Institute of Technology in Zurich, Switzerland (6 months, Erasmus), where he did his Master's thesis (Integrated Systems Laboratory, supervisor: Norbert Felber). In 2010, he joined the SEnSE group of IAIK and is currently working on his PhD thesis. He is especially interested in elliptic curve cryptography, integrated circuit design, and embedded processor design.

Index